Lecture Notes in Compu 2146

Edited by G. Goos, J. Hartmanis

Springer
Berlin
Heidelberg
New York
Barcelona
Hong Kong
London
Milan
Paris
Tokyo

Joseph H. Silverman (Ed.)

Cryptography and Lattices

International Conference, CaLC 2001
Providence, RI, USA, March 29-30, 2001
Revised Papers

Volume Editor

Joseph H. Silverman
Brown University, Mathematics Department - Box 1917
Providence, RI 02912, USA
E-mail: jhs@math.brown.edu

Cataloging-in-Publication Data applied for

Die Deutsche Bibliothek - CIP-Einheitsaufnahme

Cryptography and lattices : international conference, revised papers / CaLC 2001, Providence, RI, USA, March 29 - 30, 2001. Joseph H. Silverman (ed.). - Berlin ; Heidelberg ; New York ; Barcelona ; Hong Kong ; London ; Milan ; Paris ; Singapore ; Tokyo : Springer, 2001
(Lecture notes in computer science ; Vol. 2146)
ISBN 3-540-42488-1

CR Subject Classification (1998): E.3, F.2.1, F.2.2, G.1, I.1.2, G.2, K.4.4

ISSN 0302-9743
ISBN 3-540-42488-1 Springer-Verlag Berlin Heidelberg New York

Springer-Verlag Berlin Heidelberg New York
a member of BertelsmannSpringer Science+Business Media GmbH

http://www.springer.de

Printed in Germany

Typesetting: Camera-ready by author, date conversion by Christian Grosche, Hamburg
Printed on acid-free paper SPIN 10840224 06/3142 5 4 3 2 1 0

Preface

These are the proceedings of CaLC 2001, the first conference devoted to cryptography and lattices. We have long believed that the importance of lattices and lattice reduction in cryptography, both for cryptographic construction and cryptographic analysis, merits a gathering devoted to this topic. The enthusiastic response that we received from the program committee, the invited speakers, the many people who submitted papers, and the 90 registered participants amply confirmed the widespread interest in lattices and their cryptographic applications.

We thank everyone whose involvement made CaLC such a successful event; in particular we thank Natalie Johnson, Larry Larrivee, Doreen Pappas, and the Brown University Mathematics Department for their assistance and support.

March 2001 Jeffrey Hoffstein, Jill Pipher, Joseph Silverman

Organization

CaLC 2001 was organized by the Department of Mathematics at Brown University. The program chairs express their thanks to the program commitee and the additional external referees for their help in selecting the papers for CaLC 2001. The program chairs would also like to thank NTRU Cryptosystems for providing financial support for the conference.

Program Commitee

- Don Coppersmith <dcopper@us.ibm.com>
  IBM Research
- Jeffrey Hoffstein (co-chair) <jhoff@math.brown.edu>, <jhoff@ntru.com>
  Brown University and NTRU Cryptosystems
- Arjen Lenstra <arjen.lenstra@citicorp.com>
  Citibank, USA
- Phong Nguyen <Phong.Nguyen@ens.fr>
  ENS
- Andrew Odlyzko <amo@research.att.com>
  AT&T Labs Research
- Joseph H. Silverman (co-chair) <jhs@math.brown.edu>, <jhs@ntru.com>
  Brown University and NTRU Cryptosystems

External Referees

Ali Akhavi, Glenn Durfee, Nick Howgrave-Graham, Daniele Micciancio

Sponsoring Institutions

NTRU Cryptosystems, Inc., Burlington, MA <www.ntru.com>

Table of Contents

An Overview of the Sieve Algorithm for the Shortest Lattice Vector Problem

Miklós Ajtai, Ravi Kumar, and Dandapani Sivakumar

IBM Almaden Research Center
650 Harry Road, San Jose, CA 95120
{ajtai,ravi,siva}@almaden.ibm.com

We present an overview of a randomized $2^{O(n)}$ time algorithm to compute a shortest non-zero vector in an n-dimensional rational lattice. The complete details of this algorithm can be found in [2].

A lattice is a discrete additive subgroup of $\mathbf{R}^n$. One way to specify a lattice is through a basis. A basis $B = \{b_1, \ldots, b_n\}$ is a set of linearly independent vectors in $\mathbf{R}^n$. The lattice generated by a basis B is $L = L(B) = \{\sum_{i=1}^{n} c_i b_i \mid c_i \in \mathbf{Z}\}$. The shortest lattice vector problem (SVP) is the problem of finding a shortest non-zero vector (under some norm, usually ℓ_2) in L. The α-approximate version of SVP is to find a non-zero lattice vector whose length is at most α times the length of a shortest non-zero lattice vector.

SVP has a rich history. Gauss and Hermite studied an equivalent of SVP in the context of minimizing quadratic forms [4, 7]. Dirichlet formulated SVP under the guise of diophantine approximations. Using the convex body theorem, Minkowski gave an existential bound on the shortest vector in a lattice [13].

Though the extended Euclidean GCD algorithm can be used to solve SVP in two dimensions, the first algorithmic breakthrough in n dimensions was obtained in a celebrated result of Lenstra, Lenstra, and Lovász [10], who gave an algorithm (the LLL algorithm) that computes a $2^{n/2}$-approximate shortest vector in polynomial time. This was improved in a generalization of the LLL algorithm by Schnorr [14], who obtained a hierarchy of algorithms that provide a uniform trade-off between the running time and the approximation factor. This algorithm runs in $n^{O(1)} k^{O(k)}$ steps to solve a $k^{O(n/k)}$-approximate SVP. For instance, a polynomial-time version of this algorithm improves the approximation factor obtained by the LLL algorithm to $2^{n(\log\log n)^2/\log n}$. Kannan [8] obtained a $2^{O(n \log n)}$ time algorithm for the exact SVP. The constant in the exponent of this algorithm was improved to about 1/2 by Helfrich [6]. Recently, Kumar and Sivakumar solved the decision version of n^3-approximate SVP in $2^{O(n)}$ time [9].

On the hardness front, SVP for the L_∞ norm was shown to be NP-complete by van Emde Boas [3]. Ajtai [1] proved that SVP under the ℓ_2 norm is NP-hard under randomized reductions. Micciancio [12] showed that the α-approximate SVP remains NP-hard for any $\alpha < \sqrt{2}$. Lagarias, Lenstra, and Schnorr [11] showed that n-approximate SVP is unlikely to be NP-hard. Goldreich and Goldwasser [5] showed that $\sqrt{n/\log n}$-approximate SVP is unlikely to be NP-hard.

We sketch a randomized $2^{O(n)}$ algorithm for SVP (in ℓ_2 norm) for a lattice L in $\mathbf{R}^n$. In fact, in $2^{O(n)}$ time, our algorithm can find all α-approximate shortest vectors for any constant $\alpha \geq 1$.

J.H. Silverman (Ed.): CaLC 2001, LNCS 2146, pp. 1–3, 2001.

We make a few simplifying assumptions about the lattice L: (1) the length of shortest vector is at least 1 and at most 2 — this can be realized by appropriately scaling L; (2) the length of the longest vector in the basis is at most $2^{O(n)}$ — this can be realized by appropriate applications of the LLL algorithm.

We create a large (sides of exponential length) parallelepiped $\mathcal{P}$ that is fairly close to being a cube. Then we uniformly sample a large number of lattice points $z_1, \ldots, z_N$, $N = 2^{O(n)}$, from $\mathcal{P} \cap L$, and to each sample z_i, we add a uniform perturbation vector y_i of expected length $O(1)$ to obtain a sequence of points $x_1, \ldots, x_N$. For each perturbed lattice point x_i, we will keep track of two lattice points: its "true identity" z_i, and an "approximator" a_i, initially set to 0.

Then, we use the following sieve procedure — given sufficiently many points in $\mathbf{R}^n$ of length at most R, identify a small set of "representatives" from the set of points and a large set of "survivors" such that for every survivor point, there is a representative at distance at most $R/2$. We repeatedly apply the sieve to the vectors $x_i - a_i$; for each survivor $x_i - a_i$ with representative $x_j - a_j$, we know that the distance between $x_i - a_i$ and $x_j - a_j$ is about half the distance between x_i and a_i. Therefore, $a_i + x_j - a_j$ is a better approximation to x_i, and since x_j is close to its true identity z_j, we define the new approximator for x_i to be $a_i + z_j - a_j$. In these steps, once the true identity of a point is revealed, we will not use it in the future. We repeat this process until the distance between the points and their approximators are bounded by another constant. Finally, if x_i still survives and has an approximator a_i, output the lattice point $w_i = z_i - a_i$. Since both z_i and a_i are close to x_i, with high probability, the length of w_i is bounded by a constant. We will denote this process as the basic algorithm.

Note that if the basic algorithm stops with a non-zero w_i, we already have a constant factor approximation algorithm for SVP. To ensure that w_i is non-zero with good probability and to obtain the shortest vector, we make the following argument. Let u denote a shortest vector in L. Let $w = w_i$ be a lattice point of constant length that is output by the procedure above. Let x be a sample point from which w was obtained, and let $z \in L$ be the true identity of x. Since the perturbations are small, we can argue that the probability (conditioned on x being a sample) that one of $z \pm u$ is the true identity of x is at least $2^{-O(n)}$ times the probability that z is the true identity of x. Furthermore — and this is the crucial point — the basic algorithm is oblivious to the true identity of x. Using this fact, we will argue that for some w, $w + u$ has at least $2^{-O(n)}$ times the probability of w to be the output of the basic algorithm. Since the number of lattice points in the ball of constant radius around the origin is at most $2^{O(n)}$, we obtain that there is at least one $w \in L$ whose probability of being output is at least $2^{-O(n)}$ and $w + u$ has the probability of being output at least $2^{-O(n)}$. Therefore, by repeating the basic algorithm $2^{O(n)}$ times we can ensure that with high probability both w and $w + u$ are output. Thus, the final algorithm is the following: repeat the basic algorithm $2^{O(n)}$ times, take all possible pairwise differences of the points output by the basic algorithm, and output the shortest of these vectors.

More details of this algorithm can be found in [2].

References

1. M. Ajtai. The shortest vector problem in L_2 is NP-hard for randomized reductions. *Proc. 30th ACM Symposium on Theory of Computing*, pp. 10–19, 1998.
2. M. Ajtai, R. Kumar, and D. Sivakumar. A sieve algorithm for the shortest lattice vector problem. *Proc. 33rd ACM Symposium on Theory of Computing*, 2001. To appear.
3. P. van Emde Boas. Another NP-complete partition problem and the complexity of computing short vectors in lattices. *Mathematics Department, University of Amsterdam*, TR 81-04, 1981.
4. C. F. Gauss. *Disquisitiones Arithmeticae.* English edition, (Translated by A. A. Clarke) Springer-Verlag, 1966.
5. O. Goldreich and S. Goldwasser. On the limits of nonapproximability of lattice problems. *Journal of Computer and System Sciences*, 60(3):540–563, 2000.
6. B. Helfrich. Algorithms to construct Minkowski reduced and Hermite reduced bases. *Theoretical Computer Science*, 41:125–139, 1985.
7. C. Hermite. Second letter to Jacobi, Oeuvres, I, *Journal für Mathematik*, 40:122–135, 1905.
8. R. Kannan. Minkowski's convex body theorem and integer programming. *Mathematics of Operations Research*, 12:415–440, 1987. Preliminary version in *ACM Symposium on Theory of Computing* 1983.
9. R. Kumar and D. Sivakumar. On polynomial approximations to the shortest lattice vector length. *Proc. 12th Symposium on Discrete Algorithms*, 2001.
10. A. K. Lenstra, H. W. Lenstra, and L. Lovász. Factoring polynomials with rational coefficients. *Mathematische Annalen*, 261:515–534, 1982.
11. J. C. Lagarias, H. W. Lenstra, and C. P. Schnorr. Korkine-Zolotarev bases and successive minima of a lattice and its reciprocal lattice. *Combinatorica*, 10:333–348, 1990.
12. D. Micciancio. The shortest vector in a lattice is hard to approximate to within some constant. *Proc. 39th IEEE Symposium on Foundations of Computer Science*, pp. 92–98, 1998.
13. H. Minkowski. *Geometrie der Zahlen.* Leipzig, Teubner, 1990.
14. C. P. Schnorr. A hierarchy of polynomial time basis reduction algorithms. *Theoretical Computer Science*, 53:201–224, 1987.

Low Secret Exponent RSA Revisited

Johannes Blömer and Alexander May

Department of Mathematics and Computer Science
University of Paderborn, 33095 Paderborn, Germany
{bloemer,alexx}@uni-paderborn.de

Abstract. We present a lattice attack on low exponent RSA with short secret exponent $d = N^\delta$ for every $\delta < 0.29$. The attack is a variation of an approach by Boneh and Durfee [4] based on lattice reduction techniques and Coppersmith's method for finding small roots of modular polynomial equations. Although our results are slightly worse than the results of Boneh and Durfee they have several interesting features. We partially analyze the structure of the lattices we are using. For most $\delta < 0.29$ our method requires lattices of smaller dimension than the approach by Boneh and Durfee. Hence, we get a more practical attack on low exponent RSA. We demonstrate this by experiments, where $\delta > 0.265$.
Our method, as well as the method by Boneh and Durfee, is heuristic, since the method is based on Coppersmith's approach for bivariate polynomials. Coppersmith [6] pointed out that this heuristic must fail in some cases. We argue in this paper, that a (practically not interesting) variant of the Boneh/Durfee attack proposed in [4] always fails. Many authors have already stressed the necessity for rigorous proofs of Coppersmith's method in the multivariate case. This is even more evident in light of these results.
Keywords: Low secret exponent RSA, cryptanalysis, Coppersmith's method, lattice reduction.

1 Introduction

In this paper we consider the problem of breaking the RSA cryptosystem for short secret keys. An RSA public key is a pair (N, e) where $N = pq$ is a product of two n-bit primes. The corresponding secret key d is chosen such that it satisfies the equation

$$ed \equiv 1 \pmod{\tfrac{1}{2}\phi(N)},$$

where $\phi(N) = (p-1)(q-1)$.

The first result showing that RSA is insecure, if the secret key is too small, is due to Wiener. In 1990, Wiener [20] showed that $d < \frac{1}{3}N^{0.25}$ leads to a polynomial time attack on the RSA system. Wiener's method is based on continued fractions. Basically, Wiener showed that d is the denominator of some convergent of the continued fraction expansion of e/N. A variant of Euclid's algorithm computes the continued fraction expansion of a number. Since N, e both are public, this shows that d can be computed efficiently from the public key (N, e).

J.H. Silverman (Ed.): CaLC 2001, LNCS 2146, pp. 4–19, 2001.

Recently, Boneh and Durfee [4] proposed an attack on RSA, that shows that RSA is insecure provided $d < N^{0.292}$. Unlike Wiener's attack, the attack by Boneh and Durfee is a heuristic. It builds upon Coppersmith's result for finding small solutions of modular polynomial equations [6]. Coppersmith's method for the univariate case is rigorous but the proposed generalization for the multivariate case is a heuristic. More precisely, Boneh and Durfee show that for a small secret key d, the number $s = -\frac{p+q}{2}$ can be found as a small solution to some modular bivariate polynomial equation. Once s is known, one can immediately solve the equations $s = -\frac{p+q}{2}$ and $N = pq$ for the unknowns p and q. Using Coppersmith's method, which in turn is based on the famous L^3-lattice reduction algorithm, Boneh and Durfee reduce the problem of finding s to finding a common root of two bivariate polynomials $f(x, y), g(x, y)$ over the integers. As proposed by Coppersmith, finding a common root of f, g is done by first computing the resultant $r(y)$ of f, g with respect to the variable x. Provided $r \not\equiv 0$, the parameter s, and hence the factorization, can be found by computing the roots (over $\mathbb{Z}$) of r. Unfortunately, this method, as well as any other method based on Coppersmith's approach for multivariate polynomials[1], fails if the resultant r is identically 0. As it has never been proved that $r \not\equiv 0$, the Boneh/Durfee approach is heuristic.

In this paper we study the method by Boneh and Durfee in more detail. In Section 4, we propose a new lattice for cryptanalysing low secret exponent RSA with $d < N^{0.290}$. The new approach uses the same heuristical assumption as Boneh/Durfee. Although the new attack does not improve the bound $d < N^{0.292}$ of Boneh and Durfee [4], it has several advantages. First, the lattice dimension is reduced. Therefore, in practice we are able to get closer to the theoretical bounds. Second, the new lattice basis is triangular. This leads to rather simple proofs. Third, the new lattice basis takes advantage of special properties of the lattice vectors. We believe that some of our structural results in Section 4 can be applied to other applications of Coppersmith's method as well.

Actually, Boneh and Durfee present three different variations of the Coppersmith methodology to break RSA versions with small secret exponent d. The first one works for $d < N^{1/4}$, hence this variant basically reproduces Wiener's result. The second variation of Boneh and Durfee works for $d < N^{0.284}$. Finally they have a method that works for d up to $N^{0.292}$.

We made the experimental observation, that the first method of Boneh and Durfee, supposed to work for $d < N^{1/4}$ always failed. In fact, in all experiments the resultant r mentioned above was identically zero. Although one cannot recover the factorization by resultant computation, we show that RSA with secret key $d < \frac{1}{3}N^{1/4}$ can be broken using lattice reduction in dimension 2. In fact, we show that for an appropriately chosen lattice, a shortest vector in the lattice immediately reveals the secret key d.

Since we have not found examples where the other two variants for $d < N^{0.284}$ and $d < N^{0.292}$ described by Boneh and Durfee fail, this observation in no way invalidates the results of Boneh and Durfee. On the other hand, this is

[1] This includes among others [1, 4, 8, 12].

to our knowledge the first case mentioned in literature, that an application of Coppersmith's approach fails in general. Some authors [6, 14] already pointed out that the heuristic must fail in some cases, but no general failure has been reported for real applications of the method.

Although we are not quite able to rigorously analyze the Boneh and Durfee method for $d < N^{1/4}$, in Section 5 we prove several results that almost completely explain the behavior observed in experiments. Many authors already stressed the necessity of a rigorous analysis of methods based on Coppersmith's approach in the multivariate case. This is even more evident in light of our results.

In Section 6 we give experimental results for our new attack on RSA with short secret key d. We carried out cryptanalysis of secret keys up to $d \leq N^{0.278}$. We also compared our experimental results with the experimental results of Boneh and Durfee. In [3], they only provided examples with $d \leq N^{0.265}$. In all cases we considered, our method was faster.

2 The Boneh-Durfee Lattice

In this section we review the lattice attack by Boneh and Durfee on low exponent RSA. For an introduction into lattice theory and lattice basis reduction, we refer to the textbooks [9, 17]. Descriptions of Wiener's RSA attack and the method of Coppersmith can be found in [6, 20]. For a good overview of RSA attacks, we refer to a survey article of Boneh [2].

Let $d < e^{\delta}$. We assume that the size of e is in the order of the size of N. If e is smaller, the attack of Boneh and Durfee becomes even more effective (see [4], section 5).

All known attacks on RSA with short secret exponent focus on the identity

$$ed = 1 \bmod \frac{\phi(N)}{2} \quad \Leftrightarrow \quad ed + k\left(\frac{N+1}{2} + s\right) = 1, \tag{1}$$

where $k \in \mathbb{Z}$, $s = -\frac{p+q}{2}$ and d are unknown quantities. Since $e < \frac{\phi(N)}{2}$, we obtain $k < d$. Boneh and Durfee [4] look at equation (1) modulo e.

$$k\left(\frac{N+1}{2} + s\right) - 1 = 0 \bmod e$$

They define the polynomial

$$f(x,y) = x(A+y) - 1$$

with $A = \frac{N+1}{2}$. Let $X = e^{\delta}$ and $Y = e^{0.5}$. We know, that f has a root $(x_0, y_0) = (k, s)$ modulo e, that satisfies $|x_0| < X$ and $|y_0| < Y$. To transform the modular equation into an equation over the integers, Boneh/Durfee use a theorem of Howgrave-Graham [11]. Given a polynomial $p(x,y) = \sum_{i,j} a_{i,j} x^i y^j$, we define the norm $\|p(x,y)\|^2 = \sum_{i,j} a_{i,j}^2$.

Theorem 1 (Howgrave-Graham [11]). *Let $p(x,y)$ be a polynomial which is a sum of at most w monomials. Suppose that $p(x_0,y_0) = 0 \bmod e^m$ for some positive integer m, where $|x_0| < X$ and $|y_0| < Y$. If $\|p(xX,yY)\| < e^m/\sqrt{w}$, then $p(x_0,y_0) = 0$ holds over the integers.*

Next, Boneh and Durfee define polynomials

$$g_{i,k}(x,y) = x^i f^k(x,y)e^{m-k} \quad \text{and} \quad h_{j,k}(x,y) = y^j f^k(x,y)e^{m-k}$$

for a given positive integer m.

In the sequel, the polynomials $g_{i,k}$ are referred to as x-shifts and analogously the polynomials $h_{j,k}$ are referred to as y-shifts. By construction, the point (x_0,y_0) is a root of all these polynomials modulo e^m. Thus, we can apply Howgrave's theorem and search for a small norm linear combination of polynomials $g_{i,k}(xX,yY)$ and $h_{j,k}(xX,yY)$. This is done by using the L^3 lattice reduction algorithm. The goal is to construct a lattice that is guaranteed to contain a vector shorter than $e^m/\sqrt{w}$.

Boneh and Durfee suggest to build the lattice spanned by the coefficient vectors of the polynomials $g_{i,k}, h_{j,k}$ for certain parameters i, j and k. For each $k = 0,\dots,m$, they use the x-shifts $g_{i,k}(xX,yY)$ for $i = 0,\dots,m-k$. Additionally, they use the y-shifts $h_{j,k}$ for $j = 0,\dots,t$ for some parameter t.

In the sequel, we call the lattice constructed by Boneh and Durfee the lattice L_{BD}. The basis for L_{BD} is denoted by B_{BD}. The lattice L_{BD} is spanned by the row vectors of B_{BD}. Since the lattice depends on the parameters m and t, we sometimes refer to the parameters by $B_{BD}(m,t)$ to clarify notation.

It is easy to see, that the basis vectors of lattice L_{BD} form a triangular matrix. We give an example of the lattice basis for the parameter choice $m = 2$ and $t = 1$.

$B_{BD}(2,1) =$

	1	x	xy	x^2	x^2y	x^2y^2	y	xy^2	x^2y^3
e^2	e^2								
xe^2		e^2X							
fe	$-e$	eAX	eXY						
x^2e^2				e^2X^2					
xfe		$-eX$		eAX^2	eX^2Y				
f^2	1	$-2AX$	$-2XY$	A^2X^2	$2AX^2Y$	X^2Y^2			
ye^2							e^2Y		
yfe			$eAXY$				$-eY$	eXY^2	
yf^2			$-2AXY$		A^2X^2Y	$2AX^2Y^2$	Y	$-2XY^2$	X^2Y^3

Boneh and Durfee showed for $\delta < 0.284$, one can find m, t such that an L^3-reduced basis of L_{BD} contains vectors short enough to apply Howgrave's theorem and factor the modulus N. This was improved in the same paper to $\delta < 0.292$ by using non-triangular lattice bases. This is up to now the best bound for cryptanalysis of low secret exponent RSA. The attack works under the assumption that polynomials obtained from two sufficiently short vectors in the reduced basis have a non-vanishing resultant. Although heuristic, no failure of the method for sufficiently large δ is known.

Boneh and Durfee also argue that using $t = 0$, that is only x-shifts are used to construct a lattice basis, one obtains already an attack working for $\delta < 0.25$. This reproduces Wiener's result. However, experiments show that the method of Boneh and Durfee never works when using only x-shifts. In Section 5, we will explain why this is the case. Of course, this failure of the Boneh/Durfee method in the special case where only x-shifts are used does not affect the method in general. It only points out that one has to be careful when using Coppersmith's heuristic in the multivariate case.

3 Notations

Since the lattice L_{BD} defined in Section 2 is the starting point of our further constructions, we introduce some notations on the rows and columns of the lattice basis B_{BD}.

We refer to the coefficient vectors of the polynomials $g_{i,k}(xX, yY)$ as the X-block. The X-block is further divided into $X_l, l = 0, \dots, m$, blocks, where the block X_l consist of the $l+1$ coefficient vectors of $g_{i,k}$ with $i + k = l$. These $l+1$ vectors are called $X_{l,k}$, that is the k-th vectors in the X_l block is the coefficient vector of $g_{l-k,k}$.

The coefficient vectors of the polynomials $h_{j,k}$ form the Y-block. We define the Y_j block as the block of all $m+1$ coefficient vectors of polynomials that are shifted by y^j. The k-th vector in the block Y_j is called $Y_{j,k}$, it is identical to the coefficient vector of $h_{j,k}$.

Every column in the basis B_{BD} is labeled by a monomial $x^i y^j$. All column vectors with label $x^l y^j$, $l \geq j$, form the $X^{(l)}$ column block. Analogously, we define the $Y^{(l)}$ column block to consist of all column vectors labeled with $x^i y^{i+l}$.

In the example in Section 2, the horizontol lines divide the basis $B_{BD}(2,1)$ into the blocks X_1, X_2, X_3 and Y_1. Similarly, the vertical lines divide $B_{BD}(2,1)$ into the column blocks $X^{(1)}$, $X^{(2)}$, $X^{(3)}$ and $Y^{(1)}$. In this example, the basis entry in row $Y_{1,2}$ and column x^2y is A^2X^2Y.

4 A New Method for All $\delta < 0.290$

We introduce an alternative method for factoring the modulus N if $d < N^{0.290}$. This does not improve the bound $\delta < 0.292$ given by Boneh and Durfee. However, it has several advantages compared to their approach.

First, our method significantly reduces the lattice dimension as a function of m and t. The practical implication is that we are able to get closer to the theoretical bound. We give experimental results for $\delta > 0.265$. Second, our proofs are simple. As opposed to the Boneh/Durfee lattices for $\delta < 0.292$, the lattice bases we use in the attack for $\delta < 0.290$ remain triangular. Hence, determinant computations are simple. Third, our construction makes use of structural properties of the underlying polynomials. Thus, it should apply also to other lattice constructions using these polynomials.

Construction of the new lattice L with basis B

1. Choose lattice parameters m and t and build the Boneh-Durfee lattice basis $B_{BD}(m,t)$ as explained in Section 2.
2. In the Y_t block of the basis B_{BD} remove every vector except for the last vector $Y_{t,m}$, in the Y_{t-1} block remove every vector except for the last two vectors $Y_{t,m-1}$ and $Y_{t,m}$, and so on. Finally, in the Y_1 block remove every vector except for the last t vectors $Y_{m-t+1}, \ldots, Y_m$.
3. Remove every vector in the X-block except for the vectors in the $t+1$ blocks X_{m-t}, X_{m-t+1}, $\ldots, X_m$.
4. Delete columns in such a way that the resulting basis is again triangular. This is, remove all column blocks $X^{(0)}, X^{(1)}, \ldots, X^{(m-t-1)}$. Furthermore in the column block $Y^{(l)}$, $l = 1, \ldots, t$, remove the columns labeled with $x^i y^{i+l}$ for $0 \leq i < m - t + l$.

This construction leads to a triangular basis B of a new lattice L, which will be used in our approach. Since B depends on m and t, we sometimes write $B(m,t)$.

As opposed to Boneh and Durfee, we do not integrate more y-shifts to improve the bound $\delta < 0.284$, instead we remove some x-shifts.

Remark 1. *In our construction, we take the pattern*

$$(p_0, p_1, \ldots, p_t) = (1, 2, \ldots, t+1).$$

That is, we take the last p_i, $0 \leq i < t$ vectors from the Y_{t-i} block and the last p_t X-blocks and delete columns appropriately. The proofs in this section easily generalize to every strictly increasing pattern $(p_0, p_1, \ldots, p_t)$, $p_0 < p_1 < \cdots < p_t$. This includes among others the pattern used by Boneh/Durfee [4] to show the bound $d < N^{0.292}$. We give the proof of this generalization in the full version of the paper.

Applying the construction to the example given in Section 2, we obtain the following lattice basis of L with parameters $m = 2$ and $t = 1$.

$B(2,1) =$

	x	xy	x^2	x^2y	x^2y^2	x^2y^3
xe^2	e^2X					
fe	eAX	eXY				
x^2e^2			e^2X^2			
xfe	$-eX$		eAX^2	eX^2Y		
f^2	$-2AX$	$-2XY$	A^2X^2	$2AX^2Y$	X^2Y^2	
yf^2		$-2AXY$		A^2X^2Y	$2AX^2Y^2$	X^2Y^3

Let $\bar{B}$ be the non-triangular basis we obtain after Step 3 of the construction. That is, $\bar{B}$ consists of the remaining basis vectors of B_{BD} in the construction after removing row vectors but without removing columns. The lattice spanned by the row vectors of $\bar{B}$ is called $L_{\bar{B}}$. We adopt the notations of Section 3 for the rows and columns of B and $\bar{B}$. For example, the row vector $X_{l,k}$ of B is the coefficient vector of $g_{l-k,k}$, where we removed all the entries specified in Step 4 of the construction. In the basis $B(2,1)$ above, the row vector $X_{2,0}$ is the vector $(0,0,e^2X^2,0,0,0)$.

We call a column vector x^iy^j that appears in the basis $\bar{B}$ but not in the basis B a *removed column* of B. The bases B and $\bar{B}$ are constructed using the same coefficient vectors, where in B certain columns are removed. Having a vector $u = \sum_{b \in B} c_b b$ in the span of B, one can compute the corresponding linear combination $\bar{u} = \sum_{b \in \bar{B}} c_b b$ of vectors in $\bar{B}$ with the same coefficients c_b. Hence, the vector dimension of $\bar{u}$ is larger than the vector dimension of u. One can regard the additional vector entries in $\bar{u}$ as a reconstruction of the vector entries of u in the removed columns. Therefore, we call $\bar{u}$ the *reconstruction vector* of u.

The row vectors

$$X_{l,k}, (l = m-t, \ldots, m; k \leq l) \quad \text{and} \quad Y_{j,k}, (j = 1, \ldots, t; k = m-t+j, \ldots, m)$$

form the basis B. These vectors are no longer the coefficient vectors of the polynomials $g_{l-k,k}(xX, yY)$ and $h_{j,k}(xX, yY)$, respectively, since we remove columns in Step 4 of the construction. However in order to apply Howgrave's theorem, we must ensure that we construct a linear combination of bivariate polynomials that evaluates to zero modulo e^m at the point $(x_0, y_0) = (k, s)$. Hence, we still have to associate the rows $X_{l,k}$ and $Y_{j,k}$ with the polynomials $g_{l-k,k}$ and $h_{j,k}$. The basis vectors of $\bar{B}$ represent the coefficient vectors of these polynomials. Therefore, after finding a small vector $u = \sum_{b \in B} c_b b$ in L, we compute the reconstruction vector $\bar{u} = \sum_{b \in \bar{B}} c_b b$ in $L_{\bar{B}}$. That is, we reconstruct the entries in the removed columns. Once the reconstruction vectors of two sufficiently short vectors in L are computed, the rest of our method is the same as in the Boneh/Durfee method.

In the remainder of this section we show that short vectors u in L lead to short reconstruction vectors $\bar{u}$ in $L_{\bar{B}}$. To prove this, we first show that removed columns of B are small linear combinations of column vectors in B. We give an example for the removed column x^0y^0 in $B(2,1)$. Applying the construction in the following proof of Lemma 2, we see that this column is a linear combination of the columns x^1y^1 and x^2y^2 in B.

$$\begin{pmatrix} 0 \\ -e \\ \hline 0 \\ 0 \\ 1 \\ \hline 0 \end{pmatrix} = -\frac{1}{XY} \begin{pmatrix} 0 \\ eXY \\ \hline 0 \\ 0 \\ -2XY \\ \hline -2AXY \end{pmatrix} - \frac{1}{X^2Y^2} \begin{pmatrix} 0 \\ 0 \\ \hline 0 \\ 0 \\ X^2Y^2 \\ \hline 2AX^2Y^2 \end{pmatrix}$$

Lemma 2. *All removed columns in the column blocks $X^i, i < m - t$, are linear combinations of columns in B. Moreover, in these linear combinations, the coefficient for a column vector in $X^{(l)}, l \geq m - t$, can be bounded by $\frac{1}{(XY)^{l-i}} \cdot c$, where c depends only on m and t.*

Proof: If $x^i y^j$ is a removed column of B, we show that $x^i y^j$ is a linear combination of columns $x^{i+1} y^{j+1}$, ..., $x^m y^{m-i+j}$. If $x^{i+1} y^{i+1}$ is a removed column, we can repeat the argument to show that $x^{i+1} y^{i+1}$ is a linear combination of the remaining columns $x^{i+2} y^{j+2}$, ..., $x^m y^{m-i+j}$. Continuing in this way until all removed columns have been represented as linear combinations of columns in B, proves the lemma. Hence, it suffices to prove the following claim.

Claim 1. *If $x^i y^j$ is a removed column of B, then $x^i y^j$ is a linear combination of the columns $x^{i+1} y^{j+1}$, $x^{i+2} y^{j+2}$,..., $x^m y^{m-i+j}$, where the coefficient of column $x^{i+b} y^{j+b}, b = 1, \ldots, m - i$, is given by*

$$-\frac{1}{(XY)^b} \binom{j+b}{j}.$$

Note, that the coefficient $c_b = \binom{j+b}{j}$ depends only on m and t, since i, j depend on m and t.

We will prove Claim 1 by showing that for each row in $B(m, t)$ the entry of the column $x^i y^j$ in this row is a linear combination of the entries of the columns $x^{i+b} y^{j+b}$ in this row, with the coefficients as in the claim. We prove this for the rows in the X-block and Y-block separately.

Let $X_{l,k}$ be a row in block X_l, where $l \geq m - t$. The coefficients in this row are the coefficients of the polynomial $e^{m-k} x^{l-k} f^k(xX, yY)$. By definition of f this polynomial is

$$e^{m-k} x^{l-k} f^k(xX, yY) = e^{m-k} \sum_{p=0}^{k} \sum_{q=0}^{p} (-1)^{k+p} \binom{k}{p} \binom{p}{q} A^{p-q} X^p Y^q x^{p+l-k} y^q. \tag{2}$$

To obtain the coefficient of $x^{i+b} y^{j+b}$ in $e^{m-k} x^{l-k} f^k(xX, yY)$, we set $p = i - l + k + b$ and $q = j + b$. Hence, this coefficient is given by

$$e^{m-k} (-1)^{i-l+b} \binom{k}{i-l+k+b} \binom{i-l+k+b}{j+b} A^{i-l+k-j} X^{i-l+k+b} Y^{j+b}$$
$$= e^{m-k} A^{i-l+k-j} X^{i-l+k} Y^j (-1)^{i-l} (-1)^b \binom{k}{i-l+k+b} \binom{i-l+k+b}{j+b} (XY)^b.$$

We can ignore the factor $e^{m-k} A^{i-l+k-j} X^{i-l+k} Y^j (-1)^{i-l}$, common to all entries in row $X_{l,k}$ in the columns $x^{i+b} y^{j+b}$. Then Claim 1 restricted to row $X_{l,k}$ reads as

$$\binom{k}{i-l+k}\binom{i-l+k}{j}$$
$$= \sum_{b=1}^{m-i}(-1)^{b+1}\frac{1}{(XY)^b}\binom{j+b}{j}\binom{k}{i-l+k+b}\binom{i-l+k+b}{j+b}(XY)^b$$

Since the binomial coefficient $\binom{k}{i-l+k+b}$ is non-zero only for $k \geq i-l+k+b$, we only have to sum up to $b \leq l-i$. Substituting $i-l+k$ by i' yields

$$\binom{k}{i'}\binom{i'}{j} = \sum_{b=1}^{k-i'}(-1)^{b+1}\binom{j+b}{j}\binom{k}{i'+b}\binom{i'+b}{j+b}. \tag{3}$$

Subtracting the left-hand side, Claim 1 restricted to row $X_{l,k}$ reduces to

$$0 = \sum_{b=0}^{k-i'}(-1)^{b+1}\binom{j+b}{j}\binom{k}{i'+b}\binom{i'+b}{j+b}. \tag{4}$$

One checks that

$$\binom{j+b}{j}\binom{k}{i'+b}\binom{i'+b}{j+b} = \frac{k!}{(k-i')!j!(i'-j)!}\binom{k-i'}{b}.$$

This shows

$$\sum_{b=0}^{k-i'}(-1)^{b+1}\binom{j+b}{j}\binom{k}{i'+b}\binom{i'+b}{j+b} = -\frac{k!}{(k-i')!j!(i'-j)!}\sum_{b=0}^{k-i'}(-1)^b\binom{k-i'}{b}.$$

Since $\sum(-1)^b\binom{k-i'}{b} = (1+(-1))^b = 0$ we get equation (4).

In the same manner, Claim 1 is proved for the Y-block. Let $Y_{l,k}$ be a row in block Y_l. Analogously to equation (2), we get

$$e^{m-k}y^l f^k(xX, yY) = e^{m-k}\sum_{p=0}^{k}\sum_{q=0}^{p}(-1)^{k+p}\binom{k}{p}\binom{p}{q}A^{p-q}X^pY^qx^py^{q+l}.$$

We obtain the coefficients of $x^{i+b}y^{j+b}$ in $e^{m-k}y^l f^k(xX, yY)$ by setting $p = i+b$ and $q = j-l+b$. Again, we ignore the common factors in e, A, X and Y. Now, Claim 1 for $Y_{l,k}$ reduces to

$$\binom{k}{i}\binom{i}{j-l} = \sum_{b=1}^{m-i}(-1)^{b+1}\binom{j+b}{j}\binom{k}{i+b}\binom{i+b}{j-l+b}$$

We only have to sum up to $b \leq k-i$, because the factor $\binom{k}{i+b}$ is zero for $k < i+b$. Substituting $j-l$ by j and i by i' yields equation (3). This concludes the proof. □

Lemma 3. *Every removed column vector $x^i y^{i+l}, i < m-t+l$, is a linear combination of the columns in the column block $Y^{(l)}$ of B. In this linear combination, the coefficient for a column vector $x^k y^{k+l}, k \geq m-t+l$, can be bounded by $\frac{1}{(XY)^{k-i}} \cdot c$, where c depends only on m and t.*

Proof: Analogously to the proof of Lemma 2. Therefore we omit it. □

Theorem 2. *Let $u = \sum_{b \in B} c_b b$ be a linear combination of vectors in B with $\|u\| < e^m$. For fixed m and t and for every removed column $x^i y^j$ in the $X^{(i)}$ block $(0 \leq i < m-t)$, the entry $x^i y^j$ in the reconstruction vector $\bar{u} = \sum_{b \in \bar{B}} c_b b$ can be bounded by $O\left(\frac{e^m}{(XY)^{m-t-i}}\right)$.*

Proof: Consider a removed column $x^i y^j$. Let $v = (v_1, v_2, \ldots, v_n)^T$ be the column vector $x^i y^j$ in $\bar{B}$, where the entries are multiplied by the coefficients c_b. We want to show that $|\sum_{k=1}^n v_k| = O(\frac{e^m}{(XY)^{m-t-i}})$. This will prove the theorem.

Apply Lemma 2 and write v as a linear combination of the t+1 columns $x^{m-t} y^{m-t-i+j}, \ldots, x^m y^{m-i+j}$ in B, where again the entries in each of the $t+1$ vectors are multiplied by the coefficients c_b. Call these columns $w_i = (w_{i,1}, \ldots, w_{i,n})^T$ for $i = 0, \ldots, t$. Applying Lemma 2 yields

$$v = \frac{d_0}{(XY)^{m-t-i}} w_0 + \frac{d_1}{(XY)^{m-t-i+1}} w_1 + \cdots + \frac{d_t}{(XY)^{m-i}} w_t$$

According to Lemma 2, the d_i are constant for fixed m and t. By assumption $\|u\| < e^m$. Hence, all components of u are less than e^m. From this, we obtain $|\sum_k w_{i,k}| < e^m$. This implies

$$\begin{aligned}
\left|\sum_k v_k\right| &= \left|\frac{d_0}{(XY)^{m-t-i}} \sum_k w_{0,k} + \cdots + \frac{d_t}{(XY)^{m-i}} \sum_k w_{t,k}\right| \\
&\leq \left|\frac{d_0}{(XY)^{m-t-i}} \sum_k w_{0,k}\right| + \cdots + \left|\frac{d_t}{(XY)^{m-i}} \sum_k w_{t,k}\right| \\
&\leq \left|\frac{d_0}{(XY)^{m-t-i}} e^m\right| + \cdots + \left|\frac{d_t}{(XY)^{m-i}} e^m\right| \\
&= O\left(\frac{e^m}{(XY)^{m-t-i}}\right) + \cdots + O\left(\frac{e^m}{(XY)^{m-i}}\right)
\end{aligned}$$

Therefore, $|\sum_k v_k|$ can be bounded by $O\left(\frac{e^m}{(XY)^{m-t-i}}\right)$. □

Theorem 3. *Let $u = \sum_{b \in B} c_b b$ be a linear combination of vectors in B with $\|u\| < e^m$. For fixed m and t and for every removed column $x^i y^{i+l}$ in the $Y^{(l)}$ block $(0 \leq i < m-t+l, 1 \leq l \leq t)$, the entry $x^i y^{i+l}$ in the reconstruction vector $\bar{u} = \sum_{b \in \bar{B}} c_b b$ can be bounded by $O\left(\frac{e^m}{(XY)^{m-t+l-i}}\right)$.*

Proof: Analogously to the proof of Theorem 2. The proof is omitted. □

From Theorems 2 and 3, we can conclude that if we use the reconstruction vector $\bar{u}$ instead of the short vector u, we do not enlarge the norm significantly. This shows the correctness of our approach.

Corollary 4. *Let* $u = \sum_{b \in B} c_b b$ *with* $||u|| < e^m$ *be a vector in* L. *Then the reconstruction vector* $\bar{u} = \sum_{b \in \bar{B}} c_b b$ *satisfies* $||\bar{u}|| < e^m + O(\frac{e^m}{XY})$.

4.1 Computation of the New Bound

Since the lattice basis for L is triangular, computing the determinant of lattice L is easy. We do not carry out the computation. Manipulating the expressions for the determinant is straightforward, but requires tedious arithmetic.

The lattice dimension w equals the number of vectors in the two sets of blocks $X_{m-t}, \ldots, X_m$ and $Y_1, \ldots, Y_t$, so

$$w = \sum_{i=m-t}^{m} (i+1) + \sum_{i=1}^{t} i = (m+1)(t+1)$$

Notice that we have $w = (m+1)(m+2)/2 + t(m+1)$ for the lattice L_{BD}.

We compute the determinant $\det(L)$ as a function of e, m, t and δ. We find the optimal t as a function of m, δ by elementary calculus. Analogously to the method of Boneh/Durfee, we solve the equation

$$\det(L) < e^{mw}$$

for the maximal value of δ. This leads to the bound

$$\delta < \frac{\sqrt{6}-1}{5} \approx 0.290.$$

5 A Case Where the Heuristic Fails

As mentioned before, if L_{BD} is constructed using only x-shifted polynomials $g_{i,k}$ then the Boneh/Durfee method always failed in our experiments. More precisely, the polynomials we obtained from the two shortest vectors in an L^3-reduced basis for L_{BD} led to two polynomials whose resultant with respect to x was identically 0. We want to explain this phenomenon.

Using the construction of Section 4 in the special case of $t = 0$ and $m = l_1$, the lattice L consists only of the vectors in the block X_{l_1} with the columns in $X^{(l_1)}$. A simple determinant computation shows, that for every X_{l_1} block there is a linear combination of vectors in block X_{l_1} that is shorter than e^m provided $\delta < 0.25$.

Moreover, unless a combination of vectors in block X_{l_2} is much shorter than e^m (according to Theorem 2 it must be of size $O(\frac{e^m}{XY})$), combinations of vectors from different blocks X_{l_1}, X_{l_2} can not be shorter than vectors obtained as combinations of vectors from a single block X_{l_1}. Although not a rigorous proof, this explains the following observation. In our experiments every vector in an L^3-reduced basis for $B_{BD}(m, 0)$ was a combination of basis vectors from a single block X_{l_1}[2].

Now assume that we compute the resultant of two polynomials p_1, p_2, obtained from vectors v_1, v_2 whose length is smaller than $e^m/\sqrt{w}$ and that are linear combinations of basis vectors in X_{l_1} and X_{l_2}, respectively. By construction of L_{BD} and Howgrave's theorem (Theorem 1), p_1 and p_2 have a common root $(x_0, y_0) = (k, s)$. The following theorem shows that in this case the Boneh/Durfee attack fails.

Theorem 4. *Let $p_1(x, y)$ and $p_2(x, y)$ be polynomials that are non-zero linear combinations of the X_{l_1} and X_{l_2} block, respectively. If $p_1(x_0, y_0) = p_2(x_0, y_0) = 0$ for at least one pair $(x_0, y_0) \in \mathbb{C} \times \mathbb{C}$ then* $\mathrm{Res}_x(p_1, p_2) = 0$.

Proof: Write p_1 and p_2 as linear combinations

$$p_1(x, y) = \sum_{i=0}^{l_1} c_i x^{l_1 - i} f^i(x, y) e^{m-i} \ , \ p_2(x, y) = \sum_{i=0}^{l_2} d_i x^{l_2 - i} f^i(x, y) e^{m-i} .$$

We know $l_1, l_2 > 0$, since $p_1(x_0, y_0) = p_2(x_0, y_0) = 0$ for at least one pair (x_0, y_0). If $c_0 = d_0 = 0$, then f is a common factor of p_1 and p_2 and $\mathrm{Res}_x(p_1, p_2) = 0$. Hence, we may assume $c_0 \neq 0$ or $d_0 \neq 0$.

Let $r(y) = \mathrm{Res}_x(p_1, p_2)$ be the resultant of p_1, p_2 with respect to the variable x. Let $T = \{z \in \mathbb{C} \mid r(z) = 0\}$ be the set of roots of $r(y)$. Next, define S as

$$S = \{y \in \mathbb{C} \mid \text{ there is an } x \in \mathbb{C} \text{ such that } p_1(x, y) = p_2(x, y) = 0\}.$$

S is the projection of the common roots of p_1, p_2 onto the second coordinate. It is well-known that $S \subseteq T$ (see for example [7]). Our goal is to show, that $|S| = \infty$. Then $|T| = \infty$ as well, and $r(y) = \mathrm{Res}_x(p_1, p_2) = 0$ as stated in the theorem.

To show that $|S| = \infty$, we first perform the transformation τ defined by

$$\tau(x, y) = (x', y) \ \text{ with } \ (x', y) = \left(\frac{x}{f(x, y)}, y\right).$$

We obtain

$$x' = \frac{x}{x(A + y) - 1} = \frac{1}{A + y - \frac{1}{x}}.$$

This implies

$$\frac{1}{x'} = A + y - \frac{1}{x} \quad \text{and} \quad \frac{1}{x} = A + y - \frac{1}{x'}.$$

[2] In fact, the following was true for arbitrary L_{BD}, even those constructed using y-shifts: Every vector in an L^3-reduced basis for L_{BD} that depended only on basis vectors in the X-block was a combination of basis vectors from a single block X_{l_1}.

From the second equality we get

$$x = \frac{x'}{x'(A+y)-1} = \frac{x'}{f(x',y)}.$$

We also get

$$f(x',y) = \frac{x}{x(A+y)-1} \cdot (A+y) - 1 = \frac{1}{x(A+y)-1} = \frac{1}{f(x,y)}.$$

Applying the transformation to the polynomials p_1, p_2 gives rational polynomials q_1, q_2, where

$$q_1(x',y) = f^{-l_1}(x',y) \sum_{i=0}^{l_1} c_i x'^{l_1-i} e^{m-i},$$

$$q_2(x',y) = f^{-l_2}(x',y) \sum_{i=0}^{l_2} d_i x'^{l_2-i} e^{m-i}.$$

Hence q_1, q_2 are of the form $q_1(x',y) = \frac{1}{f^{l_1}} g_1(x'), q_2(x',y) = \frac{1}{f^{l_2}} g_2(x')$, for polynomials g_1, g_2 that depend only on x'.
Let

$$S' = \{y \in \mathbb{C} \mid \text{ there is an } x \in \mathbb{C} \text{ such that } q_1(x,y) = q_2(x,y) = 0\}.$$

If g_1, g_2 do not have a common root, then $S' = \emptyset$. On the other hand, if g_1, g_2 have a common root x then $S' = \mathbb{C} \backslash \{\frac{1}{x} - A\}$, since $y = \frac{1}{x} - A$ is the only value y for which $f(x,y) = 0$. In particular, either $S' = \emptyset$ or $|S'| = \infty$. In order to show that $|S'| = \infty$, it suffices to show that there is at least one $y \in S'$.

In order to prove the theorem, it suffices to show that the transformation τ induces a bijective mapping of the common roots of p_1, p_2 onto the common roots of q_1, q_2. By assumption, p_1 and p_2 share the common root (x_0, y_0). Then, $\tau(x_0, y_0)$ is a common root of q_1, q_2. This implies $y_0 \in S'$ and therefore $|S'| = \infty$. By the bijectivity of τ, we get $|S| = \infty$ as well.

Whenever defined, the transformation τ is the inverse of itself. This implies that τ is bijective on its domain, that is on all points (x,y) where $f(x,y) \neq 0$. Hence, a common root (x,y) of p_1, p_2 is not in the domain of τ iff (x,y) is a root of f. So assume $(x,y) \in \mathbb{C} \times \mathbb{C}$ is such that $f(x,y) = 0$. Then all term in p_1 and p_2 vanish except for $i = 0$. We get $p_1(x,y) = c_0 x^{l_1} e^m$ and $p_2(x,y) = d_0 x^{l_2} e^m$. But $f(x,y) = x(A+y) - 1 = 0$ implies $x \neq 0$. In this case, p_1 and p_2 do not have a common root because either $c_0 \neq 0$ or $d_0 \neq 0$. Hence p_1, p_2, f do not have a common root and the transformation τ induces a bijective mapping of the common roots of p_1, p_2 onto the common roots of q_1, q_2. This concludes the proof of the theorem. □

More can be said about the Boneh/Durfee attack when L_{BD} is constructed using only x-shifts. In the experiments we carried out, an L^3-reduced basis of

L_{BD} always contained a vector v depending only on the basis vectors in X_1. As usually, we denote the basis vectors in X_1 by $X_{1,0}, X_{1,1}$. The vector v was of the form $d \cdot X_{1,0} + k \cdot X_{1,1}$, where d is the secret key and k is defined by $ed = 1 - k\frac{\phi(N)}{2}$. Hence, v alone reveals the secret key.

This is explained by the following theorem. Consider the lattice L spanned by the rows of the (2×2) lattice basis

$$B(1,0) = \begin{bmatrix} eX & 0 \\ AX & XY \end{bmatrix}.$$

Theorem 5. *If we choose $Y = e^{1/2}$ and $X = 2e^{1/4}$ for the basis $B(1,0)$, then the coefficients of the shortest lattice vector equal the secret parameters k and d provided $d < \frac{1}{3}N^{1/4}$.*

Proof: We only sketch the proof idea. Details are given in the full version of the paper.

We show that for the shortest vector $u = c_1 X_{1,0} + c_2 X_{1,1}$ the quotient c_1/c_2 is a convergent in the continued fraction expansion of A/e. Furthermore, c_1/c_2 is the last convergent of A/e whose denominator is less than $e^{1/4}$.

It can be shown using Wiener's argument, that d/k is also the last convergent of A/e with denominator less than $e^{1/4}$. This implies $c_1/c_2 = d/k$. □

6 Experiments

We implemented our new method and carried out several experiments on a Linux-PC with 550 MHz. The L^3 reduction was done using Victor Shoup's NTL library [18].

In every experiment, we found two vectors with norm smaller than $\frac{e^m}{\sqrt{w}}$. Interestingly in the experiments carried out, the reduced lattice basis contained not only two sufficiently small vectors, but all vectors in the L^3-reduced basis of L had about the same norm. This is a difference to the Boneh/Durfee lattice bases. The resultant of the corresponding polynomials was computed using the Maple computer algebra system. The resultant with respect to x was always a polynomial in y, and the root delivered the factorization of N. Our results compare well to those of Boneh and Durfee in the Eurocrypt paper [3]. Boneh/Durfee ran new experiments in [4], but used additional tricks to enlarge d by a few bits:

1. Lattice reduction with Schnorr's block reduction variant [16].
2. Use of Chebychev polynomials (a trick due to Coppersmith).
3. If $||p(xX, yY)|| < c \cdot e^m/\sqrt{w}$, we know $|p(x_0, y_0)| < c \cdot e^m$ and $p(x_0, y_0) = 0 \bmod e^m$. Hence, we can guess $\gamma \in (-c, c)$ such that $p(x, y) + \gamma e^m$ satisfies $p(x_0, y_0) + \gamma e^m = 0$ over $\mathbb{Z}$.

These tricks apply to our method as well, but we did not implement them. Comparing instances with the same bitsize of p, q and the same δ as in [3], our algorithm was several times faster due to the reduced lattice dimension. Table 1 contains several running times we obtained. Where availabe, we also included the corresponding running times as provided in [3] (these running times were achieved on a 400 MHz SUN workstation).

Table 1. Results of Experiments.

p, q	δ	m	t	w	our running time	running time in [4]
1000 bits	0.265	4	2	15	6 minutes	45 minutes
3000 bits	0.265	4	2	15	100 minutes	300 minutes
3000 bits	0.269	5	2	18	8 hours	—
500 bits	0.270	6	2	21	19 minutes	—
500 bits	0.274	8	3	36	300 minutes	—
500 bits	0.2765	10	4	55	26 hours	—
500 bits	0.278	11	5	72	6 days	—

In all examples we chose d uniformly with $\delta \log(N)$ bits, until $\log_N(d)$ was δ within precision at least 10^{-4}. The running time measures only the time for L^3-reduction. With growing m and t, the time for resultant computation can take longer than reducing the lattice basis $B(m, t)$.

Acknowledgement

We want to thank Glenn Durfee for pointing out the additional tricks in Section 6 and the anonymous referees for their helpful comments.

References

1. D. Bleichenbacher, "On the Security of the KMOV public key cryptosystem", Proc. of Crypto'97
2. D. Boneh, "Twenty years of attacks on the RSA cryptosystem", Notices of the AMS, 1999
3. D. Boneh, G. Durfee, "Cryptanalysis of RSA with private key d less than $N^{0.292}$", Proc. Eurocrypt'99
4. D. Boneh, G. Durfee, "Cryptanalysis of RSA with private key d less than $N^{0.292}$", IEEE Trans. on Information Theory, vol. 46(4), 2000
5. H. Cohen, "A Course in Computational Algebraic Number Theory", Springer Verlag, 1996
6. D. Coppersmith, "Small Solutions to Polynomial Equations, and Low Exponent RSA Vulnerabilities", Journal of Cryptology 10(4), 1997
7. D. Cox, J. Little, D. O'Shea, "Ideals, Varieties and Algorithms", Springer Verlag, 1992
8. G. Durfee, P. Nguyen, "Cryptanalysis of the RSA Schemes with Short Secret Exponent from Asiacrypt'99", Proc. of Asiacrypt 2000

9. M. Gruber, C.G. Lekkerkerker, "Geometry of Numbers", North-Holland, 1987
10. G.H. Hardy, E.M. Wright, "An Introduction to the Theory of Numbers", Oxford University Press, 1979
11. N. Howgrave-Graham, "Finding small roots of univariate modular equations revisited", Proc. of Cryptography and Coding, LNCS 1355, Springer-Verlag, 1997
12. C. Jutla, "On finding small solutions of modular multivariate polynomial equations", Proc. of Eurocrypt'98
13. A. Lenstra, H. Lenstra and L. Lovasz, "Factoring polynomials with rational coefficients", Mathematische Annalen, 1982
14. P. Nguyen, J. Stern, "Lattice Reduction in Cryptology: An Update", Algorithmic Number Theory Symposium ANTS-IV, 2000
15. R. Rivest, A. Shamir and L. Adleman, "A method for obtaining digital signatures and public key cryptosystems", Communications of the ACM, volume 21, 1978
16. C.P. Schnorr, "A hierarchy of polynomial time lattice basis reduction algorithms", Theoretical Computer Science, volume 53, 1987
17. C.L. Siegel, "Lectures on the Geometry of Numbers", Springer Verlag, 1989
18. V. Shoup, Number Theory Library (NTL), `http://www.cs.wisc.edu/~shoup/ntl`
19. E. Verheul, H. van Tilborg, "Cryptanalysis of less short RSA secret exponents", Applicable Algebra in Engineering, Communication and Computing, Springer Verlag, vol. 8, 1997
20. M. Wiener, "Cryptanalysis of short RSA secret exponents", IEEE Transactions on Information Theory, vol. 36, 1990

Finding Small Solutions to Small Degree Polynomials

Don Coppersmith

IBM Research, T.J. Watson Research Center
Yorktown Heights, NY 10598, USA
copper@watson.ibm.com

Abstract. This talk is a brief survey of recent results and ideas concerning the problem of finding a small root of a univariate polynomial mod N, and the companion problem of finding a small solution to a bivariate equation over $\mathbb{Z}$. We start with the lattice-based approach from [2, 3], and speculate on directions for improvement.
Keywords: Modular polynomials, lattice reduction.

1 Univariate Modular Polynomial

Our basic setup is a univariate polynomial $p(x)$ of small degree d, and a modulus N of unknown factorization. For a suitable bound B, we wish to find all integers x_0 such that $|x_0| < B$ and $p(x_0) = 0 \bmod N$. (Call such integers "*small roots*".) Our efforts will be concentrated in increasing the bound B.

An early paper in this line of research was [2], but the author was working with an unnatural space. Here we follow the more natural presentation of Howgrave-Graham [8].

For simplicity we assume p is monic, although we really only need that the gcd of its coefficients be relatively prime to N; see Remark 1 below. We set

$$p(x) = x^d + p_{d-1}x^{d-1} + \cdots + p_2x^2 + p_1x + p_0.$$

The first approach is essentially due to Håstad [7]: Consider first the collection C_1 of polynomials:

$$C_1 = \{x^i, 0 \le i < d\} \cup \{p(x)/N\}.$$

For each polynomial $q \in C_1$, for each small root x_0, we see that $q(x_0)$ is an integer. The same will be true of any integer linear combination of polynomials in C_1.

J.H. Silverman (Ed.): CaLC 2001, LNCS 2146, pp. 20–31, 2001.

So it makes sense to consider the lattice of dimension $d+1$ generated by the *columns* of the matrix

$$L_1 = \begin{bmatrix} 1 & 0 & 0 & \cdots & 0 & 0 & p_0/N \\ 0 & B & 0 & \cdots & 0 & 0 & p_1B/N \\ 0 & 0 & B^2 & \cdots & 0 & 0 & p_2B^2/N \\ & & & \vdots & & & \\ 0 & 0 & 0 & \cdots & B^{d-2} & 0 & p_{d-2}B^{d-2}/N \\ 0 & 0 & 0 & \cdots & 0 & B^{d-1} & p_{d-1}B^{d-1}/N \\ 0 & 0 & 0 & \cdots & 0 & 0 & 1B^d/N \end{bmatrix}.$$

Each column corresponds to a polynomial $q(x)$ in C_1, expressed in the basis $\{x^i/B^i\}$. So row i corresponds to the coefficient of x^i in $q(x)$, multiplied by a scaling factor B^i.

Now apply lattice basis reduction [13]. Because L_1 has dimension $d+1$ and determinant $N^{-1}B^{d(d+1)/2}$, we will find a nonzero vector $\mathbf{v}$ of length

$$|\mathbf{v}| < c_1(d)\,(\det L_1)^{1/(d+1)} = c_1(d)N^{-1/(d+1)}B^{d/2},$$

where $c_1(d)$ is a constant depending only on the dimension.

We interpret the vector $\mathbf{v} = [v_0, v_1B, v_2B^2, \ldots, v_dB^d]$ as a polynomial $v(x) = \sum v_i x^i$, again expressing $v(x)$ in the basis $\{x^i/B^i\}$.

Suppose we know that

$$c_1(d)\,(\det L_1)^{1/(d+1)} < \frac{1}{d+1},$$

or equivalently,

$$B \le c_1'(d)N^{2/[d(d+1)]},$$

where $c_1'(d)$ is another constant depending only on d. Then we also know that $|\mathbf{v}| < 1/(d+1)$, and each entry of $\mathbf{v}$ satisfies $|v_iB^i| < 1/(d+1)$. Evaluate the polynomial $v(x)$ at a small root x_0. On one hand

$$|v(x_0)| \le \sum |v_i x_0^i| \le \sum |v_iB^i| < \sum \frac{1}{d+1} = 1,$$

so that $|v(x_0)| < 1$. On the other hand, $v(x_0)$ is an integer. These two conditions together imply that $v(x_0) = 0 \in \mathbb{Q}$.

In summary: We have computed a polynomial $v(x)$ in $\mathbb{Q}[x]$ whose roots include all "small roots" x_0, that is, all those x_0 with $|x_0| < B = c_1'(d)N^{2/d(d+1)}$ and with $p(x_0) = 0 \bmod N$.

Incidentally, we have also bounded the number of small roots, by $\dim L_1$.

Remark 1: If $p(x)$ is not monic, but its content is relatively prime to N, we augment C_1 with x^d, that is, we replace C_1 by

$$C_1' = \{x^i, 0 \le i \le d\} \cup \{p(x)/N\}.$$

One checks that the corresponding lattice L_1' still has dimension $d+1$ and determinant $N^{-1}B^{d(d+1)/2}$, and the rest of the argument goes through.

Remark 2: Using the Cauchy-Schwarz inequality, one can replace the condition

$$c_1(d)\,(\det L_1)^{1/(d+1)} < \frac{1}{d+1}$$

by the weaker condition

$$c_1(d)\,(\det L_1)^{1/(d+1)} < \frac{1}{\sqrt{d+1}}.$$

Remark 3: Sometimes we scale thing differently, using $C_1'' = \{Nx^i, 0 \le i < d\} \cup \{p(x)\}$ and using the fact that for each $q \in C_1''$ and small root x_0 we have that $q(x_0)$ is a multiple of N. The two approaches are numerically equivalent, the only difference being esthetics.

2 Improvements in the Exponent

The first improvement comes when we consider a larger collection of polynomials. Define

$$C_2 = \{x^i, 0 \le i < d\} \cup \{(p(x)/N)x^i, 0 \le i < d\}.$$

The corresponding lattice L_2 has dimension $2d$ and determinant $N^{-d}B^{2d(2d-1)/2}$. The enabling condition becomes

$$c_2(d)\left(N^{-d}B^{2d(2d-1)/2}\right)^{1/(2d)} < \frac{1}{2d},$$

or equivalently

$$B \le c_2'(d)N^{1/(2d-1)}.$$

The exponent of N has improved from $2/d(d+1)$ to $1/(2d-1)$. The improvement came about because the dimension of L increased, while its determinant decreased.

We obtain a second improvement by considering higher powers of the modulus N, along with a still larger collection of polynomials. Fix a positive integer h. Define

$$C_3 = \{(p(x)/N)^j x^i, 0 \le i < d, 0 \le j < h\}.$$

For each polynomial $q(x) \in C_3$, for each small root x_0, we see that $q(x_0)$ is an integer. Again the same holds for any integer linear combination of polynomials in C_3.

The corresponding lattice L_3 has dimension dh and determinant

$$N^{-dh(h-1)/2}B^{(dh)(dh-1)/2}.$$

(The powers of N on the diagonal consist of d each of $N^0, N^{-1}, \ldots, N^{-(h-1)}$, while the powers of B are $B^0, B^1, \ldots, B^{dh-1}$.)

Our enabling equation is now:

$$\left[N^{-dh(h-1)/2}B^{(dh)(dh-1)/2}\right]^{1/(dh)} < c_3(d,h),$$

which will be satisfied if

$$B \le c_3'(d,h) N^{(h-1)/(dh-1)}.$$

The exponent of N, namely $\frac{h-1}{dh-1}$, differs from $\frac{1}{d}$ by $\frac{d-1}{d(dh-1)} < \frac{1}{dh}$; this difference can be made arbitrarily small by choosing h larger, at the expense of computational complexity. Put another way, we achieve a bound

$$B = c_3''(d,\epsilon) N^{(1/d)-\epsilon}$$

by choosing $h = O\left(\frac{1}{d\epsilon}\right)$. The running time is polynomial in $(d, 1/\epsilon, \log N)$.

We can extend the bound to $N^{1/d}$ by breaking the interval of size $2N^{1/d}$ into N^{ϵ} intervals of size $2N^{(1/d)-\epsilon}$. This is still polynomial time, but in practice it gets much more expensive as the exponent gets closer to $1/d$.

As before, we have bounded the number of small roots, as well as showing how to compute them all in polynomial time. The existential results match those of Konyagin and Steger [12], who bound the number of small roots by

$$O\left(\frac{1+\log B}{\log(1+B^{-1}N^{1/d})}\right).$$

For $B = N^{(1/d)-\epsilon}$ their bound is essentially $O\left(\frac{1}{d\epsilon}\right)$, while for $B = N^{(1/d)+\epsilon}$ their bound becomes $O\left(\frac{N^{\epsilon}\log N}{d}\right)$.

3 Minor Improvements

We can improve the lower order factor—the $c(d)$ factor in the estimate of the bound B—by more careful consideration of the process.

One idea, due independently to Hendrik Lenstra [15] and to Nick Howgrave-Graham [9], is to recognize that, for each positive integer k, the rational polynomial $b_k(x) = x(x-1)\cdots(x-k+1)/k!$ takes on integer values for all integer arguments x. So we augment (say) the collection of polynomials

$$C_3 = \{(p(x)/N)^j x^i, 0 \le i < d, 0 \le j < h\},$$

with the polynomials $b_k(x), 0 \le k < dh$, that is

$$C_3' = \{(p(x)/N)^j x^i, 0 \le i < d, 0 \le j < h\} \cup \{b_k(x), 0 \le k < dh\}.$$

(Assume here that N is free of small factors, that is, N is relatively prime to $(dh-1)!$.) Whereas L_3 could be represented by an upper triangular matrix whose kth diagonal element is $B^k N^{-\lfloor k/d \rfloor}$, one finds that L_3' can be represented by an upper triangular matrix whose kth diagonal element is $(1/k!)B^k N^{-\lfloor k/d \rfloor}$. This decreases $\det(L_3)$ by a factor

$$\prod_{0 \le k < dh} k!.$$

This allows us to increase B in compensation, by a factor

$$\left(\prod_{0\le k<dh} k!\right)^{2/[(dh)(dh-1)]} \approx \frac{dh}{e^{3/2}} \approx \frac{dh}{4.5}.$$

Phong Nguyen [17] reports that in practice this does speed up computations, by perhaps a factor of 5. Nguyen also remarks that one could further augment C_3' with the polynomials $\{b_j(p(x)/N)b_i(x)\}$, but that one does not thereby change the lattice L_3'.

Another idea, developed here in its explicit form but related to earlier work by Boneh [1], is probably less profitable. We have used the fact that the monomials $(x/B)^i$ are bounded by 1 when $|x| < B$. The Chebyshev polynomials [18] share that property, but more efficiently. These polynomials are defined by:

$$T_k(\cos\theta) = \cos(k\theta),$$

$$T_k(x) = 2^{k-1}x^k + \text{smaller terms}(k \ge 1).$$

Where we currently use the monomial basis—row i corresponds to $(x/B)^i$—to express $q(x) \in C$ as a column of L, we can instead use the Chebyshev basis—row i corresponds to $T_i(x/B)$. This will decrease $\det(L)$ by a factor

$$2^{0+0+1+2+\cdots+(dh-2)} = 2^{(dh-1)(dh-2)/2},$$

leading to an increase in B by a factor of

$$2^{(dh-2)/(dh)} \approx 2.$$

The two ideas can be applied simultaneously, and the improvements accumulate. But they increase B by only a polynomial factor, and therefore improve running time only by that factor. The same effect could be achieved by solving several polynomials $p(x-2iB), |i| \le k/2$ over the range $|x| < B$ and using the result to solve the single polynomial $p(x)$ over the larger range $|x| < kB$.

4 Speculative Improvement of Exponent

Can we improve the asymptotic bound $B = N^{1/d}$? The bound is a natural-looking bound, and it matches the existential results of Konyagin and Steger [12]. Indeed, even in the simple case $p(x) = x^3 - A \pmod N$, we don't know an efficient way of finding roots x larger than $B = N^{1/3}$, while those smaller than $B = N^{1/3}$ are trivially found by solving $x^3 - A = 0$ over the integers.

The following example gives cause for pessimism. Set $N = q^3$ with q prime, and set $p(x) = x^3 + Dqx^2 + Eq^2x$ with $D, E \in \mathbb{Z}$. Clearly if x_0 is any multiple of q then $p(x_0) = 0 \pmod N$. So if we select a bound $B = N^{(1/3)+\epsilon}$, the number of "small roots" x_0 with $|x_0| < B$ and $p(x_0) = 0 \pmod N$ is about $2N^\epsilon$, i.e. exponentially many. (Again this essentially matches the bound of Konyagin and Steger [12].) Our lattice techniques cannot hope to find them, since in our setup all the small roots are roots of $v(x)$, so that the number of small roots needs to be bounded by $\dim(L)$.

More generally, we can expect trouble whenever $q^2|N$ and $p(x)$ has a repeated root modulo q. (We don't know whether this family contains all the polynomials with an exponentially large number of roots smaller than $N^{(1/d)+\epsilon}$.) When this happens, we know that N shares a common factor with the discriminant of $p(x)$,

$$\mathrm{Discr}(p) = \mathrm{Res}(p, p'),$$

$$\gcd(N, \mathrm{Discr}(p)) > 1.$$

So any improvement of the exponent past $1/d$ must somehow rule out this case.

With this in mind, we hypothesize a *"Discriminant Attack"*:

Suppose we can guarantee that we are never in the unfavorable situation. We can demand that $\gcd(N, \mathrm{Discr}(p)) = 1$. Equivalently, we can demand existence of $D(x), E(x) \in \mathbb{Z}[x]$ and $F \in \mathbb{Z}$ satisfying

$$D(x)p(x) + E(x)p'(x) + FN = 1;$$

if $\gcd(N, \mathrm{Discr}(p)) = 1$, then $D(x), E(x), F$ exist and are easily computed. Perhaps D, E, F can be incorporated into the construction of the lattice L, in such a way that the bound B can be improved to $N^{(1/d)+\epsilon}$. But I don't see how to do it.

A related effort is the *"Divided Difference Attack"*:

Suppose we know that there are *two* small roots x, y, which differ modulo each prime factor of N; that is, $\gcd(N, x - y) = 1$. Then besides the modular equations of degree d,

$$p(x) = 0 \pmod{N},$$

$$p(y) = 0 \pmod{N},$$

we get a third equation of total degree $d - 1$:

$$r(x, y) = \frac{p(x) - p(y)}{x - y} = 0 \pmod{N}.$$

(Despite its appearance, $r(x, y)$ is actually an integer polynomial.)

We are now dealing with *bivariate* modular polynomials. As mentioned in [2] and in Section 6 below, our techniques can sometimes handle this, but there are no guarantees. Let's try an example and see.

Select a positive integer h. Let the family of polynomials be

$$C = \{(p(x)/N)^k (r(x, y)/N)^\ell x^i y^m, k + \ell \le h, 0 \le i < d, 0 \le m < d - 1\}.$$

(We are not using $p(y)$ because it is already generated by $p(y) = p(x) + (y - x)r(x, y)$.) The polynomials of C are related to their monomial basis, which is *nearly*

$$\left\{x^a y^b;\ \ a, b \ge 0;\ \ \frac{a}{d} + \frac{b}{d-1} \le h + 1\right\}.$$

where B_x, B_y are the bounds on $|x|$ and $|y|$. The number of such monomials is roughly the area of a triangle, namely $A = [d(h+1)][(d-1)(h+1)]/2$. (We are ignoring inaccuracies due to edge effects near the hypotenuse of the triangle.) Associate with each polynomial $q \in C$ the monomial corresponding to its leading term (in reverse lexicographical order). Estimate that the average exponent x (in the monomial basis) is about $d(h+1)/3$, so that the average exponent of $p(x)$ (among $q \in C$) is about $h/3$. Similarly the average exponent of $r(x,y)$ is about $h/3$, and the average exponent of N is about $-2h/3$. Build the lattice L as before, incorporating the scaling factors B_x, B_y (the bounds on $|x_0|$ and $|y_0|$). We estimate the determinant of L as

$$B_x^{Ad(h+1)/3} B_y^{A(d-1)(h+1)/3} N^{-2Ah/3}.$$

The "enabling condition" then becomes

$$B_x^{Ad(h+1)/3} B_y^{A(d-1)(h+1)/3} N^{-2Ah/3} < c,$$

$$B_x^{d(h+1)} B_y^{(d-1)(h+1)} < c' N^{2h},$$

and in the limit of large h with d held fixed,

$$B_x^d B_y^{d-1} < c'' N^2.$$

If the enabling condition is satisfied, we will obtain an equation $v_1(x,y)$ in $\mathbb{Z}[x,y]$ relating x_0 and y_0 for all small pairs (x_0, y_0) satisfying our original modular equations p and r. But a single equation is not enough to solve for x_0 and y_0. We have to hope that two independent equations are generated. Indeed, some work by Charanjit Jutla [11] indicates that under certain conditions we can guarantee that at least two equations $v_1(x,y), v_2(x,y)$ will be generated, both with coefficients small enough that they hold in $\mathbb{Z}$. If v_1, v_2 are algebraically independent, then we can solve them by using the resultant: $u(y) = \mathrm{Res}(v1, v2; x) \in \mathbb{Z}[y]$ is a univariate equation in $\mathbb{Z}$ whose roots contain all y_0 that participate in any small pair (x_0, y_0) of interest. For each y_0 we can then easily find all the corresponding x_0.

But we cannot always guarantee that the two equations will be algebraically independent. One can be a multiple of the other, in which case the resultant will be 0 and we will learn nothing.

Let's examine the limit of the enabling equation:

$$B_x^d B_y^{d-1} < c'' N^2.$$

If either root were below $N^{1/d}$, the standard methods would find it. But the present method may work when both roots are in the narrow range:

$$N^{1/d} < |x_0|, |y_0| < N^{1/(d-1)},$$

so that we obtain a slight advantage.

We have tried to abuse this method to obtain information that should otherwise be hard to get, and we always fail. Here are some examples.

Suppose we know a root x_0 with $|x_0| < N^{1/d}$, and we want to find a second small root y_0 with $\gcd(N, x_0 - y_0) = 1$. The straightforward approach would be to divide $p(y)$ by $y - x_0 \pmod N$ to obtain a polynomial of degree $d-1$. Our usual method will solve this as long as $|y_0| < N^{1/(d-1)}$. But using the bivariate method, we might expect to be able to find y_0 as long as

$$|x_0^d y_0^{d-1}| < c'' N^2.$$

Since $|x_0| < N^{1/d}$, it seems superficially that we can allow $|y_0| > N^{1/(d-1)}$ and still satisfy our condition. But when we try it, the equations $v_i(x, y) = 0$ that we recover always involve multiples of $x - x_0 = 0$, giving no information about y.

As a second example, suppose $p(x)$ is of degree 2, and we are told there are two independent small roots x_0 and y_0, both of size about $N^{2/3}$. So

$$p(x) = x^2 + Ax + B = 0 \pmod N.$$

Since the roots satisfy

$$|x_0^d y_0^{d-1}| = |x_0^2 y_0^1| \approx N^2,$$

the present method should apply. But again the equations it gives are useless: multiples of

$$x_0 + y_0 + A = 0,$$

which we could have gotten from the original equation by inspection. Since both roots are small, A is also bounded by about $N^{2/3}$, and the equation $x_0 + y_0 = -A$ can be taken to hold in $\mathbb{Z}$.

But there may exist situations where this bivariate approach gives answers that we could not otherwise obtain.

Remark 3: One relation between the "discriminant attack" and the "divided difference attack" is the usage of $p'(x)$ in the former case and $\frac{p(x)-p(y)}{x-y}$ in the latter case. Where the divided difference attack treats the circumstance that there *exist* two (small) roots that differ modulo each prime factor of N, the discriminant attack demands that there *not exist* repeated roots, or in some sense that all roots are different mod q.

5 Bivariate Integer Polynomials

The present author [3] applied techniques similar to those of his other paper [2], to the problem of finding a small solution to a bivariate integer polynomial

$$p(x, y) = 0 \in \mathbb{Z}.$$

The primary application was to integer factorization when half the bits of one of the factors are known.

The presentation in [3] is difficult to understand. Once again Howgrave-Graham's presentation makes it more accessible, but this simplification seems to only apply to the specific equation describing integer factorization (and some related equations), and not to the general bivariate integer polynomial.

For integer factorization, suppose we have an integer N of unknown factorization $N = PQ$, where we have some approximation to the factors P and Q. We can write

$$N = (P' + x_0)(Q' + y_0)$$

where x_0, y_0 are small. If $P = N^{\beta}$ and $Q = N^{1-\beta}$, we will be able to solve this as long as

$$|x_0| < B_x = N^{\beta^2}, |y_0| < B_y = N^{(1-\beta)^2}.$$

Notice that in this setup, we know the $\beta(1-\beta)\log N$ high order bits of P, which is equivalent to knowing the $\beta(1-\beta)\log N$ high order bits of Q.

Select positive integers h, k with $h < k$. Define the family of polynomials

$$C = \{N^{h-i}(P' + x)^i, 0 \le i < h\} \cup \{x^{i-h}(P' + x)^h, h \le i < k\}.$$

These polynomials, when evaluated at $x = x_0$, are all multiples of the (unknown) integer P^h.

The corresponding lattice L has dimension k and determinant $N^{h(h+1)/2} B_x^{k(k-1)/2}$. The enabling equation is then

$$\left(N^{h(h+1)/2} B_x^{k(k-1)/2}\right)^{1/k} \le cP^h = cN^{\beta h},$$

$$B_x \le c' N^{h(2k\beta - h - 1)/[k(k-1)]}.$$

For large h, k, we optimize this by selecting $h = k\beta$, obtaining

$$B_x \approx N^{\beta^2}.$$

The rest of the development is similar to the univariate modular case. We find an equation in $\mathbb{Z}[x]$, whose roots include the root x_0 of interest.

Howgrave-Graham [10] develops these techniques even further, applied to the problem of an "approximate gcd", finding a gcd when the inputs are only approximately known.

The same technique can be applied, almost without change, to the problem of divisors in residue classes. Hendrik Lenstra [14] showed that if $0 < r < s < N$ are given positive integers, pairwise relatively prime, with $s > N^{\alpha}$ and $\alpha > 1/4$, the number of divisors of N equivalent to r mod s is bounded by a function of α, independent of r, s, N. Applying the present techniques we not only recover that existential bound but actually construct those divisors in polynomial time. [5]

For more general bivariate integer equations, the reader is referred to the author's earlier work [3], [4], where the development is less intuitive but handles a more general situation.

The reader may also enjoy the more recent work of Elkies [6], the techniques being closely related. Elkies treats a more general setting, where instead of lattice points *on* a curve, he is looking for lattice points *near* a curve. He finds much wider applicability, including non-algebraic curves.

6 Possible (and Impossible) Extensions

We can sometimes extend these techniques, to bivariate modular equations or to multivariate integer equations. But the extensions are not guaranteed to work, and in fact there are impossibility results that argue against their application to the general case.

Consider the bivariate modular case. As in the example above, we can easily build a lattice consisting of multiples of N and of $p(x,y)$ (or of their powers), and we can find a short vector in that lattice, corresponding to a polynomial in $\mathbb{Z}$ satisfied by all small roots. But this polynomial will, in general, be difficult to solve.

As in the example, we can hope to find two short vectors, corresponding to two polynomials, and we can hope that they are algebraically independent, so that taking their resultant we can recover a single polynomial in a single variable, whose roots include all those y_0 belonging to a short pair of roots (x_0, y_0). The trouble is that although we can arrange things so that two short vectors will be found, we cannot in general guarantee that the corresponding polynomials will be algebraically independent.

We start with the following theorem of Manders and Adleman [16]:

Theorem 1. *(Manders and Adleman) The (problem of accepting the) set of Diophantine equations (in a standard binary encoding) of the form*

$$\alpha x_1^2 + \beta x_2 - \gamma = 0$$

which have natural-number solutions x_1, x_2 is NP-complete.

Manders and Adleman go on to remark that the problem remains NP-complete even when β is given in fully factored form.

We need to make minor adjustments to use this theorem. Let us first center the range of x_2. Let δ approximate half of its range:

$$\delta = \left\lfloor \frac{\gamma}{2\beta} + \frac{1}{2} \right\rfloor,$$

and define

$$x = x_1,$$

$$y = x_2 - \delta,$$

$$\tau = \gamma - \beta\delta.$$

Then we are asking for existence of solutions (x_0, y_0) to

$$\alpha x^2 + \beta y - \tau = 0,$$

with $|x_0| < B_x \approx \sqrt{\gamma/\alpha}$ and $|y_0| < B_y \approx \delta$. (Clearly if we can compute all small solutions (x_0, y_0) within these bounds, we can decide whether exact solutions to the original problem exist.)

Now select N arbitrarily large, as long as N exceeds $|\alpha B_x^2| + |\beta B_y| + |\tau|$. Given the bivariate modular equation

$$\alpha x^2 + \beta y - \tau = 0 \pmod N,$$

and bounds B_x, B_y as before, it will be hard to decide whether there are small solutions (x_0, y_0); the reduction mod N is meaningless. Now, the bounds B_x, B_y stay fixed as N grows arbitrarily large.

Recall that in the univariate modular case, the allowable bound B_x grew with the $1/d$ power of N. In the bivariate modular case we cannot hope to find a similar theorem. The achievable bounds cannot grow as N grows.

Our method will derive, from the bivariate modular equation

$$\alpha x^2 + \beta y - \tau = 0 \pmod N,$$

a bivariate integer equation, namely

$$\alpha x^2 + \beta y - \tau = 0.$$

But it cannot enable us to solve either one.

Exactly the same example shows that *trivariate integer* equation

$$\alpha x^2 + \beta y - \tau - zN = 0,$$

is difficult to solve with bounds B_x, B_y as before, and $B_z = 2$. In the work on bivariate integer equations [3], the bounds grew with the coefficients of $p(x, y)$ (in a complicated way that depended on the degree of p), and because we have an arbitrarily large coefficient N here, again we cannot hope to achieve a similar theorem in the trivariate integer case.

But these negative results should not dissuade us. As Jutla and others have shown, many times one *can* use the multivariate versions of the present techniques. A tool that has been shown to be ineffective in one percent of the cases, can still be quite useful in the other 99 percent.

Acknowledgments

The author is grateful to Joe Silverman and Phong Nguyen for prodding him to actually write down the ideas that have been floating around. Conversations with Nick Howgrave-Graham are always informative and stimulating.

References

1. Dan Boneh, personal communication.
2. D. Coppersmith, Finding a small root of a univariate modular equation. *Advances in Cryptology - EUROCRYPT'96*, LNCS 1070, Springer, 1996, 155-165.
3. D. Coppersmith, Finding a small root of a bivariate integer equation; factoring with high bits known, *Advances in Cryptology - EUROCRYPT'96*, LNCS 1070, Springer, 1996, 178-189.

4. D. Coppersmith, Small solutions to polynomial equations, and low exponent RSA vulnerabilities. *J. Crypt.* **vol 10 no 4** (Autumn 1997), 233-260.
5. D. Coppersmith, N.A. Howgrave-Graham, S.V. Nagaraj, Divisors in Residue classes—Constructively. Manuscript.
6. N. Elkies, Rational points near curves and small nonzero $|x^3 - y^2|$ via lattice reduction, *ANTS-4*, LNCS vol 1838 (2000) Springer Verlag, 33-63.
7. J. Håstad, On using RSA with low exponent in a public key network, *Advances in Cryptology - CRYPTO'85*, LNCS 218, Springer-Verlag, 1986, 403-408.
8. N.A. Howgrave-Graham, Finding small solutions of univariate modular equations revisited. *Cryptography and Coding* LNCS vol 1355. (1997) Springer-Verlag. 131-142.
9. N.A. Howgrave-Graham, personal communication, 1997.
10. N.A. Howgrave-Graham, Approximate Integer Common Divisors, This volume, pp. 51–66.
11. C.S. Jutla, On finding small solutions of modular multivariate polynomial equations, *Advances in Cryptology - EUROCRYPT'98*, LNCS 1403, Springer, 1998, 158-170.
12. S.V. Konyagin and T. Steger, On polynomial congruences, *Mathematical Notes* **Vol 55** No 6 (1994), 596-600.
13. A.K. Lenstra, H.W. Lenstra, and L. Lovasz, Factoring polynomials with rational coefficients, *Math. Ann.* **261** (1982), 515-534.
14. H. W. Lenstra, Jr., "Divisors in Residue Classes," *Mathematics of Computation*, volume 42, number 165, January 1984, pages 331-340.
15. H.W. Lenstra, personal communication.
16. K.L. Manders and L.M. Adleman, NP-Complete Decision Problems for Binary Quadratics. *JCSS* **16**(2), 1978, 168-184.
17. Phong Nguyen, personal communication.
18. T.J. Rivlin, *Chebyshev Polynomials, From Approximation Theory to Algebra and Number Theory*, Wiley (1990).

Fast Reduction of Ternary Quadratic Forms

Friedrich Eisenbrand[1] and Günter Rote[2]

[1] Max-Planck-Institut für Informatik
Stuhlsatzenhausweg 85, 66123 Saarbrücken, Germany
eisen@mpi-sb.mpg.de
[2] Institut für Informatik, Freie Universität Berlin
Takustraß e 9, 14195 Berlin, Germany
rote@inf.fu-berlin.de

Abstract We show that a positive definite integral ternary form can be reduced with $O(M(s)\log^2 s)$ bit operations, where s is the binary encoding length of the form and $M(s)$ is the bit-complexity of s-bit integer multiplication.
This result is achieved in two steps. First we prove that the the classical Gaussian algorithm for ternary form reduction, in the variant of Lagarias, has this worst case running time. Then we show that, given a ternary form which is reduced in the Gaussian sense, it takes only a constant number of arithmetic operations and a constant number of binary-form reductions to fully reduce the form.
Finally we describe how this algorithm can be generalized to higher dimensions. Lattice basis reduction and shortest vector computation in fixed dimension d can be done with $O(M(s)\log^{d-1} s)$ bit-operations.

1 Introduction

A *positive definite integral quadratic form* F, or *form* for short, is a homogeneous polynomial

$$F(X_1,\dots,X_d) = (X_1,\dots,X_d)\,A\,(X_1,\dots,X_d)^{\mathrm{T}},$$

where $A \in \mathbb{Z}^{d\times d}$ is an integral positive definite matrix, i.e., $A = A^{\mathrm{T}}$ and $x^{\mathrm{T}}Ax > 0$ for all $x \neq 0$. The study of forms is a fundamental topic in the geometry of numbers (see, e.g., [2]). A basic question here is: Given a form F, what is the minimal nonzero value $\lambda(F) = \min\{\,F(x_1,\dots,x_d) \mid x \in \mathbb{Z}^d,\ x \neq 0\,\}$ of the form which is attained at an integral vector? This problem will be of central interest in this paper.

Problem 1. Given a form F, compute $\lambda(F)$.

At least since Lenstra's [9] polynomial algorithm for integer programming in fixed dimension, the study of quadratic forms has also become a major topic in theoretical computer science. Here, one is interested in the lattice variant of Problem 1, which is: Given a basis of an integral lattice, find a shortest nonzero vector of the lattice w.r.t. the ℓ_2-norm.

J.H. Silverman (Ed.): CaLC 2001, LNCS 2146, pp. 32–44, 2001.

In fixed dimension, Problem 1 can be quickly solved if F is *reduced* (see Theorem 4 in Section 5). In our setting, this shall mean that the product of the diagonal elements of A satisfies

$$\prod_{i=1}^{d} a_{ii} \leq \gamma_d \, \Delta_F \tag{1}$$

for some constant γ_d depending on the dimension d only. Here $\Delta_F = \det A$ is the *determianant* of the form F. Algorithms which transform a form F into an equivalent reduced form are called *reduction algorithms.*

In algorithmic number theory, the cost measure that is widely used in the analysis of algorithms is the number of required *bit operations.* The famous *LLL algorithm* [8] is a reduction algorithm which has polynomial running time, even in varying dimension. In fixed dimension, the LLL reduction algorithm reduces a form F of binary encoding size s with $O(s)$ arithmetic operations on integers of size $O(s)$. This amounts to $O(M(s)\, s)$ bit-operations, where $M(s)$ is the bit-complexity of s-bit integer multiplication. If one plugs in the current record for $M(s) = O(s \log s \log\log s)$ [11], this shows that a form F can be reduced with a close to quadratic amount of bit-operations.

A form in two variables is called a *binary form.* Here one has asymptotically fast reduction algorithms. It was shown by Schönhage [10] and independently by Yap [16] that a binary quadratic form can be reduced with $O(M(s) \log s)$ bit-operations, see also Eisenbrand [3] for an easier approach.

In his famous *disquisitiones arithmeticae* [4], Gauß provided a "reduction algorithm" for forms in three variables, called *ternary forms.* He showed how to compute a ternary form, equivalent to a given form, such that the first diagonal element of the coefficient matrix is at most $\frac{4}{3}\sqrt[3]{\Delta_F}$. A form which is reduced in the Gaussian sense is not necessarily reduced in the sense of (1). The Gaussian notion of reduction was modified by Seeber [13] such that a reduced form satisfies (1) with $\gamma_3 = 3$. Gauß [5] showed later that $\gamma_3 = 2$.

The "reduction algorithm" of Gauß was modified by Lagarias [7] to produce so called *quasi-reduced* forms. They satisfy the slightly weaker condition that the first diagonal element is at most twice the cubic root of the determinant. Lagarias proved that his modified ternary form algorithm runs in polynomial time. However, a quasi-reduced form is not necessarily reduced in the sense of (1).

Results. We prove that ternary forms can be reduced with a close to linear amount of bit-operations, as it is the case for binary forms. More precisely, a ternary form F of binary encoding length s can be reduced in the sense of (1) with $\gamma_3 = \frac{16}{3}$ using $O(M(s) \log^2 s)$ bit-operations. Unfortunately, the complexity of the proposed reduction procedure has still an extra ($\log s$)-factor compared to the complexity of binary form reduction. However our result largely improves on the $O(M(s)\, s)$ complexity of algorithms for ternary form reduction which are based on the LLL algorithm.

We proceed as follows. First we show that the Gaussian ternary form algorithm, in the variant of Lagarias [7], requires $O(M(s) \log^2 s)$ bit-operations.

This is achieved via a refinement of the analysis given by Lagarias. Then we prove that, given a quasi-reduced ternary form, it takes at most $O(M(s)\log s)$ bit-operations to compute an equivalent reduced form. Therefore, a ternary form can be reduced with $O(M(s)\log^2 s)$ bit-operations. This improves on the best previously known algorithms. It follows that, for ternary forms, Problem 1 can be solved with $O(M(s)\log^2 s)$ bit-operations.

Finally we generalize the described algorithm to any fixed dimension d. The resulting lattice basis reduction algorithm requires $O(M(s)\log^{d-1} s)$ bit-operations.

Related Work. Apart from the already mentioned articles, three-dimensional lattice reduction was extensively studied by various authors. Vallée [15] invented a generalization of the two-dimensional Gaussian algorithm in three dimensions. Vallée's algorithm requires $O(M(s)\,s)$ bit-operations. Semaev [14] provides an algorithm for three-dimensional lattice basis reduction which is based on pair reduction. The running time of his algorithm is $O(s^2)$ bit-operations even if one uses the naive quadratic methods for integer multiplication and division. This matches the complexity of the Euclidean algorithm for the greatest common divisor.

2 Preliminaries and Notation

The letters $\mathbb{Z}$ and $\mathbb{Q}$ denote the integers and rationals respectively. The running times of algorithms are always given in terms of the binary encoding length of the input data. The cost measure is the amount of *bit operations*. The function $M(s)$ denotes the bit-complexity of s-bit integer multiplication. All basic arithmetic operations can be done in time $O(M(s))$ [1].

We will only consider positive definite integral quadratic forms. We identify a form F with its *coefficient matrix* $M_F \in \mathbb{Z}^{d\times d}$ such that

$$F(X_1,\dots,X_d) = (X_1,\dots,X_d)\; M_F\; (X_1,\dots,X_d)^{\mathrm{T}}.$$

The function size(F) denotes the binary encoding length of M_F. Two forms F and G are *equivalent* if there exists a unimodular matrix $U \in \mathbb{Z}^{d\times d}$ with $M_G = U^{\mathrm{T}} M_F U$. We say that U *transforms* F into G. The number $\Delta_F = \det M_F$ is the determinant of the form. The determinant is invariant under equivalence. See, e.g., [2] for more on the theory of quadratic forms. The coefficient matrix $M_F \in \mathbb{Z}^{d\times d}$ has a unique $R^{\mathrm{T}}DR$ factorization, i.e, a factorization $M_F = R^{\mathrm{T}}DR$, where $R \in \mathbb{Q}^{d\times d}$ is an upper triangular matrix with ones on the diagonal and D is a diagonal matrix. The matrix R has a unique *normalization* $R' = RU$, where U is unimodular and R' is upper triangular with ones on the diagonal and elements above the diagonal in the range $(-\frac{1}{2},\frac{1}{2}]$. The corresponding matrix $R'^{\mathrm{T}}DR'$ defines a form F' which is equivalent to F. The form F' is called the *Gram-Schmidt normalization* of F. This is the normalization step of the LLL algorithm [8], translated into the language of quadratic forms. In fixed dimension, the Gram-Schmidt normalization of a form F of size s can be computed with a constant number of arithmetic operations, and hence with $O(M(s))$ bit-operations. We say that a form G is a *γ-reduction* of F, if G is equivalent to F and if the product of the diagonal elements of M_G is at most $\gamma\,\Delta_F$.

2.1 Binary Forms

A *binary form* is a form in two variables. We denote binary forms with lower case letters f or g. The binary form f is *reduced* if $M_f = \left(\begin{smallmatrix} a_{11} & a_{12} \\ a_{12} & a_{22} \end{smallmatrix}\right)$ satisfies

$$a_{11} \leq a_{22} \tag{2}$$

$$|a_{12}| \leq \tfrac{1}{2} a_{11}. \tag{3}$$

If f is reduced one has

$$\tfrac{3}{4}\, a_{11}\, a_{22} \leq \Delta_f. \tag{4}$$

The unimodular matrix $\left(\begin{smallmatrix} 1 & -r \\ 0 & 1 \end{smallmatrix}\right)$, where r is the nearest integer to $\frac{a_{12}}{a_{11}}$, transforms a binary form f to an equivalent form which is called the *normalization* of f. The normalization of f satisfies (3).

We have the following result of Schönhage [10] and Yap [16].

Theorem 1. *Given a positive definite integral binary quadratic form f of size s, one can compute with $O(M(s)\log s)$ bit-operations an equivalent reduced form g and a unimodular matrix $U \in \mathbb{Z}^{2\times 2}$ which transforms f into g.* □

2.2 Ternary Forms

Ternary forms will be denoted by capital letters F or G. Let F be given by its coefficient matrix

$$M_F = \begin{pmatrix} a_{11} & a_{12} & a_{13} \\ a_{12} & a_{22} & a_{23} \\ a_{13} & a_{23} & a_{33} \end{pmatrix}.$$

The form F defines *associated binary forms* f_{ij}, $1 \leq i, j \leq 3$, $i \neq j$ which have coefficient matrix

$$M_{f_{ij}} = \begin{pmatrix} a_{ii} & a_{ij} \\ a_{ij} & a_{jj} \end{pmatrix}.$$

By *reducing f_{ij} in F*, we mean that we compute the unimodular transformation which reduces f_{ij} and apply it to the whole coefficient matrix M_F. This changes only the i-th and j-th row and column of M_F and leaves the third diagonal element a_{kk} unchanged. It follows from Theorem 1 that such a reduction of f_{ij} in F can be done with $O(M(s)\log s)$ bit-operations on forms F of size s.

The *adjoint* F^* of F is defined by the coefficient matrix $M_{F^*} = \det M_F \cdot M_F^{-1}$ and we write

$$M_{F^*} = \begin{pmatrix} A_{11} & A_{12} & A_{13} \\ A_{12} & A_{22} & A_{23} \\ A_{13} & A_{23} & A_{33} \end{pmatrix}.$$

Clearly M_{F^*} is integral and positive definite. A unimodular matrix $S \in \mathbb{Z}^{3\times3}$ transforms F into G if and only if $(S^{\mathrm{T}})^{-1}$ transforms F^* into G^*. The associated binary forms of F^* are denoted by f^*_{ij} and by *reducing* such an associated form *in* F we mean that we apply the corresponding reduction operations on F. Notice that $\mathrm{size}(F^*) = O(\mathrm{size}(F))$ and $\mathrm{size}(F) = O(\mathrm{size}(F^*))$ and that $\Delta_{F^*} = \Delta_F^2$.

The ternary form F is *quasi-reduced* (see [7, p. 162]) if

$$a_{11} \le 2\sqrt[3]{\Delta_F} \tag{5}$$

$$A_{33} \le 2\sqrt[3]{\Delta_F^2} \tag{6}$$

$$|a_{12}| \le \tfrac{1}{2}\, a_{11} \tag{7}$$

$$|A_{13}| \le \tfrac{1}{2}\, A_{33} \tag{8}$$

$$|A_{23}| \le \tfrac{1}{2}\, A_{33}. \tag{9}$$

This notion is a relaxation of Gauß ' concept of reduction of ternary forms, which has the constant 4/3 instead of 2 in (5–6).

3 Computing a Quasi-reduced Ternary Form

The Gaussian algorithm [4, Arts. 272–275] for ternary form "reduction" proceeds by iteratively reducing the associated binary forms f_{12} and f^*_{32} in F. Lagarias [7] modified the algorithm by keeping the entries above and below the diagonal of the intermediate forms small so that (7–9) are fulfilled after every iteration. So we only have to see that (5) and (6) are fulfilled. One iterates until

$$A_{33} < 2\sqrt[3]{\Delta_F^2}. \tag{10}$$

In the following we prove that the number of iterations until a ternary form F of size s satisfies (10) is $O(\log s)$. For F and its adjoint F^* one has

$$\begin{aligned} A_{33} &= \Delta_{f_{12}} \\ a_{11}\Delta_F &= \Delta_{f^*_{32}}. \end{aligned} \tag{11}$$

Thus reducing f_{12} in F leaves A_{33} unchanged and reducing f^*_{32} in F leaves a_{11} unchanged. Furthermore, after reducing f_{12} in F one has

$$a_{11} \le \sqrt{\tfrac{4}{3}\, A_{33}} \tag{12}$$

by (11), (2) and (4). Similarly, after reducing f^*_{32} in F one has

$$A_{33} \le \sqrt{\tfrac{4}{3}\, a_{11}\Delta_F}. \tag{13}$$

This shows that each iteration decreases the binary encoding length of A_{33} by roughly a factor of 4 as long as A_{33} exceeds $\sqrt[3]{\Delta_F^2}$ by a large amount. We make this observation more precise.

Let $A_{kl}^{(i)}$ denote the coefficients of F^* after the i-th iteration of this procedure. By combining (12) and (13) we get the following relation (see [7, p. 166, (4.65)])

$$A_{33}^{(i+1)} \leq (\tfrac{4}{3})^{(3/4)} \sqrt{\Delta_F}\, (A_{33}^{(i)})^{1/4}. \tag{14}$$

Lagarias then remarks that, if $A_{33}^{(i)} \geq 2\sqrt[3]{\Delta_F^2}$, then

$$A_{33}^{(i+1)} \leq (\tfrac{2}{3})^{3/4} A_{33}^{(i)} \tag{15}$$

and it follows that the number of iterations is bounded by $O(s)$. Lagarias does not take full advantage of (14). By rewriting (14) in the form

$$\frac{A_{33}^{(i+1)}}{\frac{4}{3}\Delta_F^{2/3}} \leq \sqrt[4]{\frac{A_{33}^{(i)}}{\frac{4}{3}\Delta_F^{2/3}}},$$

we see that we can achieve

$$\frac{A_{33}^{(i+1)}}{\frac{4}{3}\Delta_F^{2/3}} \leq 2$$

in at most

$$i = \log_4 \log_2 \left[A_{33}^{(0)}/(\tfrac{4}{3}\Delta_F^{2/3})\right] \leq \log_4 \log_2 A_{33}^{(0)} = O(\log s)$$

iterations. After we have achieved $A_{33}^{(i)} \leq \frac{8}{3}\sqrt[3]{\Delta_F^2}$, then, by (15), the modified ternary form algorithm requires at most one additional iteration to obtain an equivalent quasi-reduced form.

This shows that the modified ternary form algorithm requires $O(\log s)$ iterations to quasi-reduce a ternary form of size s. If one iteration of the reduction algorithm is performed with the fast reduction algorithm for binary forms one obtains the following result.

Theorem 2. *The modified ternary form reduction method reduces a ternary form of size s in $O(M(s)\log^2 s)$ bit-operations.*

Proof. Lagarias proves that the sizes of the intermediate ternary forms are $O(s)$. We have seen that the number of iterations is $O(\log s)$. One iteration requires $O(M(s)\log s)$ bit-operations if one uses the fast reduction for binary forms. □

4 From Quasi-reduced to Reduced

A quasi-reduced form (or a form which is reduced in the sense of Gauß) is not necessarily reduced. For example, the form F given by

$$M_F = \begin{pmatrix} 4x & 2x & 0 \\ 2x & x+1 & 0 \\ 0 & 0 & 2x^2 \end{pmatrix}, \quad M_{F^*} = \begin{pmatrix} 2x^3+2x^2 & -4x^2 & 0 \\ -4x^2 & 8x^3 & 0 \\ 0 & 0 & 4x \end{pmatrix}$$

with $\Delta_F = 8x^3$ is quasi-reduced, but it is far from being reduced, for $x \to \infty$.

In this section we show that we can compute a $\frac{16}{3}$-reduction of a quasi-reduced ternary form F with $O(M(s) \log s)$ bit-operations.

The following lemma states that, if F has two small entries on the diagonal which belong to an associated reduced binary form, then the Gram-Schmidt normalization of F is reduced.

Lemma 1. *Let F be a ternary form such that f_{12} is reduced and $a_{11}, a_{22} \leq \kappa \sqrt[3]{\Delta_F}$ for some κ. Then one has*

$$a'_{11} a'_{22} a'_{33} \leq \left(\tfrac{4}{3} + \tfrac{1}{2}\kappa^3\right) \Delta_F,$$

for the Gram-Schmidt normalization F' of F.

Proof. Let

$$\begin{pmatrix} a_{11} & a_{12} & a_{13} \\ a_{12} & a_{22} & a_{23} \\ a_{13} & a_{23} & a_{33} \end{pmatrix} = \begin{pmatrix} 1 & 0 & 0 \\ r_{12} & 1 & 0 \\ r_{13} & r_{23} & 1 \end{pmatrix} \begin{pmatrix} d_1 & 0 & 0 \\ 0 & d_2 & 0 \\ 0 & 0 & d_3 \end{pmatrix} \begin{pmatrix} 1 & r_{12} & r_{13} \\ 0 & 1 & r_{23} \\ 0 & 0 & 1 \end{pmatrix}$$

be the $R^{\mathrm{T}} D R$ factorization of the coefficient matrix of F. Since $\Delta_{f_{12}} = d_1 d_2$, f_{12} is reduced, and $d_1 = a_{11}$, it follows that

$$d_2 \geq \tfrac{3}{4}\, a_{22}. \tag{16}$$

Now $\Delta_F = d_1 d_2 d_3$ and (16) imply

$$d_3 \leq \frac{4}{3} \frac{\Delta_F}{a_{11}\, a_{22}}. \tag{17}$$

Let $F' = R'^{\mathrm{T}} D R'$ be the Gram-Schmidt normalization of F, then

$$\begin{aligned} a'_{33} &= d_3 + (r'_{23})^2\, d_2 + (r'_{13})^2\, d_1 \leq d_3 + (r'_{23})^2\, a_{22} + (r'_{13})^2\, a_{11} \\ &\leq d_3 + \tfrac{1}{2}\kappa \sqrt[3]{\Delta_F}. \end{aligned} \tag{18}$$

Since f_{12} is reduced we have not only $a'_{11} = a_{11}$ but also $a'_{22} = a_{22}$ since $|r_{12}| \leq \frac{1}{2}$. By combining (17) and (18) and the assumption that a_{11}, $a_{22} \leq \kappa \sqrt[3]{\Delta_F}$, one obtains

$$\begin{aligned} a'_{11}\, a'_{22}\, a'_{33} &= a_{11}\, a_{22}\, a'_{33} \\ &\leq a_{11}\, a_{22} \left(d_3 + \tfrac{1}{2}\kappa \sqrt[3]{\Delta_F}\right) \\ &\leq \tfrac{4}{3}\Delta_F + \tfrac{1}{2}\kappa^3\, \Delta_F = \left(\tfrac{4}{3} + \tfrac{1}{2}\kappa^3\right) \Delta_F. \end{aligned}$$

Now we are ready to prove that, given a quasi-reduced ternary form F, an equivalent γ-reduction is readily available, for $\gamma = \frac{16}{3}$.

Proposition 1. *Given a quasi-reduced ternary form F of size s, one can compute with $O(M(s) \log s)$ bit-operations a $\frac{16}{3}$-reduction G of F.*

Proof. Let F be quasi reduced and let F^* be the adjoint of F. First reduce f^*_{32} in F. This leaves a_{11} unchanged and maybe decreases A_{33}. Recall that $a_{11} \leq 2\sqrt[3]{\Delta_F}$. It follows from (4) that

$$\tfrac{3}{4} A_{33} A_{22} \leq \det f^*_{32} = a_{11} \Delta_F. \tag{19}$$

We normalize f_{12} in F. This leaves the form f_{13} unchanged. Also normalizing f_{13} in F leaves f_{12} unchanged. Therefore normalizing f_{12} and f_{13} in F leaves $A_{33} = \Delta_{f_{12}}$ and $A_{22} = \Delta_{f_{13}}$ unchanged. If, after these normalizations, f_{12} or f_{13} is not reduced, (2) must be violated and we have two diagonal elements of value at most $2\sqrt[3]{\Delta}$. By one more binary form reduction step performed on f_{12} or f_{13} in F, we are in the situation of Lemma 1 with $\kappa = 2$ after swapping the second and third row and column if necessary. It is clear that the computations in the proof of Lemma 1 can be carried out in $O(M(s))$ bit operations. In this case we compute a γ-reduction of F with $\gamma \leq \frac{4}{3} + 4 = \frac{16}{3}$.

If f_{12} and f_{13} are reduced then (4) implies

$$A_{33} = \det f_{12} \geq \tfrac{3}{4} a_{11} a_{22}$$
$$A_{22} = \det f_{13} \geq \tfrac{3}{4} a_{11} a_{33}$$

We conclude from (19) that

$$a_{11} \Delta_F \geq \left(\tfrac{3}{4}\right)^3 a_{11}^2\, a_{22}\, a_{33}$$

and thus that

$$\Delta_F \geq \left(\tfrac{3}{4}\right)^3 a_{11}\, a_{22}\, a_{33} \geq \tfrac{3}{16}\, a_{11}\, a_{22}\, a_{33},$$

and we have a $\frac{16}{3}$-reduction of F. The overall amount of bit operations is $O(M(s) \log s)$, where the factor $\log s$ is required for the binary reduction steps that may be necessary. □

By combining Theorem 2 and Proposition 1 we have our main result.

Theorem 3. *Given an integral positive definite ternary form F of size s, one can compute with $O(M(s) \log^2 s)$ bit-operations a $\frac{16}{3}$-reduction of F.* □

5 Finding the Minimum of a Ternary Form

The following theorem is well known.

Theorem 4. *If F is a form in d variables with coefficient matrix $M_F = (a_{ij})$ such that $\prod_{i=1}^{d} a_{ii} \leq \gamma\, \Delta_F$, then*

$$\lambda(F) = \min\{\, F(x_1, \ldots, x_d) \mid |x_i| \leq \sqrt{\gamma},\, x_i \in \mathbb{Z},\, i = 1, \ldots d \,\}. \qquad \square$$

If the dimension is fixed and F is reduced, then Theorem 4 states that $\lambda(F)$ can be quickly computed from a constant number of candidates. This gives rise to the next theorem.

Theorem 5. *The minimum $\lambda(F)$ of a positive definite integral ternary form F of binary encoding length s can be computed with $O(M(s)\log^2 s)$ bit-operations, where $M(s)$ is the bit-complexity of s-bit integer multiplication.*

Proof. Given a ternary form F of size s, we first compute a $\frac{16}{3}$-reduction G of F. Now $\lambda(F) = \lambda(G)$ and by Theorem 4, the minimum of G is attained at an integral vector $x \in \mathbb{Z}^3$ with $|x_i| \leq \frac{16}{3}$, $i = 1, \dots, 3$. By Theorem 3, all this can be done with $O(M(s)\log^2 s)$ bit-operations. □

6 Fast Reduction in Any Fixed Dimension

In this section we sketch how the previous technique can be generalized to any fixed dimension. It is more convenient to describe this in the language of lattices. For this we review some terminology. A *(rational) lattice* $\Lambda \subseteq \mathbb{Q}^d$ is a set of the form $\Lambda = \Lambda(A) = \{Ax \mid x \in \mathbb{Z}^k\}$, where $A \in \mathbb{Q}^{d\times k}$ is a rational matrix of full column rank. The matrix A is a *basis* of the lattice Λ and its columns are the *basis vectors.* The lattice Λ is *integral* if $A \in \mathbb{Z}^{d\times k}$. The number k is the *dimension* of the lattice. If $k = d$, then Λ is *full-dimensional.* Let F be the quadratic form with coefficient matrix $A^{\mathrm{T}}A$. The *lattice determinant* of Λ is the number $\det \Lambda = \sqrt{\Delta_F}$ and the lattice basis $A = (x_1, \dots, x_k)$ is *reduced* if the form F is reduced. More explicitly, this means that

$$\prod_{i=1}^{k} \|x_i\| \leq \gamma \det \Lambda \tag{20}$$

for some constant γ. The *Lattice Reduction Problem* is the problem of computing a reduced basis for a given lattice.

The *dual lattice* of a full-dimensional lattice Λ is the lattice $\Lambda^* = \{\, y \in \mathbb{Q}^d \mid y^{\mathrm{T}}x \in \mathbb{Z}, \forall x \in \Lambda \,\}$. Clearly $\Lambda^* = \Lambda(A^{\mathrm{T}^{-1}})$ and $\det \Lambda^* = 1/\det \Lambda$.

6.1 Lattice Reduction, Shortest Vectors, and Short Vectors

The *Shortest Vector Problem* is the problem of finding a shortest nonzero vector of a given lattice. This is just the translation of Problem 1 into lattice terminology. Hermite [6] proved that a d-dimensional lattice Λ always contains a (shortest) vector x with $\|x\| \leq (4/3)^{(d-1)/4}(\det \Lambda)^{1/d}$. We call the problem of computing a vector x with

$$\|x\| \leq \kappa \cdot (\det \Lambda)^{1/d},$$

where κ is an arbitrary constant, the *SHORT Vector Problem.*

Clearly, every shortest vector is also a short vector. If a reduced lattice basis is available, a shortest vector can be computed fast, as mentioned above in Section 5 (Theorem 4). The availability of a reduced lattice bases also implies an easy solution of the Short Vector Problem, either directly by (20) or via the Shortest Vector Problem.

So, the Short Vector Problem is apparently the easiest problem among the three problems Lattice Reduction, Shortest Vector, and Short Vector. We will show in Section 6.3 that Lattice Reduction (and hence the Shortest Vector Problem) can be reduced to Short Vector. In Section 6.2, we will first describe a solution of the Short Vector Problem which proceeds by induction on the dimension, analogously to the procedure of Section 3.

6.2 Finding a Short Vector

First we describe how one can find a lattice vector $x \in \Lambda$ of a d-dimensional integral lattice $\Lambda \subseteq \mathbb{Z}^d$ with $\|x\| \leq \alpha\,(4/3)^{(d-1)/4}\sqrt[d]{\det\Lambda}$, for any constant $\alpha > 1$. The procedure mimicks the proof of Hermite [6] who showed that such a vector (with $\alpha = 1$) exists, see also [12, p. 79].

The idea is to compute a sequence of lattice vectors $x_0, x_1, x_2, \ldots$ which satisfy the relation

$$\|x_{i+1}\| \leq (\kappa_{d-1})^{d/(d-1)}(\det\Lambda)^{(d-2)/(d-1)^2}\|x_i\|^{1/(d-1)^2}, \tag{21}$$

for a certain constant κ_{d-1}. This is the generalization of (14) to higher dimensions. We rewrite (21) as

$$\frac{\|x_{i+1}\|}{(\kappa_{d-1})^{(d-1)/(d-2)}\sqrt[d]{\det\Lambda}} \leq \left[\frac{\|x_i\|}{(\kappa_{d-1})^{(d-1)/(d-2)}\sqrt[d]{\det\Lambda}}\right]^{1/(d-1)^2}.$$

Arguing as in Section 3, we can obtain $\|x_i\| \leq \kappa_d\cdot(\det\Lambda)^{1/d}$ in $i = O(\log\log\|x_0\|)$ steps, if we choose the constant $\kappa_d > (\kappa_{d-1})^{(d-1)/(d-2)}$.

We now describe how the successor of x_i is computed. Let x_i be given. Consider the $(d-1)$-dimensional sublattice Ω^* of Λ^* defined by

$$\Omega^* = \{\, y \in \Lambda^* \mid y^{\mathrm{T}}x_i = 0 \,\}.$$

The lattice Ω^* has determinant

$$\det\Omega^* \leq \|x_i\|\,\det\Lambda^* = \|x_i\|\,(\det\Lambda)^{-1}.$$

We find a short vector $\tilde{y}$ in Ω^* with

$$\|\tilde{y}\| \leq \kappa_{d-1}(\|x_i\|\,(\det\Lambda)^{-1})^{1/(d-1)}.$$

This is a Short Vector Problem in $d-1$ dimensions, which is solved inductively. Now we repeat the same procedure, going from the dual lattice back to the original lattice: consider the $(d-1)$-dimensional sublattice Γ of Λ defined by

$$\Gamma = \{\, x \in \Lambda \mid \tilde{y}^{\mathrm{T}}x = 0 \,\},$$

whose determinant satisfies

$$\det\Gamma \leq \|\tilde{y}\|\cdot\det\Lambda \leq \kappa_{d-1}(\det\Lambda)^{(d-2)/(d-1)}\|x_i\|^{1/(d-1)}.$$

We find a short vector x_{i+1} of Γ with $\|x_{i+1}\| \leq \kappa_{d-1}\,(\det\Gamma)^{1/(d-1)}$, which immediately yields (21). □

As a consequence one obtains the following proposition which generalizes Theorem 2.

Proposition 2. *Let $d \in \mathbb{N}$, $d \geq 3$, and let κ_{d-1} be some constant. Suppose that, in an integral lattice Γ of dimension $d-1$ with binary encoding length s, a short vector x with*

$$\|x\| \leq \kappa_{d-1} \, (\det \Gamma)^{1/(d-1)}$$

can be found in $T_{d-1}(s)$ bit-operations. Then, for an integral lattice basis $A \in \mathbb{Z}^{d \times d}$ with binary encoding length s, we can compute a basis $B \in \mathbb{Z}^{d \times d}$ of the generated lattice Λ such that the first column vector x of B satisfies

$$\|x\| \leq \kappa_d \, (\det \Lambda)^{1/d},$$

in $T_d(s) = O(T_{d-1}(s) \log s + M(s) \log s)$ bit-operations, for any constant κ_d with $\kappa_d > (\kappa_{d-1})^{(d-1)/(d-2)}$.

Proof. We start the sequence $x_0, \ldots, x_k$ with an arbitrary vector x_0 out of the basis A. The successors are computed as described above. The computation of $\tilde{y}$ can be done with $O(T_{d-1}(s) + M(s))$ bit-operations, since this involves only one $(d-1)$-dimensional shortest vector problem and basic linear algebra. The same time bound holds for the computation of x_{i+1}. These computations have to be repeated at most $O(\log \log \|x_0\|)$ times and we arrive at a lattice vector x with $\|x\| \leq \kappa_d \, (\det \Lambda)^{1/d}$. Now we determine an integral vector $y \in \mathbb{Z}^d$ with $Ay = x$. With the extended euclidean algorithm one can find a unimodular matrix $U \in \mathbb{Z}^{d \times d}$ with first column $y / \gcd(y_1, \ldots, y_d)$. The matrix $B = AU$ is as claimed. □

We can use this proposition inductively, starting with $\kappa_2 = \sqrt[4]{4/3}$ and $T_2(s) = O(M(s) \log s)$. We see that we can choose κ_d as close to $(4/3)^{(d-1)/4}$ as we like. So we obtain:

Corollary 1. *In a d-dimensional integral lattice $\Lambda \subseteq \mathbb{Z}^d$, a lattice vector x with $\|x\| \leq \kappa \sqrt[d]{\det \Lambda}$ can be found in $O(M(s) \log^{d-1} s)$ time, for any constant $\kappa > (\frac{4}{3})^{(d-1)/4}$.* □

6.3 Augmenting the Number of Short Vectors in the Basis

Now we generalize the approach of Section 4 to get a reduced basis. Suppose we have a basis $v_1, \ldots, v_d$ of the d-dimensional lattice Λ which is not reduced and such that the first $k \geq 1$ basis vectors satisfy $\|v_i\| \leq \alpha \sqrt[d]{\det \Lambda}$, $1 \leq i \leq k$ for some constant α depending on d and k only. We describe a procedure that computes a new basis $v'_1, \ldots, v'_d$ which satisfies one of the following.

(a) $v'_1, \ldots, v'_d$ is reduced, or
(b) for all $1 \leq j \leq k+1$ one has $v'_j \leq \alpha^* \sqrt[d]{\det \Lambda}$ for some constant α^* depending on d and $k+1$ only.

Let L be the subspace of $\mathbb{R}^d$ which is generated by the vectors $v_1, \ldots, v_k$ and denote its orthogonal complement by $L^\perp$. Let $\bar{v}_j$ denote the projection of v_j into $L^\perp$. Let $\Lambda^{(1)}$ be the k-dimensional lattice generated by $v_1, \ldots, v_k$ and let $\Lambda^{(2)}$

be the $(d-k)$-dimensional lattice generated by the vectors $\bar{v}_{k+1}, \ldots, \bar{v}_d$. Clearly $\det \Lambda^{(1)} \det \Lambda^{(2)} = \det \Lambda$. Let

$$\bar{u}_{k+1}, \ldots, \bar{u}_d$$

be a reduced basis of $\Lambda^{(2)}$ and suppose that $\bar{u}_{k+1}$ is the shortest among these basis vectors. Let $U \in \mathbb{Z}^{(d-k)\times(d-k)}$ denote the unimodular matrix which transforms $(\bar{v}_{k+1}, \ldots, \bar{v}_d)$ into $(\bar{u}_{k+1}, \ldots, \bar{u}_d)$. The vectors $v_j^* \in \Lambda$ defined by $(v_{k+1}^*, \ldots, v_d^*)$ $= (v_{k+1}, \ldots, v_d)\, U$ are of the form

$$v_j^* = \bar{u}_j + \sum_{i=1}^{k} \mu_{ij}\, v_i,$$

with some real coefficients μ_{ij}. It follows that

$$v_j' = \bar{u}_{k+1} + \sum_{i=1}^{k} \{\mu_{ij}\}\, v_i \in \Lambda,$$

where $\{x\}$ denotes the fractional part of x. Clearly

$$v_1, \ldots, v_k, v_{k+1}', \ldots, v_d'$$

is a basis of Λ and

$$\|v_j'\| \leq \|\bar{u}_j\| + k\alpha \sqrt[d]{\det \Lambda}.$$

There are two cases. If $\|\bar{u}_{k+1}\| > \sqrt[d]{\det \Lambda}$, then for all $j = k+1, \ldots, d$,

$$\|v_j'\| \leq (k\alpha + 1)\, \|\bar{u}_j\|.$$

Thus we get $\|v_{k+1}'\| \cdots \|v_d'\| \leq \alpha_2 \det \Lambda^{(2)}$ for some constant α_2 since $\bar{u}_{k+1}, \ldots, \bar{u}_d$ is reduced. Now let $v_1', \ldots, v_k'$ be a reduced basis of $\Lambda^{(1)}$. Then

$$\|v_1'\| \cdots \|v_d'\| \leq \alpha_1 \det \Lambda^{(1)}\, \alpha_2 \det \Lambda^{(2)} = \alpha_1\, \alpha_2 \det \Lambda,$$

which means that $v_1', \ldots, v_d'$ is reduced and thus (a) holds.

If, on the other hand, $\|\bar{u}_{k+1}\| \leq \sqrt[d]{\det \Lambda}$, then the basis $v_1, \ldots, v_k, v_{k+1}', \ldots, v_d'$ satisfies (b). □

Now it is clear how to proceed. We find the first short basis vector by Proposition 2, and we iterate the above procedure as long as case (b) prevails, increasing k. We must eventually end up with a reduced basis, because as soon as k reaches d, we have $\|v_i\| \leq \alpha \sqrt[d]{\det \Lambda}$ for *all* basis vectors v_i, and this implies that the basis is reduced.

In this way, we have reduced the Lattice Reduction Problem in dimension d to one d-dimensional Short Vector Problem and a constant number (fewer than $2d$) of lower-dimensional lattice reduction problems, plus some linear algebra which can be done in $O(M(n))$ time. Thus we obtain the following theorem by induction on the dimension.

Theorem 6. *Let $d \in \mathbb{N}$, $d \geq 2$, $A \in \mathbb{Z}^{d\times d}$ be a lattice basis generating Λ and suppose that the binary encoding length of A is s. Then one can compute with $O(M(s)\log^{d-1} s)$ bit-operations a reduced basis of Λ or a shortest vector of Λ.* □

References

1. A.V. Aho, J.E. Hopcroft, and J.D. Ullman. *The Design and Analysis of Computer Algorithms.* Addison-Wesley, Reading, 1974.
2. J.W.S. Cassels. *Rational quadratic forms.* Academic Press, 1978.
3. F. Eisenbrand. Short vectors of planar lattices via continued fractions. *Information Processing Letters*, 2001, to appear. `http://www.mpi-sb.mpg.de/~eisen/report_lattice.ps.gz`
4. C.F. Gauß. *Disquisitiones arithmeticae.* Gerh. Fleischer Iun., 1801.
5. C.F. Gauß. Recension der "Untersuchungen über die Eigenschaften der positiven ternären quadratischen Formen von Ludwig August Seeber." Reprinted in *Journal für die reine und angewandte Mathematik*, 20:312–320, 1840.
6. Ch. Hermite. Extraits de lettres de M. Ch. Hermite à M. Jacobi sur différents objets de la théorie des nombres. *Journal für die reine und angewandte Mathematik*, 40, 1850.
7. J. C. Lagarias. Worst-case complexity bounds for algorithms in the theory of integral quadratic forms. *Journal of Algorithms*, 1:142–186, 1980.
8. A.K. Lenstra, H.W. Lenstra, and L. Lovász. Factoring polynomials with rational coefficients. *Math. Annalen*, 261:515–534, 1982.
9. H.W. Lenstra. Integer programming with a fixed number of variables. *Mathematics of Operations Research*, 8(4):538–548, 1983.
10. A. Schönhage. Fast reduction and composition of binary quadratic forms. In *International Symposium on Symbolic and Algebraic Computation, ISSAC'91*, pages 128–133. ACM Press, 1991.
11. A. Schönhage and V. Strassen. Schnelle Multiplikation grosser Zahlen (Fast multiplication of large numbers). *Computing*, 7:281–292, 1971.
12. A. Schrijver. *Theory of Linear and Integer Programming.* John Wiley, 1986.
13. L.A. Seeber. *Untersuchung über die Eigenschaften der positiven ternären quadratischen Formen.* Loeffler, Mannheim, 1831.
14. I. Semaev. A 3-dimensional lattice reduction algorithm. In *Cryptography and Lattices Conference, CALC 2001.* This volume, pp. 181–193, 2001.
15. B. Vallée. An affine point of view on minima finding in integer lattices of lower dimensions. In *Proceedings of the European Conference on Computer Algebra, EUROCAL'87*, volume 378 of *Lecture Notes in Computer Science*, pp. 376–378. Springer, Berlin, 1989.
16. C.K. Yap. Fast unimodular reduction: Planar integer lattices. In *Proceedings of the 33rd Annual Symposium on Foundations of Computer Science*, pages 437–446, Pittsburgh, 1992. IEEE Computer Society Press.

Factoring Polynomials and 0–1 Vectors

Mark van Hoeij

Department of Mathematics, Florida State University
Tallahassee, FL 32306-3027
hoeij@math.fsu.edu

Abstract. A summary is given of an algorithm, published in [4], that uses lattice reduction to handle the combinatorial problem in the factoring algorithm of Zassenhaus. Contrary to Lenstra, Lenstra and Lovász, the lattice reduction is not used to calculate coefficients of a factor but is only used to solve the combinatorial problem, which is a problem with much smaller coefficients and dimension. The factors are then constructed efficiently in the same way as in Zassenhaus' algorithm.

1 Comparison of Three Factoring Algorithms

Let $f \in \mathbb{Q}[x]$ be a polynomial of degree N. Write

$$f = \sum_{i=0}^{N} a_i x^i \qquad \text{with } a_i \in \mathbb{Q}.$$

Suppose that f is monic (i.e. $a_N = 1$) and square-free (i.e. $\gcd(f, f') = 1$). Assume that the coefficients a_i do not have more than D digits.

Let p be a prime number and let $\mathbb{Q}_p$ be the field of p-adic numbers. Let

$$f = \prod_{i=1}^{n} f_i$$

where $f_i \in \mathbb{Q}_p[x]$ are the monic irreducible factors of f in $\mathbb{Q}_p[x]$. Let v be a 0–1 vector, i.e. $v = (v_1 \ldots v_n) \in \{0,1\}^n$. Let $g_v = \prod f_i^{v_i}$. Then

$$\{g_v | v \in \{0,1\}^n\}$$

is the set of all monic factors of f in $\mathbb{Q}_p[x]$. Let

$$V = \{v \in \{0,1\}^n | g_v \in \mathbb{Q}[x]\}$$

so $\{g_v | v \in V\}$ is the set of all monic factors of f in $\mathbb{Q}[x]$. Let

$$B = \{v \in V | g_v \text{ irreducible in } \mathbb{Q}[x]\}.$$

Because $\mathbb{Q}[x]$ has unique factorization it follows that V is the set of all $\{0,1\}$-linear combinations of B.

J.H. Silverman (Ed.): CaLC 2001, LNCS 2146, pp. 45–50, 2001.

The algorithm of Zassenhaus [1] to factor f in $\mathbb{Q}[x]$ works as follows. First the f_i are computed modulo a power of p using Hensel lifting. The computation time is bounded by $P_Z(N, D)$ which is a polynomial in N and D. Then, for $v \in \{0,1\}^n$, a method is given to decide if $g_v \in \mathbb{Q}[x]$ or not, and if so, to calculate $g_v \in \mathbb{Q}[x]$. If g_v is not in $\mathbb{Q}[x]$, then the time c to verity that is very small, and is almost independent of N and D. For all practical purposes we can assume that this cost c is constant. Calculating g_v for all $v \in B$ gives the set of all irreducible factors of f. But no method is given to calculate the set B, other than to try all v. Denote $|v| = \sum v_i$, so g_v is a product of $|v|$ p-adic factors. First one tries all v with $|v| = 1$, then $|v| = 2$, etc. Whenever a $g_v \in \mathbb{Q}[x]$ is found, the corresponding f_i are removed, and n decreases. The complexity depends on

$$M = \max\{|v|, v \in B\}.$$

The worst case is $M = n$, i.e. f is irreducible, because then all 2^n vectors v will be be tried (or 2^{n-1} vectors, by skipping the complements of the v's that were already tried). So the total cost is at most

$$P_Z(N, D) + c2^{n-1},$$

where c is a very small constant. If $M = n/2$ then the cost is essentially the same. If $M < n/2$ then the cost is lower, we can bound the cost by

$$P_Z(N, D) + cE_Z(n, M),$$

where $E_Z(n, M) \leq 2^{n-1}$ depends exponentially on M. After trying all $|v| \leq 3$, which can be done quickly, we may assume that $M > 3$ (if there are still any f_i left).

In most examples (even with large N) the number M will be small. Then P_Z dominates the computation time and the algorithm works fast. However, in some examples M can be large, in which case cE_Z can dominate the computation time. This happens when f has few factors in $\mathbb{Q}[x]$ but many factors in $\mathbb{Q}_p[x]$ for every p. Such polynomials have a very special Galois group; order$(\sigma) \ll N$ for every σ in the Galois group. Extreme examples are the Swinnerton-Dyer polynomials, where order$(\sigma) \leq 2$ for all σ. Other examples are resolvent polynomials, which tend to be polynomials of high degree with small Galois groups. For these polynomials, the computation time is dominated by cE_Z, and the algorithm of Zassenhaus will be exponentially slow.

The first polynomial time algorithm was given by Lenstra, Lenstra and Lovász. In their paper [2] they give a lattice reduction algorithm (the LLL algorithm). Many combinatorial problems can be solved in polynomial time with LLL by encoding the solutions of the problem as short vectors in a lattice. The LLL algorithm can find the set S of short vectors, provided that all vectors outside of span(S) are sufficiently long in comparison.

In [2] the LLL algorithm is used in the following way. Take one factor $f_1 \in \mathbb{Q}_p[x]$ of f. The problem to be solved with LLL is: Find, if it exists, a non-zero $g \in \mathbb{Q}[x]$ of degree $< N$ such that f_1 divides g. If such g exists, then

$$\gcd(f, g) \in \mathbb{Q}[x]$$

is a non-trivial factor of f. This problem is reduced to lattice reduction as follows. First calculate f_1 modulo a sufficiently high power of p by Hensel lifting. The cost of Hensel lifting can be bounded by $P_{L_1}(N, D)$ which is a polynomial in N, D. From that, a lattice can be constructed that (if f is reducible) contains a vector U_g whose entries are the coefficients of g. Then the LLL algorithm can find this vector in a time bounded by $P_{L_2}(N, D)$ which is also a polynomial in N and D. So the total computation time is bounded by $P_L = P_{L_1} + P_{L_2}$, which is a polynomial in N, D. However, $P_{L_1} > P_Z$ because one must Hensel lift up to a substantially higher power of p. Furthermore, $P_{L_2} \gg P_Z$, in other words, the algorithm of Zassenhaus is much faster, except when $cE_Z \gg P_Z$ which only happens for polynomials with special Galois groups.

So, there exists a polynomial time algorithm [2], and an exponential time algorithm [1] that is faster most of the time (except when M is large). How can the advantages of both algorithms be combined?

Suppose that $g_v \in \mathbb{Q}[x]$. The algorithm in [2] would find such $g = g_v$ by computing a vector U_g in an N-dimensional lattice whose entries are the coefficients of g. Suppose that g is large (say degree five hundred and coefficients with thousands of digits). Lattice reduction is a very general method that can be applied to solve many combinatorial problems. So it is to be expected that if it is used to construct a large expression such as U_g that it will take a long time. To have a faster computation, we must use LLL to construct smaller vectors instead.

The vectors in B are much smaller than the vector U_g, in two ways. The entries have only 1 digit whereas the entries of U_g can have many digits. And the number of entries is only n, usually n is much smaller than N (except for Swinnerton-Dyer polynomials where $n = N/2$, which is smaller but not much smaller than N).

Because of the much smaller size, LLL can calculate the elements of B much faster than the vector U_g. We need to design an input lattice for LLL in such a way, that span(B) and hence B can be obtained from the short vectors found by LLL. To keep the LLL cost to a minimum, we must make sure that the short vectors found by LLL do not contain any information (other than the set B) about the coefficients of a factor g, so that the LLL cost will not depend on the size of g. The LLL cost will then be bounded by a polynomial $P(n)$ that depends only on n, and not on N or D. The cost $cE_Z(n, M)$ of finding the set B in Zassenhaus' algorithm is now replaced by $P(n)$, and the total cost of factoring is now:

$$P_Z(N, D) + P(n).$$

So the resulting algorithm is faster than Zassenhaus' algorithm whenever

$$P(n) < cE_Z(n, M).$$

It turns out in experiments that the cut-off point is low. That means that when $P(n)$ is not smaller than $cE_Z(n, M)$, then $P(n)$ and $cE_Z(n, M)$ are both small, so then the computation time is close to $P_Z(N, D)$ for both algorithms. However, when n is larger, then $P(n)$ can be much smaller than $cE_Z(n, M)$. Experiments show that polynomials with $n > 400$, $N, D > 2000$ can be handled, which is far beyond the reach of [1, 2].

2 How to Construct the Lattice to Find B?

To find linear conditions on v we can not use the coefficients of the polynomial g_v, because they do not depend linearly on v, $g_{u+v} = g_u g_v$. In order to find linear conditions, a vector $T_A(g)$ with s entries (s will be small compared to n) will be defined, that has the following property:

$$T_A(g_1 g_2) = T_A(g_1) + T_A(g_2) \in \mathbb{Q}_p^s \qquad \text{for all non-zero } g_1, g_2 \in \mathbb{Q}_p[x].$$

This $T_A(g)$ will be constructed in such a way that the entries of $T_A(g_v)$ are p-adic integers for all 0–1 vectors v, and if furthermore $g_v \in \mathbb{Q}[x]$ then the entries are integers, bounded in absolute value by $\frac{1}{2}p^b$ for some integer b.

Now $T_A(g_v) = \sum v_i T_A(f_i)$ is a linear combination of the $T_A(f_i)$, and when $g_v \in \mathbb{Q}[x]$ then the entries of this linear combination are integers. However, the entries of $T_A(f_i)$ are not yet suitable for use in the LLL input vectors for two reasons. First, these entries of $T_A(f_i)$ are p-adic integers, which are not finite expressions. Second, if $g_v \in \mathbb{Q}[x]$, then the entries of $T_A(g_v)$ are integers bounded by $\frac{1}{2}p^b$, and these integers give some partial information about the coefficients of g_v. It would be inefficient to have this information in the lattice. Both problems are solved by cutting each entry of $T_A(f_i)$ on two sides. If t is such an entry, then the p-adic integer t can be written as

$$t = \sum_{i=0}^{\infty} t_i p^i \qquad \text{with } t_i \text{ integers and } -\frac{p}{2} < t_i \leq \frac{p}{2}.$$

Choose $a > b$, then cut such t by replacing it with

$$\sum_{i=b}^{a-1} t_i p^{i-b}.$$

So the powers $\geq a$ and the powers $< b$ of p in t are removed. The first causes the expression to be finite, and the second removes unnecessary information about the coefficients of g from the lattice. Denote the result of this cutting by $T_A^{b,a}(f_i)$. Let $e_1, \ldots, e_n$ be the standard basis vectors, so $g_{e_i} = f_i$. Denote

$$V_i = (e_i, T_A^{b,a}(f_i))$$

as the concatenation of the vectors e_i and $T_A^{b,a}(f_i)$. The number of entries is $n + s$ which is a little more than n. Denote E_j, $j \in \{1, \ldots, s\}$ as the vector with $n + s$ entries, all 0 except for the $n + j$'th entry which is p^{a-b}.

The $n + s$ vectors V_i and E_j are now the input vectors for LLL. To find the set B, calculate the short vectors in the lattice spanned by the V_i and E_j, then take the projection on the first n entries, and then reduce those vectors to echelon form. The resulting vectors form the set B, provided that a, b and the other parameters in the algorithm were set properly. This can be verified; the polynomials g_v for v in the calculated set B are automatically irreducible if they are in $\mathbb{Q}[x]$, so to verify the correctness of B one only needs to check that $g_v \in \mathbb{Q}[x]$ for all $v \in B$.

The reason that the LLL algorithm lets us find the vectors v for which $g_v \in \mathbb{Q}[x]$ is because for those v, the entries of $T_A(g_v)$ are integers bounded by $\frac{1}{2}p^b$, hence $T_A^{b,a}(g_v) = 0$. Now $\sum v_i T_A^{b,a}(f_i)$ is almost the same as $T_A^{b,a}(g_v) = 0$, except for some round-off errors (caused by cutting the p-adic numbers) which must be of the form $\epsilon_1 + \epsilon_2 p^{a-b}$, where ϵ_1, ϵ_2 are small. The vector

$$\sum v_i V_i = (v, \sum v_i T_A^{b,a}(f_i))$$

is in the lattice. After reducing with the vectors E_i, all entries are small; the first n entries are all 0 or 1, and the last s entries are small as well. So $\sum v_i V_i$ is a short vector. When this vector is found by LLL, v can be read off by taking the first n entries.

There is a lot of freedom in the choice of the numbers p^{a-b} and s so one can choose the size of the LLL input vectors. For efficiency, the size should be not too large, but the size should also not be too small because LLL can only find the short vectors if the remaining vectors are sufficiently long in comparison. The number $s(a-b)\log(p)$ should be $O(n^2)$, and should be independent of N and D.

3 The i'th Trace

Definition 1. *The i'th trace* $\text{Tr}_i(g)$ *of a polynomial g is defined as the sum of the i'th powers of the roots (counted with multiplicity) of g.*

It is clear that

$$\text{Tr}_i(f_1) + \text{Tr}_i(f_2) = \text{Tr}_i(f_1 f_2)$$

for any two polynomials f_1, f_2. Suppose $g = \sum c_i x^i$ is monic of degree d. Then $\text{Tr}_i(g)$ for $i = 1, \ldots, k$ can be determined from c_{d-i} for $i = 1, \ldots, k$ with the Newton relations, and vice versa.

Now choose some $i_1, i_2, \ldots, i_s$, and for $g \in \mathbb{Q}_p[x]$ let $T_A(g)$ be a vector whose entries are the p-adic numbers

$$\text{Tr}_{i_1}(g), \ldots, \text{Tr}_{i_s}(g)$$

multiplied by some integer m. The integer m is chosen in such a way that if g is a factor of f in $\mathbb{Q}[x]$, then the entries of $T_A(g)$ are integers. If f is monic and has integer coefficients then $m = 1$.

The number s is normally chosen much smaller than n or degree(g), so $T_A(g)$ will only contain partial information about the coefficients of g. By computing a bound for the absolute value of the complex roots of f, it is easy to bound $\mathrm{Tr}_i(g)$ for any factor $g \in \mathbb{Q}[x]$ of f, so it is easy to calculate a number b such that the absolute values of the entries of $T_A(g)$ are bounded by $\frac{1}{2}p^b$.

We use $T_A(g)$ to fix the problem that g_v does not depend linearly on v. There is also another way to fix this problem: In [3] Victor Miller uses idempotents to factor in $\mathbb{Q}[x]$. The main difference with our algorithm is that our algorithm is closer to Zassenhaus' algorithm, because only the 0–1 vectors are computed with integer-relation-finding or LLL; the factors g themselves are constructed like in [1]. Millers algorithm is less similar to [1] but is closer to [2] in the sense that everything is computed with integer-relation-finding, more precisely: it calculates the 0–1 vector and the idempotent e simultaneously, and if e is a non-trivial idempotent then

$$\gcd(e, f) \in \mathbb{Q}[x]$$

is a non-trivial factor of f. In our algorithm, besides the 0–1 vector, no information about g is calculated with LLL, because the value of $T_A(g)$ which contains information about g is precisely what is being cut away when all powers $< b$ of p were removed.

References

1. H. Zassenhaus, *On Hensel Factorization, I.*, Journal of Number Theory **1**, 291–311 (1969).
2. A.K. Lenstra, H.W. Lenstra and L. Lovász, *Factoring polynomials with rational coefficients*, Math. Ann. **261**, 515–534 (1982).
3. V. Miller, *Factoring polynomials via Relation-Finding.* ISTCS'92, Springer Lecture Notes in Computer Science **601**, 115–121 (1992).
4. M. van Hoeij, *Factoring polynomials and the knapsack problem*, preprint available from `http://www.math.fsu.edu/~hoeij/` accepted for Journal of Number Theory (2000).

Approximate Integer Common Divisors

Nick Howgrave-Graham

IBM T.J.Watson Research Center
PO Box 704, Yorktown Heights, New York 10598, USA
nahg@watson.ibm.com

Abstract. We show that recent results of Coppersmith, Boneh, Durfee and Howgrave-Graham actually apply in the more general setting of (partially) approximate common divisors. This leads us to consider the question of "fully" approximate common divisors, i.e. where both integers are only known by approximations. We explain the lattice techniques in both the partial and general cases. As an application of the partial approximate common divisor algorithm we show that a cryptosystem proposed by Okamoto actually leaks the private information directly from the public information in polynomial time. In contrast to the partial setting, our technique with respect to the general setting can only be considered heuristic, since we encounter the same "proof of algebraic independence" problem as a subset of the above authors have in previous papers. This problem is generally considered a (hard) problem in lattice *theory*, since in our case, as in previous cases, the method still works extremely reliably in practice; indeed no counter examples have been obtained. The results in both the partial and general settings are far stronger than might be supposed from a continued-fraction standpoint (the way in which the problems were attacked in the past), and the determinant calculations admit a reasonably neat analysis.
Keywords: Greatest common divisor, approximations, Coppersmith's method, continued fractions, lattice attacks.

1 Introduction

When given new mathematical techniques, as is the case in [1] and [5], it is important to know the full extent to which the result can be used, mathematically, even if this generalisation does not have immediate applications (to cryptography, say). In this paper we will start by describing approximate common divisor problems. Later we will show that the above results can be seen as special instances of this problem, and we will describe a lattice based solution to (a version of) the general problem. For now let us just concentrate on explaining approximate common divisor problems.

As an example, we explain this in the more specific and familiar case of *greatest* common divisors. If we are given two integers a and b we can clearly find their gcd, d say, in polynomial time. If d is in some sense large then it may be possible to incur some additive "error" on either of the inputs a and b, or both, and still recover this gcd. This is what we refer to as an approximate common

J.H. Silverman (Ed.): CaLC 2001, LNCS 2146, pp. 51–66, 2001.

divisor problem (ACDP); although we delay its rigorous definition to later. Of course if there is too much error incurred on the inputs, the algorithm may well not be able to discern the gcd d we had initially over some other approximate divisors d' (e.g. they may all leave residues of similar magnitude when dividing a and b). In this sense, the problem is similar to those found in error correcting codes.

Continuing this error correcting code analogy we can state the problem from the standpoint of the design of the decoding algorithm, i.e. we wish to create an algorithm which is given two inputs a_0 and b_0, and bounds X, Y and M for which one is assured that $d|(a_0 + x_0)$ and $d|(b_0 + y_0)$ for some $d > M$ and x_0, y_0 satisfying $|x_0| \leq X$, $|y_0| \leq Y$. The output of the algorithm should be the common divisor d, or all of the possible ones if more than one exists. We explore the following questions: under what conditions on these variables and bounds

- is d uniquely defined, or more generally limited to polynomially many solutions?
- does an algorithm exist to recover these d in polynomial time?

Without loss of generality let us assume that our inputs are ordered so that $X \geq Y$. If one of the inputs is known exactly, i.e. $Y = 0$, then we call this a *partially approximate common divisor problem* (PACDP), and we refer to an algorithm for its solution as a PACDP algorithm. If however neither input is known exactly, i.e. $Y > 0$, then we refer to the problem as a *general approximate common divisor problem* (GACDP), and an algorithm for its solution is called a GACDP algorithm.

In section 3 we show that the results given in [1] and [5] may be seen as special instances of a PACDP, and the techniques used therein effectively give a PACDP algorithm.

As a motivating example of this, consider the widely appreciated result in [1], which states that if $N = pq$ where $p \sim q \sim \sqrt{N}$, and we are given the top half of the bits of p then one can recover the bottom half of the bits of p in polynomial time. Notice that in effect we are given p_0, such that $p_0 = p + x_0$, for some x_0 satisfying $|x_0| < N^{1/4}$, from which we can recover the whole of p. It is not so widely known that the result also holds if we are given any integer p'_0 such that $p'_0 = kp + x_0$ for some integer k (and the same bound on x_0 as before), i.e. we can still recover all the bits of p from this information too.

This immediately shows that Okamoto's cryptosystem [11] leaks the private information (the factorisation of $n = p^2q$) from the public information (n and $u = a + bpq$ where $a < (1/2)\sqrt{pq}$) in polynomial time[1], so this result can be considered an even stonger break than that given in [14], which recovers all plaintexts, but does not recover the secret information.

In section 4 we go on to consider the GACDP, and produce some new and interesting bounds for polynomial time algorithms which (heuristically) solve this. For ease of analysis we do restrict our attention to the case when $a_0 \sim b_0$

[1] In fact the size of a is much smaller than it need be for this attack to work.

and $X \sim Y$ (we call such input "equi-sized"), though similar techniques can brought to bear in more general situations.

This general problem is thought to be very interesting to study from a mathematical point of view, and the lattice analysis is also considered to be quite neat and interesting from a theoretical standpoint, however this generalisation admittedly lacks an obvious (useful) application in either cryptography or coding theory. It is hoped that with this presentation of the problem, a use will subsequently be found.

In section 7 we conclude with some open and fundamental questions associated with these general lattice techniques.

1.1 Presentation and Algorithm Definitions

In the remainder of this paper a and b will always denote integers which do have a "large" common divisor, and d will be used to represent this common divisor. The (known) approximations to a and b will be denoted by a_0 and b_0 respectively, and their differences by $x_0 = a - a_0$, $y_0 = y - y_0$. A real number $\alpha \in (0 \ldots 1)$ is used to indicate the quantity $\log_b d$, and α_0 is used to indicate a (known) lower bound for this quantity.

Given two integers u, v we will write $u \sim_{\varepsilon_0} v$ if $|\log_2 \log_2 u - \log_2 \log_2 v| < \varepsilon_0$, though we frequently drop the ε_0 subscript, when it is clear what we mean.

We now define the algorithms to solve the ACDPs presented in this paper. We include their bound information to fully specify the algorithms, though the proof of these bounds, and the proof that their output is polynomially bounded, is left to their respective sections. We start with the PACDP algorithms described in sections 2 and 3.

Algorithm 11. *The (continued fraction based)* partially approximate common divisor algorithm PACD_CF, *is defined thus: Its input is two integers a_0, b_0 such that $a_0 < b_0$. The algorithm should output all integers $d = b_0^{\alpha}$, $\alpha > 1/2$, such that there exists an x_0 with $|x_0| < X = b_0^{2\alpha-1}$, and d divides both $a_0 + x_0$ and b_0, or report that no such d exists (under the condition on X we are assured that there are only polynomially many solutions for d).*

Algorithm 12. *The (lattice based)* partially approximate common divisor algorithm PACD_L, *is defined thus: Its input is two integers a_0, b_0, $a_0 < b_0$ and two real numbers $\varepsilon, \alpha_0 \in (0 \ldots 1)$. Let us define $M = b_0^{\alpha_0}$ and $X = b_0^{\beta_0}$ where $\beta_0 = \alpha_0^2 - \varepsilon$. The algorithm should output all integers $d > M$ such that there exists an x_0 with $|x_0| < X$, and d divides both $a_0 + x_0$ and b_0, or report that no such d exists (under the conditions on M and X we are assured that there are only polynomially many solutions for d).*

Firstly notice that the condition $a_0 < b_0$ is not a limitation at all. Since we know b_0 exactly we may subtract any multiple of it from a_0 to ensure this.

Secondly (and far more importantly) notice that there is a subtle distinction between algorithms 11 and 12 in that in the continued fraction based algorithm

α is not an input, but rather is defined in terms of any common divisor d (this is true of the GACDP algorithms defined below too). This is preferential to the situation with the lattice algorithms, in which it is presently necessary to state in advance an $\alpha_0 < \alpha$, and then the bound X is defined in terms of α_0 rather than α (so one would wish α_0 to be very close to α to ensure X was as large as possible).

Thirdly notice the requirement for $\alpha > 1/2$ in the continued fraction techniques. A major contribution of this paper is in showing that we may solve the ACDPs for $\alpha < 1/2$ too, using the lattice based methods.

Lastly notice the appearance of ε in the lattice based variants. This is because the bound on X in these algorithms is defined asymptotically. In order to know X explicitly we allow $\varepsilon \ll \log_{b_0} X$ to be given as input to the algorithm.

We give two equi-sized (i.e. $a_0 \sim b_0$) GACDP algorithms in the paper, one in section 2 and one in section 4. The lattice based approach is only defined for $\alpha < 2/3$. Refer to figure 61 to see a graphical representation of the bounds from each of the algorithms.

Algorithm 13. *The* (continued fraction based) equi-sized GACDP algorithm `GACD_CF`, *is defined thus: Its input is two integers a_0, b_0 subject to $a_0 \sim b_0$, $a_0 < b_0$. The algorithm should output all integers $d = b_0^\alpha$, $\alpha > 1/2$ such that there exist integers x_0, y_0 with $|x_0|, |y_0| < X = b_0^\beta$ where $\beta = \max(2\alpha - 1, 1 - \alpha)$, and d divides both $a_0 + x_0$ and $b_0 + y_0$, or report that no such d exists (under the condition on X we are assured that there are only polynomially many solutions for d).*

Algorithm 14. *The* (lattice based) equi-sized GACDP algorithm `GACD_L`, *is defined thus: Its input is two integers a_0, b_0 subject to $a_0 \sim b_0$, $a_0 < b_0$, and two real numbers $\varepsilon, \alpha_0 \in (0 \ldots 2/3)$. Let us define $M = b_0^{\alpha_0}$ and $X = b_0^{\beta_0}$ where $\beta_0 = 1 - (1/2)\alpha_0 - \sqrt{1 - \alpha_0 - (1/2)\alpha_0^2} - \varepsilon$. The algorithm should output all integers $d > M$ such that there exist integers x_0, y_0 with $|x_0|, |y_0| < X$, and d divides both $a_0 + x_0$ and $b_0 + y_0$, or report that it is unlikely that such a d exists (under the conditions on M and X we are assured that there are only polynomially many solutions for d).*

As the above definitions mention, the number of common divisors d is polynomially bounded when the conditions of the algorithms are met. If one wishes to use the algorithms as encoding/decoding algorithms, and so require a *unique* output from the algorithms, then one should ensure that the "aimed for" solution is substantially below the algorithm bounds, meaning that it is highly (exponentially) unlikely that any of the other (polynomially many) common divisors will be confused with it.

Notice that since algorithm `GACD_L` is heuristic, it is possible that a d exists which meets the conditions of the algorithm, but it is not found. In order to give an indication of the probability of this event happening one can refer to the results in Table 1. As can be seen there, no such occurances were detected, and so such events are considered to be extremely rare.

2 A Continued Fraction Approach

One way to approach solving the ACDPs is to consider the sensitivity of the Euclidean algorithm to additive error of its inputs. This can be studied via the use of coninued fractions, as explained in [6]. One of the many places such an analysis was found useful was in [15], when attacking RSA with a small decrypting exponent, and we look at this further in section 5.

The main results that are useful in this analysis are the following:

Theorem 21. *Let ρ be any real number, and let g_i/h_i, $i = 1 \ldots m$ denote the (poynomially many) approximants to ρ during the continued fraction approximation.*

For all $i = 1 \ldots m$ we have that

$$\left|\rho - \frac{g_i}{h_i}\right| < \frac{1}{h_i^2}.$$

Moreover for every pair of integers s, t such that

$$\left|\rho - \frac{s}{t}\right| < \frac{1}{2t^2}$$

then the ratio s/t will occur as one of the approximants g_j/h_j for some $j \in (1 \ldots m)$.

To see how this applies to creating an ACDP algorithm, let us recall the input to the continued fraction based algorithms, namely two integers a_0 and b_0, $a_0 < b_0$, and we search for common divisors $d = b_0^\alpha$ that divide both $a_0 + x_0$ and $b_0 + y_0$ ($y_0 = 0$ in the PACDP).

In this section we will assume that x_0 and y_0 are such that $|x_0|, |y_0| < b_0^\beta$, and show the dependence of β on α so that the continued fraction approach is assured of finding the common divisors d. We will see that the basic algorithm is essentially the same for PACD_CF and GACD_CF, i.e. the fact that we know $y_0 = 0$ will only help limit the number of d, not the approach used to find them.

Let a' and b' denote $(a_0 + x_0)/d$ and $(b_0 + y_0)/d$ respectively, so clearly the sizes of a' and b' are bounded by $|a'|, |b'| < b_0^{1-\alpha}$. Also notice that

$$\frac{a_0}{b_0} = \frac{a_0 + x_0}{b_0 + y_0} + \frac{a_0 y_0 - b_0 x_0}{b_0(b_0 + y_0)} = \frac{a'}{b'} + \frac{a_0 y_0 - b_0 x_0}{b_0(b_0 + y_0)}$$

so we have

$$\left|\frac{a_0}{b_0} - \frac{a'}{b'}\right| = \frac{|a_0 y_0 - b_0 x_0|}{b_0(b_0 + y_0)} < b_0^{\beta - 1}$$

This means that by producing the continued fraction approximation of $\rho = a_0/b_0$, we will obtain a'/b' whenever $b_0^{\beta-1} < 1/(2(b')^2)$. Since we are primarily concerned with large a_0 and b_0 we choose to ignore constant terms like 2, and by using the fact that $|b'| < b_0^{1-\alpha}$ we see that this inequality holds whenever $\beta - 1 < 2(\alpha - 1)$, or $\beta < 2\alpha - 1$ as anticipated by the algorithm definitions.

Thus for both PACD_CF and GACD_CF the algorithm essentially comes down to calculating the continued fractions approximation of a_0/b_0; the only difference between these two algorithms is what is then outputted as the common divisors d.

Let us first concentrate on PACD_CF, and let g_i/h_i denote the (polynomially many) approximants in the continued fraction approximation of a_0/b_0. If $d = b_0^\alpha$ divides both $a_0 + x_0$ and b_0 and $|x_0| < X = b_0^{2\alpha-1}$ we know a_0/b_0 is one of these approximants. It remains to test each of them to see if h_i divides b_0; if it does then we output $d = b_0/h_i$ as a common divisor. Note that for all such approximants we are assured that $|x_0| < X$ since $x_0 = a_0 - dg < b_0/h^2 = b_0^{2\alpha-1}$.

This proves the existence of algorithm PACD_CF.

We now turn our attention to algorithm GACD_CF. The first stage is exactly the same, i.e. we again consider the (polynomially many) approximants g_i/h_i, however now h_i need not divide b_0. Instead we find the integer k (which will correspond to our common divisor d) which minimises the max-norm of the vector $k(g_i, h_i) - (a_0, b_0) = (x_0, y_0)$. Again we are assured that for all approximants $|x_0|, |y_0| < X = b_0^{2\alpha-1}$.

However consider $(x_0', y_0') = (k + l)(g_i, h_i) - (a_0, b_0) = (x_0, y_0) + l(g_i, h_i)$ for some integer l. If it is the case that $|lg_i|, |lh_i| < X$ as well, then this will mean $(k + i)$ for all $i = 0 \ldots l$ satisfy the properties of the common divisors d, and shoud be outputted by the algorithm. Since $h_i = b_0^{1-\alpha}$ there are $b_0^{3\alpha-2}$ choices for l, which becomes exponential for $\alpha > 2/3$.

In order to limit ourselves to polynomially many d we define $X = b_0^\beta$ where $\beta = \min(2\alpha - 1, 1 - \alpha)$, which ensures there is (exactly) one common divisor d associated with each continued fraction approximant g_i/h_i. This proves the existence of algorithm GACD_CF.

Readers who are reminded of Wiener's attack on RSA by this analysis should consult section 5.

Even though it is known that the continued fraction analysis is optimal for the Euclidean algorithm, we do obtain better bounds for the PACDP and the GACDP in the next sections. This is because we are not restricted to using the Euclidean algorithm to find the common divisors d, but rather we can make use of higher dimensional lattices, as originally done in [1].

3 Using Lattices to Solve PACDP

Essentially there is nothing new to do in this section, except point out that the techniques previously used to solve divisibility problems, e.g. [1] and [5], actually apply to the more general case of PACDPs.

To do this we will concentrate on the approach taken in [5] (because of the simpler analysis), which we sum up briefly below (primarily so that we can draw similarities with the method described in section 4). The reader is encouraged to consult the original paper for complete details of this method.

The algorithm was originally used to factor integers of the form $N = p^r q$. In this section we will restrict our attention to $r = 1$, though we note that the techniques still work for larger r (which would be analagous to ACDPs such that one of the inputs is near a *power* of the common divisor d).

A key observation in [5] is that if we have m polynomials $p_i(x) \in \mathbf{Z}[x]$, and we are assured that for some integer x_0 we have $p_i(x_0) = 0 \bmod t$ for all $i = 1 \dots m$, then any (polynomial) linear combination of these polynomials, i.e. $r(x) = \sum_{i=1}^{m} u_i(x)p_i(x)$ also has the property that $r(x_0) = 0 \bmod t$.

The trick is to find a polynomial $r(x)$ of this form, which is "small" when evaluated at all the "small" integers. Let $r(x) = r_0 + r_1 x + \dots + r_h x^h$. One way to ensure that $|r(x)| \leq hX^h$ when evaluated at any x such that $|x| < X$ is to make the r_i satisfy $|r_i|X^i \leq X^h$. This is the approach we shall take, and we shall use lattice reduction algorithms to generate these coefficients.

Notice that if t were such that $t > hX^h$ and x_0 were such that $|x_0| < X$, then the integer $r(x_0)$ which must be a multiple of t, but cannot be as large as t, must therefore equal 0. By finding all the roots of the equation $r(x) = 0$ over the integers we therefore find all possible x_0 for which the $p_i(x_0) = 0 \bmod t$ when $|x_0| < (t/h)^{1/h}$.

To see the relevance of this with the PACDP, notice that in the PACDP we are effectively given two polynomials[2] $q_1(x) = a_0 + x$ and $q_2(x) = b_0$, and told that $d > M = b_0^{\alpha_0}$ divides both $q_1(x_0)$ and $q_2(x_0)$ for some $|x_0| < X = b_0^{\beta_0}$. As in the previous section we will work out the conditions on β_0 (in terms of α_0) for us to be assured of finding such common divisors d.

Rather than just consider $q_1(x)$ and $q_2(x)$ directly we will calculate $r(x)$ by considering the polynomials $p_i(x) = q_1(x)^{u-i}q_2(x)^i$ for some fixed integer u and $i = 0 \dots u$. Let h be the degree of the polynomial $r(x)$ we are trying to produce. We will see later how the optimal choice of h is related to u, but for the moment just notice that $p_i(x_0) = 0 \bmod d^u$ for $i = 0 \dots u$.

In order to find the desired polynomial $r(x)$ and to work out the size of its coefficients we must describe the lattice we build. To this end we show an example below with $h = 4$ and $u = 2$.

$$\begin{pmatrix} b_0^2 & 0 & 0 & 0 & 0 \\ b_0 a_0 & b_0 X & 0 & 0 & 0 \\ a_0^2 & 2a_0 X & X^2 & 0 & 0 \\ 0 & a_0^2 X & 2a_0 X^2 & X^3 & 0 \\ 0 & 0 & a_0^2 X^2 & 2a_0 X^3 & X^4 \end{pmatrix}$$

The reason that we build the lattice generated by the rows of this type of matrix, is that each of the first $u+1$ rows corresponds to one of the polynomials $p_i(x)$, where the coefficient of x^j is placed into the $(j+1)^{\text{th}}$ column multiplied with $X^j = b_0^{j\beta_0}$. Notice that with this representation the integer matrix operations respect polynomial addition.

The remaining $h-u-1$ rows effectively allow for a *polynomial* multiple of the relation $(a_0+x)^u$ (polynomial multiples of the other relations are ignored because

[2] Notice that $q_2(x)$ is simply a constant, but we still refer to it as a (constant) polynomial.

of their obvious linear dependence). If these choices seem a little "plucked from the air", their justification can be seen in the determinant analysis.

We plan to reduce this lattice to obtain a small vector $\mathbf{r} = (r_0, r_1X, \ldots, r_hX^h)$ from which we obtain the polynomial $r(x) = \sum r_i x^i$.

It remains to see how small the entries of $\mathbf{r}$ will be. The dimension of the lattice is clearly $(h+1)$ and the determinant of the lattice can be seen to be

$$\Delta = X^{h(h+1)/2} b_0^{u(u+1)/2} = b_0^{u(u+1)/2 + \beta_0 h(h+1)/2},$$

which means we are assured of finding a vector $\mathbf{r}$ such that $|\mathbf{r}| < c\Delta^{1/(h+1)}$ for some c which (as in the previous section) is asymptotically small enough (compared to $\Delta^{1/(h+1)}$) for us to ignore in the following analysis. To actually find $\mathbf{r}$ one would use the LLL algorithm (see [8]) or one of its variants (e.g. [12]).

Notice that, as mentioed above, this bound on $\mathbf{r}$ is also (approximately) a bound on the integers $|r(x')|$, where $r(x) = \sum r_i x^i$, and x' is any integer such that $|x'| < X$.

We therefore wish $\Delta^{1/(h+1)}$ to be less than $d^u > b_0^{\alpha_0 u}$, so that the roots of $r(x) = 0$ over the integers must contain all solutions x_0 such that $d > b_0^{\alpha_0}$ divides both $q_1(x_0)$ and $q_2(x_0)$, and $|x_0| < X$.

For this to happen we require that $(u(u+1) + \beta_0 h(h+1))/(2(h+1)) < \alpha_0 u$, i.e.

$$\beta_0 < \frac{u(2(h+1)\alpha_0 - (u+1))}{h(h+1)}.$$

For a given α_0 the optimum choice of h turns out to be at approximately u/α_0; in which case we obtain that whenever

$$\beta_0 < \alpha_0^2 - \frac{\alpha_0(1-\alpha_0)}{h+1},$$

we will find all the possible x_0.

This proves the existence of algorithm GACD_L. One simply reduces the relevant lattice with $h = \lceil(\alpha_0(1-\alpha_0)/\varepsilon\rceil - 1$ and $u = \lceil h\alpha_0 \rceil$, and finds the roots of the resulting polyomial equation over the integers. The fact that there are only a polynomial number of possible d follows from the fact that the number of x_0 is bounded by h.

4 Using Lattices to Solve GACDP

In this section we attempt to solve GACDP by recovering the x and y such that "a paricularly large" d divides both $a_0 + x$ and $b_0 + y$. As mentioned in section 2 there are exponentially many x and y for $d > b_0^{2/3}$ which implies that our lattice technique cannot work above this bound. However we will show that the method does work (heuristically) right up to this bound.

The approach taken is similar to that described in section 3, which we will make frequent reference to, but the analysis must now deal with bivariate polynomials.

A particularly interesting question is whether one can make what follows completely rigorous. Such an argument is presently evading author, and as shown in [9] one cannot hope for a argument for a general bivariate modular equation, but it does not rule out that in in this case (and other specific cases) it may be possible to *prove* what follows.

In the GACDP we are given two polynomials[3] $q_1(x,y) = a_0 + x$ and $q_2(x,y) = b_0 + y$, where $a_0 \sim b_0$, and told that $d > M = b_0^{\alpha}$ divides $q_1(x_0,y_0)$ and $q_2(x_0,y_0)$ for some $|x_0|, |y_0| < X = b_0^{\beta}$.

The first thing to notice is that we can extend the general approach of the last section easily, since if we have m polynomials $p_i(x,y) \in \mathbf{Z}[x,y]$, and we are assured that for some integers x_0, y_0 we have $p_i(x_0,y_0) = 0 \bmod t$ for all $i = 1 \ldots m$, then any (polynomial) linear combination of these polynomials, i.e. $r(x,y) = \sum_{i=1}^{m} u_i(x,y) p_i(x,y)$ also has the property that $r(x_0,y_0) = 0 \bmod t$.

Again rather than just considering $q_1(x,y)$ and $q_2(x,y)$ directly we will calculate $r(x,y)$ by considering the polynomials $p_i(x,y) = q_1(x,y)^{u-i} q_2(x,y)^i$ for some fixed integer u. In this new case the role of the variable h is as a bound on the total degree of the polynomials. Notice that we still have $p_i(x_0,y_0) = 0 \bmod d^u$.

As an example of the lattice we create we show below the matrix whose rows generate the lattice, for $h = 4$ and $u = 2$. Note that we use the symbol Y to denote bound on $|y_0|$, though in fact $Y = X = b_0^{\beta_0}$ in our case (this is done so that the entries of the matrix may be more easily understood).

$$\begin{pmatrix}
a_0^2 & 2a_0X & X^2 & 0 & 0 & 0 & 0 & 0 & 0 & 0 & 0 & 0 & 0 & 0 & 0 \\
0 & a_0^2X & 2a_0X^2 & X^3 & 0 & 0 & 0 & 0 & 0 & 0 & 0 & 0 & 0 & 0 & 0 \\
0 & 0 & a_0^2X^2 & 2a_0X^3 & X^4 & 0 & 0 & 0 & 0 & 0 & 0 & 0 & 0 & 0 & 0 \\
a_0b_0 & b_0X & 0 & 0 & 0 & a_0Y & XY & 0 & 0 & 0 & 0 & 0 & 0 & 0 & 0 \\
0 & a_0b_0X & b_0X^2 & 0 & 0 & 0 & a_0XY & X^2Y & 0 & 0 & 0 & 0 & 0 & 0 & 0 \\
0 & 0 & a_0b_0X^2 & b_0X^3 & 0 & 0 & 0 & a_0X^2Y & X^3Y & 0 & 0 & 0 & 0 & 0 & 0 \\
b_0^2 & 0 & 0 & 0 & 0 & 2b_0Y & 0 & 0 & 0 & Y^2 & 0 & 0 & 0 & 0 & 0 \\
0 & b_0^2X & 0 & 0 & 0 & 0 & 2b_0XY & 0 & 0 & 0 & XY^2 & 0 & 0 & 0 & 0 \\
0 & 0 & b_0^2X^2 & 0 & 0 & 0 & 0 & 2b_0X^2Y & 0 & 0 & 0 & X^2Y^2 & 0 & 0 & 0 \\
0 & 0 & 0 & 0 & 0 & b_0^2Y & 0 & 0 & 0 & 2b_0Y^2 & 0 & 0 & Y^3 & 0 & 0 \\
0 & 0 & 0 & 0 & 0 & 0 & b_0^2XY & 0 & 0 & 0 & 2b_0XY^2 & 0 & 0 & XY^3 & 0 \\
0 & 0 & 0 & 0 & 0 & 0 & 0 & 0 & 0 & b_0^2Y^2 & 0 & 0 & 2b_0Y^3 & 0 & Y^4
\end{pmatrix}$$

Again, since we are proving the bounds given in algorithm GACD_L, we must treat β_0 as a variable for the time being.

As can be seen the situation is slightly different to that previously obtained in section 3, mainly due to the differences in the linear dependencies of polynomial multiples of the $p_i(x,y)$. As we see we can multiply each $p_i(x,y)$ by x^j for all $j = 0 \ldots h-u$ now, without incurring any linear dependencies. The down side is that only $(h-u)(h-u+1)/2$ other rows corresponding to multiples of $(b_0+y)^u$ may be added.

Alternatively seen, one can think of the $p_i(x,y)$ as $(a_0+x)^s(b_0+y)^t$ for $u \le s+t \le h$, since these generate the same basis as the above ones. We consider these because they admit an easier analysis, though in practice we would probably reduce the above ones because they are slightly more orthogonal.

[3] Notice that $q_1(x,y)$ is not dependent on y, and $q_2(x,y)$ is not dependent on x, but we cannot keep the univariate nature of them for long.

Again we wish to reduce this lattice, and recover a polynomial $r_1(x, y)$ which is small for all $|x|, |y| < X$. However knowing that all the x, y solutions to the GACDP are solutions to $r_1(x, y) = 0$ does not assure us of finding the x_0, y_0 we require; in general we cannot solve a bivariate equation over the integers.

The common (heuristic) argument at this stage, is to find another small vector in the basis, from which one produces another polynomial $r_2(x, y) = 0$ over the integers. If we are fortunate in that $r_1(x, y)$ and $r_2(x, y)$ are algebraically independent, then this does give us enough information to solve for all roots (x, y), e.g. using resultant techniques. Unfortunately we cannot *prove* that in general we can always find two small vectors which give rise to *algebraically* independent polyomials, although this would seem likely, and indeed is borne out by practical results (see Table 1).

Assuming that we can find two algebraically independent polynomials, we still have the problem of working out what bounds this lattice method implies. Working out the determinant of a lattice given by a non-square matrix can be a major piece of work; see for example [3]. Fortunately there is a trick in our example, which may prove useful in more general situations.

It is well known that the determinant of the lattice is bounded above by the product of all the norms of the rows. Our strategy for analysing the determinant of this lattice will be to perform some non-integral row operations[4] and then use the product of the norms of the rows.

As an example of this if we consider the fourth row of the above lattice, then this clearly corresponds to the polynomial $(a_0 + x)(b_0 + y)$. By writing $(b_0 + y)$ as $y_0 - (b_0/a_0)x + (b_0/a_0)(a_0 + x)$ we see that

$$(a_0 + x)(b_0 + y) = (a_0 + x)(y - \frac{a_0}{b_0}x) + \frac{a_0}{b_0}(n + x)^2.$$

This means that taking a_0/b_0 times the first row, away from the fourth row, will leave us with a polynomial representing $(a_0 + x)(y - (a_0/b_0)x)$. Since $a_0 \sim b_0$, if we look at each of the entries of the vector corresponding to this polynomial then we see that they are each less than b_0X (since $a_0 < b_0$), so this a bound for the norm of the vector too (again we ignore the aymptotically small factor of the difference of these two norms). This bound is a vast improvement on the b_0^2 we may have naïvely bounded the vector by.

In a similar way, we also see that the norms of the vectors corresponding to the polynomials $(a_0 + x)^s$ can be bounded by $X^{s-u}b_0^u$, since we can subtract suitable multiples of the vectors corresponding to the polynomials $(a_0 + x)^i$ for $i < s$.

In general the vector corresponding to a polynomial $p_i(x, y) = (a_0 + x)^s(b_0 + y)^t$ will contribute $X^{s+t-u}Y^t b_0^{u-t}$ for $0 \le t \le u$ and X^sY^t for $u < t \le h$. We can write the contribution of the polynomials in a table indexed by s and t, e.g. for the above example with $h = 4$ and $u = 2$

[4] These non-integral row operations will mean that the resulting matrix will no longer be a basis for our lattice, but it will have the same determinant which is all we care about estimating.

Y^4				
Y^3	XY^3			
Y^2	XY^2	X^2Y^2		
	b_0Y	b_0XY	b_0X^2Y	
		b_0^2	b_0^2X	$b_0^2X^2$

This representation allows us to count the dimension of the lattice and its determinant easily. The dimension is clearly $m = (h+1-u)(h+u+2)/2$, and the determinant is

$$\Delta = Y^{(h+1-u)(h(h+u+2)+u(u+1))/6} b_0^{u(u+1)(h+1-u)/2} X^{(h-u)(h+1-u)(h+2u+2)/6} = b_0^\delta$$

where $\delta = u(u+1)(h+1-u)/2 + \beta_0(h+1-u)(2h^2+4h+2uh-u^2-u)/6$.

As in section 3 we require that $\delta < mu\alpha_0$, so that the vectors LLL finds[5] will be valid over the integers. Representing this in terms of β_0 we obtain

$$\begin{aligned}\beta_0 &< \frac{3u(h+u+2)\alpha_0 - 3u(u+1)}{2h(h+u+2)-u(u+1)} \\ &= \frac{3u(h+u)\alpha_0 - 3u^2}{2h(h+u)-u^2} - \frac{3u(h-u)(2h-u\alpha_0)}{(2h(h+u)-u^2)(2h(h+u+2)-u(u+1))}.\end{aligned}$$

The optimal setting for γ such that $h = u/\gamma$ is now

$$\gamma = \frac{2 - 2\alpha_0 - \sqrt{4 - 4\alpha_0 - 2\alpha_0^2}}{3\alpha_0 - 2}$$

for which we obtain the bound

$$\beta_0 < 1 - \frac{1}{2}\alpha_0 - \sqrt{1 - \alpha_0 - \frac{1}{2}\alpha_0^2} - \varepsilon(h, \alpha_0)$$

where for completeness we give the precise $\varepsilon(h,\alpha_0) = \varepsilon_1/\varepsilon_2$; namely

$$\begin{aligned}\varepsilon_1 &= 3(3\alpha_0-2)\Big((\alpha_0-1)(11\alpha_0^2+8\alpha_0-12)\sqrt{4-4\alpha_0-2\alpha_0^2} \\ &\quad +(\alpha_0^2-16\alpha_0+12)(\alpha_0^2+2\alpha_0-2)\Big) \\ \varepsilon_2 &= 4h(\alpha_0^2+2\alpha_0-2)\left((10\alpha_0-8)\sqrt{4-4\alpha_0-2\alpha_0^2}+23\alpha_0^2-44\alpha_0+20\right) \\ &\quad +2(3\alpha_0-2)\Big((34\alpha_0^2-55\alpha_0+22)\sqrt{4-4\alpha_0-2\alpha_0^2} \\ &\quad +(19\alpha_0-14)(\alpha_0^2+2\alpha_0-2)\Big).\end{aligned}$$

Notice that $\lim_{h\to\infty} \varepsilon(h,\alpha_0) = 0$, and so this proves the existence of the algorithm `GACD_L`.

[5] Actually there are a few minor errors that can creep in due to: underestimating the determinant (since $a_0 \sim b_0$, but $a_0 \neq b_0$), transferring between the L_1 norm and L_2 norms, the fact that there is a $2^{(m-1)/4}$ LLL approximation factor and the common divisor may be overestimated (again since $a_0 \sim b_0$, but $a_0 \neq b_0$).

5 An Equivalent Problem?

In this paper we have been considering equations where d divides both $a_0 + x$ and $b_0 + y$, and d is particularly large with respect to the sizes of x and y. Notice that we may write this as

$$b'(a_0 + x) - a'(b_0 + y) = 0$$

where now $a' = (a_0 + x)/d$ and $b' = (b_0 + y)/d$ are particularly small.

A very related problem to this is finding small a' and b' such that there exists small x, y which satisfy

$$b'(a_0 + x) - a'(b_0 + y) = 1.$$

The author knows of no *reduction* between these two problems, but as we shall the techniques used to solve them are remarkably similar.

The *partial* variant of the second problem is very related to cryptography; indeed it has recently been named "the small inverse problem" in [3]. The connection with cryptography was first realised in [15] where it was shown (via a continued fraction method) that the use of a low decrypting exponent in RSA (i.e. one less than a quarter of the bits of the modulus N) is highly insecure; the public information of N and e effectively leaks the factorisation of N in polynomial time!

The problem was re-addressed in [3], where the new attack was now based on Coppersmith's lattice techniques, and the bound for which the decrypting exponent likely reveals the factorisation of N (in polynomial time) was increased to $1 - 1/\sqrt{2}$ of the bits of N. It is a heuristic method, but seems to work very reliably in practice.

In [3] it was hypothesised that this bound might be increased to $1/2$ the bits of N, because the bound of $1 - 1/\sqrt{2}$ seemed unnatural. However the author would like to point out that this state of affairs is unlikely, and in fact the bound of $1 - 1/\sqrt{2}$ *is* natural in that it is exactly what one would expect from the related PACDP.

Indeed if one had the bound $\beta_0 = 1/2 - \varepsilon$ for $\alpha_0 = 1/2$ for the PACDP (as one might expect the two problems to have the same bound), then this would imply a polynomial time factoring algorithm for RSA moduli[6]!

We note that this does not *prove* the falsity of the conjecture (even assuming factoring is not possible in polynomial time) because there is no known reduction between the small inverse problem and the PACDP (in either direction). However it does mean that for the conjecture to be true, the fact that the r.h.s. of the above equation is 1 rather than 0 must be of key importance to the algorithm which finds a', b' (which is not the case in any of the known attacks).

[6] One would just need to guess a constant amount of the top order bits of a factor of p, and use the algorithm to find the remaining bits.

6 Results

We sum up the results in the last sections by showing their bounds graphically. The variables $\alpha = \log_{b_0} M$ and $\beta = \log_{b_0} X$ are represented below by the $x-$ and $y-$ axes respectively. The lines show the maximum β for which the common divisor d can still be recovered for any given $\alpha \in (0 \ldots 1)$.

Figure 61. *The Bounds Implied by the Approximate Common Divisor Algorithms.*

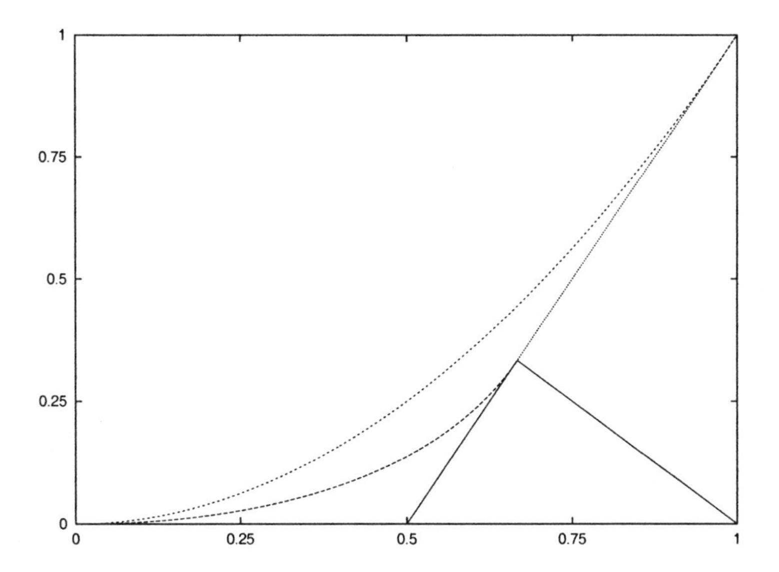

The top line is $\beta = \alpha^2$ which is the bound implied by PACD_L. It is responsible for theorems such as "factoring with high bits known", and (with a slight extension) "factoring integers of the form $N = p^r q$"; see [1] and [5].

The straight line given by $\beta = 2\alpha - 1$ is the bound implied by PACD_CF, and is equivalent to Wiener's attack on low decrypting exponent RSA. The line for GACD_CF starts off the same, but after $\alpha = 2/3$ it then becomes the line $\beta = 1 - \alpha$ to ensure a polynomial sized output.

The curved line between the previous two is

$$\beta = 1 - \frac{1}{2}\alpha - \sqrt{1 - \alpha - \frac{1}{2}\alpha^2}.$$

This is a new (but heuristic) lattice based result, explained in section 4, which also applies to GACDP. It interestingly shows that the problem may be solved even when $\alpha < 1/2$.

We now give some practical results to demonstrate the last of these methods working in practice. We show that indeed we can recover the common divisor d

in reasonable time, and that the size of the short vectors are not significantly smaller than predicted by the Minkowski bound, which impies that no sublattice of the one we reduce will reveal a better determinant analysis (this ratio is given logarithmically below by δ).

The results were carried out on a Pentium II 700MHz machine, running Redhat Linux. The software was written in C++ and used the NTL Library [13]. Notice that for consistency the total number of bits of a and b was kept near 1024.

Table 1. Table of Practical Results.

# bits of divisor	# bits of a,b	α	β_{max}	h	u	$time(s)$	δ	# bits of error	β
205	1025	0.2	0.016	10	1	731	0.941	5	0.004
307	1023	0.3	0.041	5	1	8	0.998	15	0.012
				6	1	20	0.972	20	0.019
				7	1	43	0.965	22	0.019
410	1025	0.4	0.079	4	1	2	0.984	43	0.041
				5	2	28	0.990	41	0.040
				6	2	83	0.987	51	0.049
				7	2	197	0.990	56	0.055
512	1024	0.5	0.137	4	2	9	0.994	103	0.100
				5	2	34	0.993	107	0.104
				6	3	261	0.996	113	0.110
614	1023	0.6	0.231	4	3	18	0.998	213	0.208
				5	3	100	0.990	207	0.204
				6	4	507	0.997	216	0.211

(We bring to the readers attention the fact that some of the above choices for u are sub-optimal. This effect can be noticed in the bound on β).

7 Conclusions and Open Problems

The first problem is to find more uses for approximate integer common divisor problems in cryptography, or any other part of computational mathematics (especially for the general case). Without this the techniques described herein will find it hard to reach a large target audience.

Another interesting question is to see if one can find reductions between the "= 0" and "= 1" variants given in section 5, i.e. the small inverse problem and PACDP.

The next obvious thing to do is to generalise ACDPs even more. For instance one could complete the analysis (aluded to in section 4) of when d^r divides a number close to one of the inputs. Alternatively one could work out the bounds when one has many inputs, all close to numbers which share a large common divisor d. For completeness these extensions will appear in the final version of this paper.

The last problems are of a more fundamental nature to do with lattice analysis. The proof of algebraic independence of the two bivariate polynomials still remains as a key result to prove in this field. A proof would turn both this method and the ones described in [3] and [4] in to "fully fledged" polynomial time algorithms, rather than the heuristic methods they are currently written up as. Of course as shown by [9] it is not possible to hope for a general solution to bivariate modular equations, but in this particular case (and others) one may hope to find a rigorous proof.

Finally we ask if the bounds implied by section 4 are the best possible, in polynomial time? One way in which this result may not be optimal, is that our determinant analysis may be too pessimistic, or alternatively there may be a sublattice of this lattice, for which the reduced dimension and determinant imply better bounds on the GACDP (for example this type of effect was seen in the first lattice built in [3]).

Evidence against both of these states of affairs is given by the results in Table 1, i.e. they do show that the vectors obtained are approximately what one would expect from the Minkowski bound (though for smaller α there is a small observed discrepancy), which means that neither of these above situations is likely with our method.

This being said, it still remains a very interesting and open theoretical question to be able to identify, in advance, which lattices are built "incorrectly", i.e. those for which some of the rows actually hinder the determinant analysis (e.g. like the first one in [3]).

A third way to improve the results is that one could build an entirely different lattice, whose analyis simply admitted better bounds than the ones reached here. Nothing is presently known about the existence or non-existence of such an algorithm.

Acknowledgements

The author would particularly like to thank Don Coppersmith, Yuval Ishai and Phong Nguyen for interesting conversations relating to this subject, and is also very grateful for the period of time spent on this problem at HP Labs in Bristol.

References

1. D. Coppersmith. Finding a small root of a bivariate integer equation *Proc. of Eurocrypt'96* Lecture Notes in Computer Science, Vol. 1233, Springer-Verlag, 1996
2. D. Boneh. Twenty years of attacks on the RSA cryptosystem. *Notices of the American Mathematical Society (AMS)* Vol. 46, No. 2, pp. 203–213, 1999.
3. D. Boneh and G. Durfee. Cryptanalysis of RSA with private key d less than $N^{0.292}$ *IEEE Transactions on Information Theory*, Vol 46, No. 4, pp. 1339–1349, July 2000.

4. D. Boneh, G. Durfee and Y. Frankel. An attack on RSA given a small fraction of the private key bits. *In proceedings AsiaCrypt'98*, Lecture Notes in Computer Science, Vol. 1514, Springer-Verlag, pp. 25–34, 1998.
5. D. Boneh, G. Durfee and N. Howgrave-Graham Factoring $N = p^r q$ for large r. *In Proceedings Crypto '99*, Lecture Notes in Computer Science, Vol. 1666, Springer-Verlag, pp. 326–337, 1999.
6. G.H. Hardy and E.M. Wright. An introduction to the theory of numbers, 5'th edition. Oxford University press, 1979.
7. N.A. Howgrave-Graham. Computational mathematics inspired by RSA. Ph.D. Thesis, Bath University, 1999.
8. A.K. Lenstra, H.W. Lenstra and L. Lovász. Factoring polynomials with integer coefficients *Mathematische Annalen*, Vol. 261, pp. 513–534, 1982.
9. K.L. Manders and L.M. Adleman. NP-Complete decision problems for binary quadratics *JCSS* Vol. 16(2), pp. 168–184, 1978.
10. P. Nguyen and J. Stern. Lattice reduction in cryptology: An update", *Algorithmic Number Theory – Proc. of ANTS-IV*, volume 1838 of LNCS. Springer-Verlag, 2000.
11. T. Okamoto. Fast public–key cryptosystem using congruent polynomial equations *Electronic letters*, Vol. 22, No. 11, pp. 581–582, 1986.
12. C–P. Schnorr. A hierarchy of polynomial time lattice bases reduction algorithms *Theoretical computer science*, Vol. 53, pp. 201–224, 1987.
13. V. Shoup. NTL: A Library for doing Number Theory (version 4.2) `http://www.shoup.net`
14. B. Vallée, M. Girault and P. Toffin. *Proceedings of Eurocrypt '88* LNCS vol. 330, pp. 281–291, 1988.
15. M. Wiener. Cryptanalysis of short RSA secret exponents *IEEE Transactions of Information Theory* volume **36**, pages 553-558, 1990.

Segment LLL-Reduction of Lattice Bases

Henrik Koy[1] and Claus Peter Schnorr[2]

[1] Deutsche Bank AG, Frankfurt am Main
henrik.koy@db.com
[2] Fachbereiche Mathematik und Informatik
Universität Frankfurt, PSF 111932
D-60054 Frankfurt am Main, Germany
schnorr@cs.uni-frankfurt.de

Abstract. We present an efficient variant of LLL-reduction of lattice bases in the sense of LENSTRA, LENSTRA, LOVÁSZ. We organize LLL-reduction in segments of size k. Local LLL-reduction of segments is done using local coordinates of dimension k.
We introduce *segment* LLL-*reduced bases*, a variant of LLL-reduced bases achieving a slightly weaker notion of reducedness, but speeding up the reduction time of lattices of dimension n by a factor n. We also introduce a variant of LLL-reduction using *iterated segments*. The resulting reduction algorithm runs in $O(n^3 \log_2 n)$ arithmetic steps for integer lattices of dimension n with basis vectors of length 2^n.
Keywords: LLL-reduction, shortest lattice vector, segments, iterated segments, local coordinates, local LLL-reduction, divide and conquer.

1 Introduction

The famous algorithm for LLL-reduction of lattice bases of LENSTRA, LENSTRA, LOVÁSZ [LLL82] is a basic technique for solving important problems in algorithmic number theory, integer optimization, diophantine approximation and cryptography. Of the many possible applications we refer to a few recent ones [BN00,Bo00,Co97,NS00]. Present codes for LLL-reduction merely perform for lattices up to dimension 350, our new contributions lift this barrier beyond dimension 1000. In this paper we present a theoretic reduction algorithm together with a rigorous analysis counting the number of arithmetic steps on integers of bounded length. In the companion paper [KS01] we introduce an orthogonalization via scaling using floating point arithmetic that is stable in very high dimension. The resulting segment LLL-reduction with floating point orthogonalization is a highly practical reduction algorithm that is in practice much faster than previous codes for LLL-reduction. In practice it finds a lattice basis that is as good as a truly LLL-reduced basis.

In this paper we propose the concept of segment LLL-reduction in which a basis $b_1, \ldots, b_n$ of dimension $n = km$ is partitioned into m segments $b_{kl+1}, \ldots, b_{kl+k}$ of k consecutive basis vectors. Segment LLL-reduction is designed to do most of the LLL-exchanges within the segments using local coordinates of dimension

J.H. Silverman (Ed.): CaLC 2001, LNCS 2146, pp. 67–80, 2001.

k. There is a double advantage. Firstly, local LLL-vector operations cost merely $O(k)$ arithmetic steps, whereas global vector operations require $O(n)$ steps. Secondly, as local operations are for lattices of small dimension they can be done in single precision whereas global operations require multi-precision steps.

First we introduce k-segment reduced bases, a variant of LLL-reduced bases that is designed to minimize the global overhead associated to the local LLL-reductions. Segment LLL-reduction saves a factor n in the running time compared to standard LLL-reduction of lattices of dimension n. Using *iterated segments* we present an even faster theoretic reduction algorithm that runs in $O(n^3 \log_2 n)$ arithmetic steps for integer lattices of dimension n with basis vectors of length $O(2^n)$. In this paper we analyse a theoretic version of the novel reduction algorithms in the model of integer arithmetic. In the companion paper [KS01] we propose a practical implementation using floating point orthogonalization. Our present code reduces lattice bases of dimension 1000 consisting of integers of bit length 400 in 10 hours on a 800 MHz PC. Even for dimension $n < 100$ the new code is in practice much faster than previous codes. We did not yet implement reduction using iterated segments. The use of iterated segments should further speed up the reduction in high dimensions.

Previously, Schönhage [Sc84] has used local coordinates to speed-up LLL-reduction. His concept of semi-reduction approximates the length of the shortest lattice vector up to a factor 2^n whereas we get close to a factor $(4/3)^{n/2}$. We use the [Sc84] analysis of the size of integers occuring during the reduction.

Future work. We expect that the practical reduction algorithm of [KS01] will greatly improve the present codes for finding very short lattice vectors, in particular for lattices of high dimension. LLL-reduction is the basis for the more powerful reduction algorithms of Kannan [K84] and of block reduction of Schnorr [S87,S94]. Combining the segment LLL-reduction with the techniques of [S87,S94,SH95] will speed up the search for very short basis vectors. The novel concept of segment LLL-reduction and the block reduction of [S87,S94] are based on similar structures and can be easily combined. Moreover, the search for very short lattice vectors can be enhanced by the new concept of primal-dual segment reduction proposed by Koy [K01].

2 LLL-Reduction of Lattice Bases

Let $\mathbf{R}^d$ be the d–dimensional real vector space with the Euclidean inner product $\langle \cdot, \cdot \rangle$ and the Euclidean norm, called the length, $\|y\| = \langle y, y \rangle^{1/2}$. An integer *lattice* $L \subset \mathbf{Z}^d$ is an additive subgroup of $\mathbf{Z}^d$. Its *dimension* is the dimension of the minimal linear subspace that contains L. Every lattice L of dimension n has a *basis*, i.e., a set $b_1, \ldots, b_n$ of linearly independent vectors satisfying

$$L = \{t_1 b_1 + \ldots + t_n b_n \,|\, t_1, \ldots, t_n \in \mathbf{Z}\}.$$

Let $L(b_1, \ldots, b_n)$ denote the lattice with basis $b_1, \ldots, b_n$.

With an ordered lattice basis $b_1, \ldots, b_n \in \mathbf{Z}^d$ we associate the *Gram–Schmidt orthogonalization* $\widehat{b}_1, \ldots, \widehat{b}_n \in \mathbf{R}^d$ which can be computed together with the

Gram–Schmidt coefficients $\mu_{j,i} = \langle b_j, \widehat{b}_i \rangle / \langle \widehat{b}_i, \widehat{b}_i \rangle$ by the recursion

$$\widehat{b}_1 = b_1, \; \widehat{b}_j = b_j - \sum_{i=1}^{j-1} \mu_{j,i} \widehat{b}_i \qquad \text{for } j = 2, \ldots, n.$$

We have $\mu_{j,j} = 1$ and $\mu_{j,i} = 0$ for $i > j$. From the above equations we have the Gram–Schmidt decomposition

$$[b_1, \ldots, b_n] = [\widehat{b}_1, \ldots, \widehat{b}_n] \; [\mu_{j,i}]^{\mathsf{T}}_{1 \le i,j \le n},$$

where $[b_1, \ldots, b_n]$ denotes the matrix with column vectors $b_1, \ldots, b_n$ and $[\mu_{j,i}]^{\mathsf{T}}$ is the transpose of the matrix $[\mu_{j,i}]$. The *determinant* of the lattice $L(b_1, ..., b_n)$ is defined

$$\det L = \det([b_1, ..., b_n][b_1, ..., b_n]^{\mathsf{T}})^{1/2}.$$

Definition 1. *An ordered basis* $b_1, \ldots, b_n \in \mathbf{Z}^d$ *of the lattice* L *is* LLL-reduced *with* $\delta \in]\frac{1}{4}, 1]$ *if it has properties* **1.** *and* **2.**:

1. $\quad |\mu_{j,i}| \le 1/2 \qquad$ *for* $1 \le i < j \le n$,

2. $\quad \delta \, \|\widehat{b}_i\|^2 \;\le\; \mu_{i+1,i}^2 \, \|\widehat{b}_i\|^2 + \|\widehat{b}_{i+1}\|^2 \qquad$ *for* $i = 1, \ldots, n-1$.

LLL-reduced bases have been introduced by A.K. Lenstra, H.W. Lenstra, Jr. and L. Lovász [LLL82] who focused on $\delta = 3/4$. A basis with property **1.** is called *size-reduced.* A basis $b_1, ..., b_n$ is good if the values $\|b_i\|$ are good approximations to the successive minima. The i-th *successive minimum* λ_i of a lattice L, relative to the Euclidean norm, is the smallest real number r such that there are i linearly independent vectors in L of length at most r. Extending [LLL82] to arbitrary $\delta \in]\frac{1}{4}, 1]$ and $\alpha := 1/(\delta - \frac{1}{4})$ yields

Theorem 1. *A basis* $b_1, \ldots, b_n$ *of lattice* L *that is* LLL*-reduced with* δ *satisfies*:

1. $\|b_i\|^2 \le \alpha^{n-1} \, \lambda_i^2$ *and* $\|b_1\|^2 \le \alpha^{i-1} \, \|\widehat{b}_i\|^2 \qquad$ *for* $i = 1, \ldots, n$,

2. $\|b_1\|^2 \le \alpha^{\frac{n-1}{2}} (\det L)^{\frac{2}{n}}$ *and* $\|\widehat{b}_n\|^2 \ge \alpha^{-\frac{n-1}{2}} (\det L)^{\frac{2}{n}}$.

Consider the QR-factorization $B = QR$ of the basis matrix $B = [b_1, \ldots, b_n] \in \mathbf{Z}^{d \times n}$, where $Q \in \mathbf{R}^{d \times d}$ is an orthogonal matrix and $R = [r_{i,j}] = [\mathbf{r}_1, ..., \mathbf{r}_n] \in \mathbf{R}^{d \times n}$ is an upper triangular matrix, $r_{i,j} = 0$ for $i > j$. We have $\mu_{j,i} = r_{i,j}/r_{i,i}$ and $|r_{i,i}| = \|\widehat{b}_i\|$. We present the core of the LLL-reduction algorithm using the coefficients $r_{i,j}$ of the matrix R. The vector $\mathbf{r}_l$ is the orthogonal transform of b_l.

LLL

INPUT $\quad b_1, \ldots, b_n \in \mathbf{Z}^d, \; \delta$

OUTPUT $\; b_1, \ldots, b_n \quad$ LLL-reduced basis

1. $\; l := 1 \qquad$ *(l is the stage)*

2. **while** $l \le n$ **do**

$\qquad$ compute $\mathbf{r}_l = (r_{1,l}, ..., r_{l,l}, 0, ..., 0)^{\mathsf{T}}$,

size-reduce b_l against $b_{l-1}, ..., b_1$.
if $l \neq 1$ **and** $\delta r_{l-1,l-1}^2 > r_{l-1,l}^2 + r_{l,l}^2$
then swap b_{l-1}, b_l, $l := l-1$ **else** $l := l+1$. *end*

The LLL-algorithm locally reduces the 2×2-diagonal submatrices of R by successively decreasing the length of the first column vector in a 2×2-matrix. While the coefficients $r_{i,j}$ of the matrix R are not rational, rational arithmetic must use the rational numbers $\mu_{j,i}$, $\|\widehat{b}_i\|^2$ satisfying $r_{i,j} = \mu_{j,i}\|\widehat{b}_i\|$, $r_{i,i} = \|\widehat{b}_i\|$.

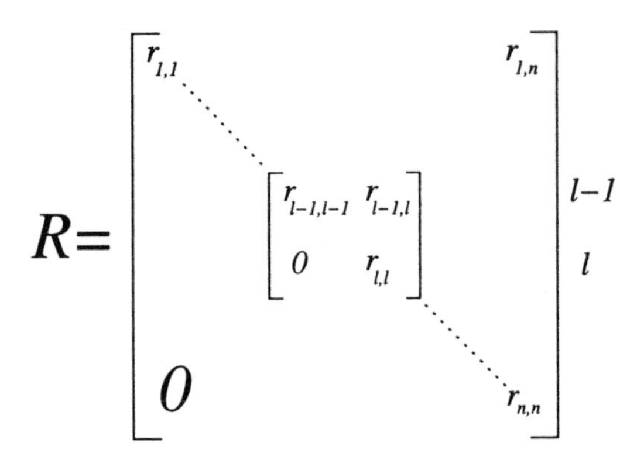

Fig. 1. The 2×2-Diagonal Submatrices of R.

LLL-time bound. Following [Sc84] we measure the size of the input in terms of

$$M_{Sc} =_{\text{def}} \max_{i=1,\ldots,n} (2^n, \|b_i\|^2, D_i), \quad \text{where} \quad D_i = \|\widehat{b}_1\|^2 \cdots \|\widehat{b}_i\|^2$$

is the *Gramian determinant* of the sublattice with basis $b_1, ..., b_i$. The M_{Sc}-measure differentiates large and small lattice determinants. It relates to the [LLL82]-measure $M = \max_i \|b_i\|^2$ by $M \leq M_{Sc} \leq M^n$.

By the Lovász volume argument, the number of LLL-exchanges — swaps of b_{l-1}, b_l — is at most $(n-1)\log_{1/\delta} M_{Sc}$. Size-reduction of b_l requires $O(dl)$ arithmetic steps. **LLL** performs at most $O(n^2 d \log_{1/\delta} M_{Sc})$ arithmetic steps. These steps operate on rational integers where numerator and denominator have at most $O(\log_2 M_{Sc})$ bits, see [LLL82,Sc84].

3 Segment LLL-Reduction

Segments and Local Coordinates. Let the basis $b_1, \ldots, b_n \in \mathbf{Z}^d$ have dimension $n = k \cdot m$ and the QR-factorization $[b_1, \ldots, b_n] = QR$. We partition the basis matrix B into m *segments* $B_l = [b_{k(l-1)+1}, \ldots, b_{kl}]$ for $l = 1, ..., m$. Local reduction of two consecutive segments uses the $2k \times 2k$-submatrix $R_l := [r_{kl+i,kl+j}]_{-k<i,j\leq k} \in \mathbf{R}^{2k \times 2k}$ of $R \in \mathbf{R}^{d \times n}$, corresponding to two consecutive segments B_{l-1}, B_l. More precisely, local reduction uses the rational representation of R_l by the coefficients $\mu_{kl+i,kl+j}$ and $\|\widehat{b}_{kl+i}\|^2$. We want to do most of

the LLL-exchanges and the corresponding size-reduction in local coordinates of some R_l. Extra global transformations are required after local LLL-reduction. The novel concept of *k-segment reduced bases* intends to minimize these global costs. We let $D(l) = \|\widehat{b}_{k(l-1)+1}\|^2 \cdots \|\widehat{b}_{kl}\|^2$ denote the *local Gramian determinant* of segment B_l. We have that $D_{kl} = D(1) \cdots D(l)$.

Definition 2. *We call an ordered basis* $b_1, \dots, b_n \in \mathbf{Z}^d, n = km$, *k*-segment LLL-reduced *with* $\delta \in]\frac{1}{4}, 1]$ *if it is size-reduced and satisfies for* $\alpha = 1/(\delta - \frac{1}{4})$:

1. $\delta \|\widehat{b}_i\|^2 \leq \mu_{i+1,i}^2 \|\widehat{b}_i\|^2 + \|\widehat{b}_{i+1}\|^2$ *for* $i \neq 0 \mod k$,
2. $D(l) \leq (\alpha/\delta)^{k^2} D(l+1)$ *for* $l = 1, \dots, m-1$,
3. $\delta^{k^2} \|\widehat{b}_{kl}\|^2 \leq \alpha \|\widehat{b}_{kl+1}\|^2$ *for* $l = 1, \dots, m-1$.

We use Inequality **3.** to bound in Theorem 2 $\|b_i\|$ by the successive minimum λ_i. Without Inequality **3.** we bound in Theorem 5 $\|b_1\|$ by $(\det L)^{\frac{1}{n}}$ as well as by $\|\widehat{b_n}\|$. The large exponent k^2 of δ^{k^2} in **3.** is impractical for $k > \sqrt{n}$, we drop Inequality **3.** of the definition 2 in Section 4.

Lemma 1. *A k-segment* LLL-*reduced basis* $b_1, \dots, b_n$ *satisfies*

1. $\delta^{2k^2+j-i} \|\widehat{b}_i\|^2 \leq \alpha^{j-i} \|\widehat{b}_j\|^2$ *for* $1 \leq i < j \leq n$,
2. $\delta^{n-1} \|b_1\|^2 \leq \alpha^{n-1} \|\widehat{b}_n\|^2$ *and* $\delta^{k^2+j-1} \|\widehat{b}_1\|^2 \leq \alpha^{j-1} \|\widehat{b}_j\|^2$ *for* $1 < j \leq n$.

Proof. The inequalities **2.** of Definition 2 imply for $l \leq l'$ that

$$D(l) \leq (\alpha/\delta)^{k^2(l'-l)} D(l').$$

As $D(l) = \|\widehat{b}_{k(l-1)+1}\|^2 \cdots \|\widehat{b}_{kl}\|^2$, there exists s, $1 \leq s \leq k$ such that

$$\|\widehat{b}_{k(l-1)+s}\|^2 \leq (\alpha/\delta)^{k(l'-l)} \|\widehat{b}_{k(l'-1)+s}\|^2.$$

Inequality **1.** follows by combining the latter inequality — at each end $k(l-1)+s$ and $k(l'-1)+s$ — with some inequalities $\|\widehat{b}_\nu\|^2 \leq \alpha \|\widehat{b}_{\nu+1}\|^2$, which hold within the segments, and possibly with an inequality **3.** of Definition 2 that bridges two consecutive segments. In particular, we can choose l, l' so that

$$i \leq k(l-1) + s \leq k(l'-1) + s \leq j$$

and that each pair $\{i, k(l-1)+s\}$ and $\{j, k(l'-1)+s\}$ are indices either of the same or of two consecutive segments.

Inequalities **2.** If $i = 1, j = n$ we can choose $l = 1, l' = m$, and thus each pair $\{1, k(l-1)+s\}, \{n, k(l'-1)+s\}$ is in a single segment. Consequently, we have that $\delta^{n-1} \|b_1\|^2 \leq \alpha^{n-1} \|\widehat{b}_n\|^2$. If $i = 1$ the pair $\{i, k(l-1)+s\}$ is in one segment and thus $\delta^{k^2+j-1} \|b_1\|^2 \leq \alpha^{j-1} \|\widehat{b}_j\|^2$. □

Theorem 2. *Let $b_1, \dots, b_n$ be a basis that is k-segment* LLL*-reduced with δ. Then we have for $i = 1, \dots, n$:*

$$\delta^{2k^2+n-1}\, \|b_i\|^2 \le \alpha^{n-1}\lambda_i^2 \qquad \textit{and} \qquad \delta^{k^2+i-1}\, \|b_1\|^2 \le \alpha^{i-1}\|\widehat{b}_i\|^2,$$

where $\lambda_1 \le \dots \le \lambda_n$ are the successive minima of the lattice.

The proof of Theorem 2 follows from Lemma 1 by standard arguments. Comparison of Theorems 1 and 2 shows that k-segment reduced bases are close to LLL-reduced bases.

Algorithm for Segment LLL*-Reduction.* The algorithm **segment LLL** transforms a given basis into a k-segment reduced basis. It iterates local LLL-reduction of two segments $[B_{l-1}, B_l] = [b_{kl-k+1}, \dots, b_{kl+k}]$ via: *The procedure* **loc-LLL**(l).

Given the orthogonalization of a k-segment reduced basis $b_1, \dots, b_{kl-k}$ the procedure **loc-LLL**(l) computes the orthogonalization and size-reduction of the segments B_{l-1}, B_l. In particular it provides the submatrix $R_l \in \mathbf{R}^{2k\times 2k}$ of $R \in \mathbf{R}^{d\times n}$ corresponding to the segments B_{l-1}, B_l. Thereafter it performs a local LLL–reduction of R_l and stores the LLL-transformation in the matrix $H \in \mathbf{Z}^{2k\times 2k}$. Finally, it transforms $[B_{l-1}, B_l]$ into the locally reduced segments $[B_{l-1}, B_l]H$ and size-reduces $[B_{l-1}, B_l]$ globally.

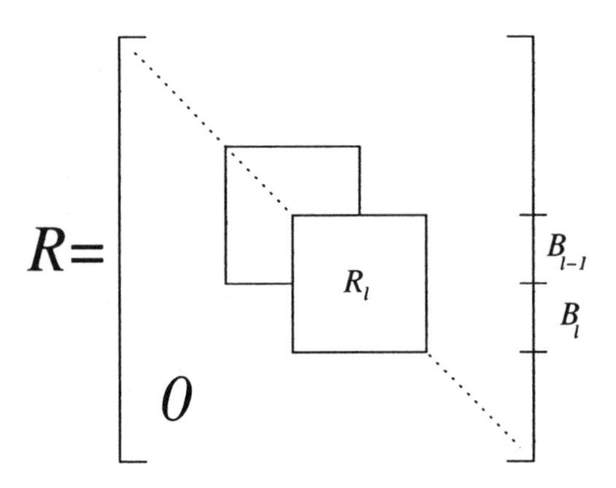

Fig. 2. Areas of Subsequent Local LLL-Reductions.

Each execution of **loc-LLL**(l) induces a global overhead of $O(ndk)$ arithmetic steps for global size-reduction, orthogonalization and segment transformation via H. The efficiency relies on the fast local LLL-reduction of R_l. Here each LLL-exchange of two consecutive basis vectors costs merely $O(k^2)$ arithmetic steps, local size-reduction included. Compare this to the $O(nd)$ arithmetic steps for an LLL-exchange in global coordinates. Here is our segment LLL-reduction algorithm.

Segment LLL

INPUT $b_1, \dots, b_n \in \mathbf{Z}^d, k, m, n = km, \ \delta$

OUTPUT $b_1, \dots, b_n$ k-segment LLL-reduced basis

1. $l := 1$

2. **while** $l \leq m - 1$ **do**

loc-LLL(l)

if $l \neq 1$ **and**

$(D(l-1) > (\alpha/\delta)^{k^2} D(l)$ **or** $\delta^{k^2} \|\widehat{b}_{k(l-1)}\|^2 > \alpha \|\widehat{b}_{k(l-1)+1}\|^2)$

then $l := l - 1$ **else** $l := l + 1$. *end*

The original LLL-algorithm — with δ replaced by δ^2 — essentially coincides with the case $k = 1$ of **segment LLL**. [1]

Segment LLL proceeds like the original LLL-algorithm replacing vectors by segments of size k. Obviously, **segment LLL** results in a k-segment LLL-reduced basis.

Time Analysis. The dominant work of **segment LLL** consists of the global overhead of the executions of **loc-LLL**(l). Initial and final global size-reduction of the segments B_{l-1}, B_l cost $O(ndk)$ arithmetic steps per execution. These costs also cover the initial computation of the column vectors $\mathbf{r}_{kl-k+1}, \dots, \mathbf{r}_{kl+k} \in \mathbf{R}^d$ of the matrix R and the global transform of $[B_{l-1}, B_l]$ into $[B_{l-1}, B_l]H$. Note that the overhead $O(ndk)$ of **loc-LLL**(l) is linear in the segment size k. Next we show by the Lovász volume argument that the number of executions of **loc-LLL** decreases cubically in k.

We let decr denote the number of times that the condition

$$D(l-1) > (\alpha/\delta)^{k^2} D(l) \quad \textbf{or} \quad \delta^{k^2} \|\widehat{b}_{kl}\|^2 > \alpha \|\widehat{b}_{kl+1}\|^2$$

holds and l is decreased for some l. The number of iterations of the while loop and the number of executions of **loc-LLL** is $m - 1 + 2 \cdot \mathrm{decr}$.

Theorem 3. $\mathrm{decr} \leq 2\frac{m-1}{k^2} \log_{1/\delta} M_{Sc} < 2\frac{n}{k^3} \log_{1/\delta} M_{Sc}$.

Remarks. 1. For $k = 1, m = n$ the bound of Theorem 3 shows that there are at most $(n-1)\log_{1/\delta} M_{Sc}$ LLL-exchanges of two consecutive basis vectors during reduction. The factor 2 in Theorem 3 disappears as **segment LLL** with $k = 1$ corresponds to **LLL** using δ^2.

2. If the reduction reverses the order of the basis $b_1, \dots, b_n$ we have $\mathrm{decr} \geq \binom{m}{2} = \frac{m(m-1)}{2}$. Then, by Theorem 3 we must have that $\log_{1/\delta} M_{Sc} \geq nk/4$, $M_{Sc} \geq \delta^{-nk/4}$. Thus, the bound M_{Sc} must be rather large for interesting bases.

[1] The inequality $D(l-1) > (\alpha/\delta)^{k^2} D(l)$ holds for $k = 1$ if and only if $\delta\, r_{l-1,l-1}^2 > \alpha\, r_{l,l}^2$. Multiplying the latter inequality by $\alpha = 1/(\delta - \frac{1}{4})$ implies that $\delta^2 r_{l-1,l-1}^2 > \frac{1}{4} r_{l-1,l-1}^2 + r_{l,l}^2$ which violates an Inequality **2.** of Definition 1 provided that δ is replaced by δ^2.

Proof. The Gramian determinant D_{kl} is the product $D_{kl} = D(1)\cdots D(l)$ of the first l local determinants. **loc-LLL**(l) performs a local LLL-reduction of two segments B_{l-1}, B_l, it merely changes $D_{k(l-1)}$, the Gramian determinant of $b_1, ..., b_{kl-k}$, and leaves $D_{kl'}$ for $l' \neq l-1$ unchanged.

Consider an execution of **loc-LLL**(l) performed after a decrease of l. We show that it decreases $D(l-1)$ by the factor $\delta^{k^2/2}$. First consider the case that $D(l-1) > (\alpha/\delta)^{k^2} D(l)$ holds upon entry of **loc-LLL**. As $D^{ter}(l-1) \leq \alpha^{k^2} D^{ter}(l)$ holds upon termination, and since the product $D(l-1)D(l)$ does not change we have that

$$\begin{aligned} D^{ter}(l-1) &\leq \alpha^{k^2} D^{ter}(l) \\ &= \alpha^{k^2} D(l-1)D(l)/D^{ter}(l-1) \\ &\leq \delta^{k^2} D(l-1)^2/D^{ter}(l-1), \end{aligned}$$

and thus $D^{ter}(l-1) \leq \delta^{k^2/2} D(l-1)$.

If **loc-LLL**(l) is performed in the case $\delta^{k^2} \|\widehat{b}_{kl}\|^2 > \alpha \|\widehat{b}_{kl+1}\|^2$ the previous argument shows again that $D^{ter}(l-1) \leq \delta^{k^2/2} D(l-1)$. Hence, **loc-LLL**$(l)$ decreases $\mathbf{D} =_{\text{def}} \prod_{j=1}^{m-1} D_{jk}$ by the factor $\delta^{k^2/2}$. As $\mathbf{D}$ is a positive integer, $\mathbf{D} \leq M_{Sc}^{m-1}$ this implies

$$\text{decr} \leq \log_{1/\delta^{k^2/2}} M_{Sc}^{m-1} \leq 2\tfrac{m-1}{k^2} \log_{1/\delta} M_{Sc}. \qquad \square$$

Theorem 4. *For $k = \Theta(m) = \Theta(\sqrt{n})$* **segment LLL** *performs*

$$O(nd \log_{1/\delta} M_{Sc})$$

arithmetic steps using integers of bit length $O(\log_2 M_{Sc})$.

Proof. Time Bound. There are at most $(n \log_{1/\delta} M_{Sc})$ LLL-exchanges — each requiring $O(k^2)$ steps for local size-reduction. There are decr $\leq 2\frac{m}{k^2} \log_{1/\delta} M_{Sc}$ calls of **loc-LLL** — each requiring $O(ndk)$ arithmetic steps for global size-reduction, global orthogonalization and global transformation of two consecutive segments. The choice $k, m = \Theta(\sqrt{n})$ equalizes for $d = O(n)$ the theoretic time bounds $O(k^2 n \log_{1/\delta} M_{Sc})$ for the LLL-exchanges in local coordinates and $O(n^2 k \frac{n}{k^3} \log_{1/\delta} M_{Sc})$ for the global overhead.

We need that $M_{Sc} \geq 2^n$, otherwise the $O(nd \log_{1/\delta} M_{Sc})$ bound does not cover the $O(n^2 d)$ steps required for QR-factorization, the $m-1$ calls **loc-LLL**(l) for $l = 1, ..., m-1$ also require $O(n^2 d)$ steps for global size-reduction.

Size of the integers. We first show that the initial bound

$$M_{Sc} = \max_{i=1,\ldots,n} (2^n, \|b_i\|^2, D_i)$$

can temporarily increase not more than by a factor $2\,\alpha^{n-1}$ for $\delta \geq \frac{3}{4}$. Recall that the determinants D_i do not increase during LLL-reduction. In particular, we always have that $1 \leq D_i \leq M_{Sc}$, and $\|\widehat{b}_i\|^2 = D_i/D_{i+1}$ is a rational integer, $M_{Sc}^{-1} \leq \|\widehat{b}_i\|^2 \leq M_{Sc}$ with numerator and denominator bounded by M_{Sc}.

The length $||b_i||^2$ can only temporarily increase during size-reduction of b_i according to $b_i := b_i - \lceil \mu_{i,h} \rfloor b_h$ for $h = i-1, ..., 1$. Assuming that $b_1, ..., b_{i-1}$ is already LLL-reduced we have that

$$|\mu_{i,h}| = \left| \frac{\langle b_i, \widehat{b}_h \rangle}{\langle \widehat{b}_h, \widehat{b}_h \rangle} \right| \leq \frac{||b_i||}{||\widehat{b}_h||} \leq \sqrt{M_{Sc} \alpha^{i-1}}.$$

We see that, during size-reduction of b_i, the value $\max(||b_1||^2, ..., ||b_i||^2)$ can temporarily increase not more than by a factor $2\,\alpha^{i-1}$ for $\delta \geq \frac{3}{4}$.

Consider the coefficients of the matrix $H \in \mathbf{Z}^{2k \times 2k}$ representing the local LLL-reduction of the segments B_{l-1}, B_l so that local LLL-reduction of B_{l-1}, B_l transforms $[B_{l-1}, B_l] := [B_{l-1}, B_l]H$. We let $b'_j, \widehat{b}'_j, \mu'_{j,i}$ denote the values corresponding to the transformed segments $[b'_{kl-k+1}, ..., b'_{kl+k}] = [B_{l-1}, B_l]H$. We let $||H||_1$ denote the maximal $||\ ||_1$-norm of the columns of H.

Lemma 2. [Sc84, Inequality (3.3)] *We have that*

1. $H = ([\mu_{j,i}]^{\top})^{-1}\ [\langle \widehat{b}_i, \widehat{b}'_j \rangle ||\widehat{b}_i||^{-2}]_{kl-k<i,j\leq kl+k}\ [\mu'_{j,i}]^{\top}$,

2. $||H||_1 \leq (2k)^2 (\frac{3}{2})^{2k-1} M_{Sc} \leq M_{Sc}^2$.

Proof. Equality **1.** follows from the equations

$$\begin{aligned}
[b_{kl-k+1}, ..., b_{kl+k}] &= [\widehat{b}_{kl-k+1}, ..., \widehat{b}_{kl+k}][\mu_{j,i}]^{\top}, \\
[b'_{kl-k+1}, ..., b'_{kl+k}] &= [\widehat{b}'_{kl-k+1}, ..., \widehat{b}'_{kl+k}][\mu'_{j,i}]^{\top} \\
&= [\widehat{b}_{kl-k+1}, ..., \widehat{b}_{kl+k}][\mu_{j,i}]^{\top} H.
\end{aligned}$$

When starting the local LLL-reduction of R_l the segments are already size-reduced, i.e., $|\mu_{j,i}| \leq \frac{1}{2}$ for $kl - k < i < j \leq kl + k$. Then the coefficients $\nu_{j,i}$ of the inverse matrix $[\nu_{j,i}] = [\mu_{j,i}]^{-1}$ satisfy $|\nu_{j,i}| \leq (\frac{3}{2})^{|j-i|}$. Inequality **2.** follows from **1.** as $|\langle \widehat{b}_i, \widehat{b}'_j \rangle ||\widehat{b}_i||^{-2}| \leq ||\widehat{b}'_j|| / ||\widehat{b}_i|| \leq M_{Sc}$ and $|\mu_{j,i}| \leq \frac{1}{2}$ for $i < j$. □

Conclusion. All integers arising during the reduction are bounded in absolute value by

$$\max(M_{Sc}^2, 2\,\alpha^{n-1} M_{Sc}) \leq M_{Sc}^2 \qquad \text{for } \delta \geq \frac{3}{4}.$$

The algorithm **segment LLL** improves the LLL-time bound — for the case $n = km$, $k = \Theta(\sqrt{n})$ — from $O(n^2 d \log_{1/\delta} M_{Sc})$ to $O(nd \log_{1/\delta} M_{Sc})$ arithmetic steps, saving a time factor n. □

Optimal Segment Size. For $k < m$ the dominant costs are for the global overhead of the executions of **loc-LLL** requiring $O(k^{-2} n^2 d \log_{1/\delta} M_{Sc})$ arithmetic steps. The LLL-exchanges require $O(k^2 n \log_{1/\delta} M_{Sc})$ arithmetic steps using local coordinates. The latter should become dominant for $k > m$. In practice, the crossover point of the two costs is for $k \gg m$. This is because the steps for the LLL-exchanges are mostly on small integers. In the [KS01] implemenation, these steps are even in floating point arithmetic. Large segment sizes as $k = n/4$ yield good results.

4 Divide and Conquer Using Iterated Segments

There is a natural way to iterate the concept of segments to subsegments, subsubsegments, etc. Let $n = k_1 \cdots k_s$ be a product of integers $k_1, ..., k_s \geq 2$, $s \leq \log_2 n$. We consider segments of size n/k_s, subsegments of size $n/(k_s k_{s-1})$, subsubsegments of size $n/(k_s k_{s-1} k_{s-2})$ and so on. We denote $\mathbf{k}(\sigma) := (k_1, ..., k_\sigma)$, $\mathbf{k}_\sigma := k_1 \cdots k_\sigma$ for $\sigma = 1, ..., s$. There are $s - 1$ levels of segments, we have $\mathbf{k}_\sigma$-segments $[b_{\mathbf{k}_\sigma(l-1)+1}, ..., b_{\mathbf{k}_\sigma l}]$ of size $\mathbf{k}_\sigma$ for $\sigma = 1, ..., s-1$. For $n = k_1 \cdots k_s = \mathbf{k}_s$ we let

$$D(l, \mathbf{k}(\sigma)) = \|\widehat{b}_{\mathbf{k}_\sigma(l-1)+1}\|^2 \cdots \|\widehat{b}_{\mathbf{k}_\sigma l}\|^2$$

denote the *local determinant* of the l-th $\mathbf{k}_\sigma$-segment. Also, let

$$\mathbf{k}_0 := 1 \qquad \text{and} \qquad D(l, \mathbf{k}(0)) := \|\widehat{b}_l\|^2.$$

For $n = 2^s$ it is natural to choose $k_1 = k_2 = ... = k_s = 2$.

Recall that Inequality **3.** of Definition 2 is impractical for segments of size $\mathbf{k}_\sigma$ greater than $\sqrt{n}$. We want to use large $\mathbf{k}_\sigma$-segments, where the exponent $\mathbf{k}_\sigma^2$ of $\delta^{\mathbf{k}_\sigma^2}$ in **3.** of Definition 2 gets impractical. Theorem 5 shows that the Inequalities **1.** and **2.** of Definition 2 — without Inequality **3.** — describe a sufficiently strong reduction. In the following we drop Inequality **3.** to render possible a divide and conquer approach.

Theorem 5. *Let the basis $b_1, \ldots, b_n$, $n = k \cdot m$ of lattice L satisfy the inequalities* **1.** *and* **2.** *of Definition 2. Then we have that*

1. $\|\widehat{b}_{kl+1}\|^2 \leq (\alpha/\delta)^{k(l'-l)-1} \|\widehat{b}_{kl'}\|^2$ *for* $1 \leq l < l' \leq n$,

2. $\|b_1\|^2 \leq (\alpha/\delta)^{\frac{n-1}{2}} (\det L)^{\frac{2}{n}}$

3. $\|\widehat{b}_n\|^2 \geq (\delta/\alpha)^{\frac{n-1}{2}} (\det L)^{\frac{2}{n}}$.

For comparison LLL-reduced bases $b_1, \ldots, b_n$ satisfy $\|b_1\|^2 \leq \alpha^{\frac{n-1}{2}} (\det L)^{\frac{2}{n}}$. Inequality **2.** is a bit weaker than the inequality $\|b_1\|^2 \leq \alpha^{n-1} \lambda_1^2$ of Theorem 1, but it is sufficient for most applications, in particular if δ is close to 1. The dual Inequality **3.** is useful in applying the method of [Co97] to find small integer solutions of polynomial equations.

Proof. Inequality **1.** follows from the inequalities

$$\|\hat{b}_{kl+1}\|^2 \leq (\alpha/\delta)^{i-1} \|\hat{b}_{kl+i}\|^2 \qquad \text{for } 1 \leq i \leq k$$

that hold within the segments, and an inequality

$$\|\hat{b}_{k(l-1)+s}\|^2 \leq (\alpha/\delta)^{k(l'-l)} \|b_{k(l'-1)+s}\|^2 \qquad \text{for some } 1 \leq s \leq k$$

that bridges the segments B_l and $B_{l'}$. The latter inequality holds as $D(l) = \|\widehat{b}_{k(l-1)+1}\|^2 \cdots \|\widehat{b}_{kl}\|^2$ satisfies $D(l) \leq (\alpha/\delta)^{k^2(l'-l)} D(l')$.

To prove Inequality **2.** we note that

$$D(1) = \|\hat{b}_1\|^2 \cdots \|\hat{b}_k\|^2 \qquad \text{and} \qquad \|b_1\|^2 \leq \alpha^{i-1} \|\hat{b}_i\|^2 \quad \text{for } i = 1, ..., k$$

imply that

$$\|b_1\|^2 \le \alpha^{\frac{k-1}{2}} D(1)^{\frac{1}{k}}.$$

Moreover, $D(1)\cdots D(m) = (\det L)^2$ and $D(1) \le (\alpha/\delta)^{k^2(i-1)} D(i)$ imply that

$$D(1) \le (\alpha/\delta)^{k^2 \frac{m-1}{2}} (\det L)^{\frac{2}{m}}.$$

Combining the two inequalities yields Inequality **2.**

We get Inequality **3.** by applying Inequality **2.** to the *dual* basis $b_1^*, ..., b_n^*$ satifying $\langle b_i^*, b_j\rangle = \delta_{i,j}$, $\|b_1^*\| = \|\widehat{b}_n\|^{-1}$ and $\det(L^*) = (\det L)^{-1}$. □

Definition 3. *A basis* $b_1, \ldots, b_n \in \mathbf{Z}^d, n = k_1 \cdots k_s = \mathbf{k}_s$ *is called* $\mathbf{k}(s)$-segment LLL-reduced *with* $\delta \in]\frac{1}{4}, 1]$ *if it is size-reduced and satisfies for* $\alpha := 1/(\delta - \frac{1}{4})$*:*

$$D(l, \mathbf{k}(\sigma)) \le (\alpha/\delta^\sigma)^{(\mathbf{k}_\sigma)^2} D(l+1, \mathbf{k}(\sigma)) \tag{1}$$

for $l \ne 0 \mod k_{\sigma+1}$, *for* $\sigma = 0, ..., s-1$, *and* $l = 1, ..., n/\mathbf{k}_\sigma - 1$.

The exponent σ of δ^σ is used for the time bound of Theorem 7. As δ can be chosen very close to 1, the factor $\delta^{-\sigma}$ is still close to 1.

Due to the restriction $l \ne 0 \mod k_{\sigma+1}$ the inequalities (1) only hold within $\mathbf{k}_{\sigma+1}$-segments, they cannot bridge distinct $\mathbf{k}_{\sigma+1}$-segments. The inequalities (1) get weaker and weaker as the size $\mathbf{k}_\sigma$ of the segments increases and $\delta^{\mathbf{k}_\sigma^2}$ decreases. For $\sigma = 0, k_0 = \mathbf{k}_0 = 1$, the inequalities (1) mean that

$$\|\widehat{b}_l\|^2 \le \alpha \|\widehat{b}_{l+1}\|^2 \qquad \text{for } l = 1, ..., n-1, l \ne 0 \mod k_1,$$

slightly weakening Clause **1.** of Definition 2.

For $n = k_1 \cdot k_2 = k \cdot m$, $s = 2$, Definition 3 recites clauses **1.** and **2.** of Definition 2 slightly weakening **1.** and dropping **3.**.

We next extend Lemma 3 to iterated segments.

Theorem 6. *A* $\mathbf{k}(s)$*-segment LLL-reduced basis* $b_1, \ldots, b_n$, $n = k_1 \cdots k_s = \mathbf{k}_s$ *satisfies*

$$\|b_1\|^2 \le (\alpha/\delta^{s-1})^{\frac{n-1}{2}} (\det L)^{\frac{2}{n}} \qquad \textit{and} \qquad \|\widehat{b}_n\|^2 \ge (\delta^{s-1}/\alpha)^{\frac{n-1}{2}} (\det L)^{\frac{2}{n}}.$$

Proof. We prove by induction on σ that

$$\|b_1\|^2 \le (\alpha/\delta^{\sigma-1})^{\frac{\mathbf{k}_\sigma - 1}{2}} D(1, \mathbf{k}(\sigma))^{\frac{1}{\mathbf{k}_\sigma}}.$$

The first claim of the theorem follows from $D(1, \mathbf{k}(s)) = (\det L)^2$, $\mathbf{k}_s = n$ for $\sigma = s$. The second claim follows by duality.

The induction hypothesis holds for $\sigma = 1$ as we have $\|b_1\|^2 \le \alpha^{i-1}\|\hat{b}_i\|^2$ for $i = 1, \ldots, k_1$ and $D(1, \mathbf{k}(1)) = \|\hat{b}_1\|^2 \cdots \|\hat{b}_{k_1}\|^2$.

The induction hypothesis extends from σ to $\sigma + 1$ by the inequalities

$$D(1, \mathbf{k}(\sigma)) \le (\alpha/\delta^\sigma)^{(\mathbf{k}_\sigma)^2 (l-1)} D(l, \mathbf{k}(\sigma)) \qquad \text{for } l = 1, \ldots, k_{\sigma+1}$$

and the equation

$$D(1, \mathbf{k}(\sigma+1)) = \prod_{l=1}^{k_{\sigma+1}} D(l, \mathbf{k}(\sigma)). \qquad \square$$

$\mathbf{k}_s$-segment LLL-reduction. The algorithm **$\mathbf{k}_s$-segment LLL** transforms a given basis into a $\mathbf{k}_s$-segment LLL-reduced basis. It iterates $\mathbf{k}_{s-1}$-segment LLL-reduction of two $\mathbf{k}_{s-1}$-segments

$$[B_{l-1}, B_l] = [b_{\mathbf{k}_{s-1}(l-1)+1}, \dots, b_{\mathbf{k}_{s-1}(l+1)}]$$

via the procedure **$\mathbf{k}_{s-1}$-segment LLL**. For $\sigma \geq 1$ the procedure **$\mathbf{k}_\sigma$- -segment LLL(l)**

- computes the orthogonal transform of the $\mathbf{k}_\sigma$-segments $[B_{l-1}, B_l]$ providing the local R-matrix $R_l \in \mathbf{R}^{2\mathbf{k}_\sigma \times 2\mathbf{k}_\sigma}$,
- performs a local $\mathbf{k}_{\sigma-1}$-segment LLL-reduction on $[B_{l-1}, B_l]$ by iterating **$\mathbf{k}_{\sigma-1}$-segment LLL**,
- stores the corresponding transformation matrix $H_l \in \mathbf{Z}^{2\mathbf{k}_\sigma \times 2\mathbf{k}_\sigma}$,
- upon termination, it transports H_l to the matrix $H_{l'}$ of the $\mathbf{k}_{\sigma+1}$-segment $B_{l'}$ that contains B_{l-1}, B_l. The $2\mathbf{k}_\sigma$ columns of $H_{l'}$ corresponding to B_{l-1}, B_l are multiplied from the right by H_l. Thereafter, H_l is reset to the identity matrix.

While procedure **$\mathbf{k}_\sigma$-segment LLL** for $\sigma \geq 1$ recursively calls **$\mathbf{k}_{\sigma-1}$-segment LLL** , the procedure **$\mathbf{k}_0$-segment LLL(l)** exchanges b_{l-1} and b_l in case that $\|\widehat{b}_{l-1}\|^2 > \alpha \|\widehat{b}_l\|^2$.

$\mathbf{k}_s$-segment LLL

INPUT $\quad b_1, \dots, b_n \in \mathbf{Z}^d,\ n = k_1 \cdots k_s = \mathbf{k}_s,\ \delta$

OUTPUT $\quad b_1, \dots, b_n \quad \mathbf{k}_s$-segment LLL-reduced basis

1. $l := 1$

2. **while** $l \leq k_s - 1$ **do**

$\mathbf{k}_{s-1}$-segment LLL(l)
if $l \neq 1$ **and** $D(l-1, \mathbf{k}(s-1)) > (\alpha/\delta^\sigma)^{(\mathbf{k}_{s-1})^2} D(l, \mathbf{k}(s-1))$
then $l := l - 1$ **else** $l := l + 1$. *end*

Theorem 7. *Given a basis $b_1, \dots, b_n \in \mathbf{Z}^d$, $n = k_1 \cdots k_s = \mathbf{k}_s$ the algorithm* **$\mathbf{k}_s$-segment LLL** *produces a $\mathbf{k}_s$-segment LLL-reduced basis. It performs at most*

$$O\left(dn^2 + d \sum_{\sigma=1}^{s} k_\sigma^2 \log_{1/\delta} M_{Sc}\right)$$

arithmetic steps. If

$$\max_\sigma k_\sigma = O(1) \qquad \text{and} \qquad \log_{1/\delta} M_{Sc} = O(n^2),$$

the number of arithmetic steps is $O(n^2 d \log_2 n)$.

Proof. Consider for $\sigma = 1, \dots, s-1$ the number of executions of **$\mathbf{k}_\sigma$-segment LLL** as sub$\cdots$subroutine of **$\mathbf{k}_s$-segment LLL**. The number of these executions is $n/\mathbf{k}_\sigma - 1 + 2 \cdot \mathrm{decr}(\mathbf{k}(\sigma))$, where $\mathrm{decr}(\mathbf{k}(\sigma))$ is the number of times that

$\mathbf{k}_\sigma$**-segment LLL** is called as sub$\cdots$subroutine of $\mathbf{k}_s$**-segment LLL** due to a violated inequality (1), where $D(l-1, \mathbf{k}(\sigma)) > (\alpha/\delta)^{(\mathbf{k}_\sigma)^2} D(l, \mathbf{k}(\sigma))$ for some l.

Formally, $\mathbf{k}_\sigma$**-segment LLL** is also executed after a decrease of l on some level σ' where $\sigma' > \sigma$. However, that execution induces no costs except for the case of a violated inequality (1) at level σ. The reason is that the $\mathbf{k}_s$-segments are already orthogonalized and size-reduced by previous calls of $\mathbf{k}_\sigma$**-segment LLL**.

Consider the product of the Gramian determinants

$$\mathbf{D}(\mathbf{k}(\sigma)) =_{\text{def}} \prod_{l=1}^{n/\mathbf{k}_\sigma} D_{\mathbf{k}_\sigma l} = \prod_{l=1}^{n/\mathbf{k}_\sigma} \left(D(1, \mathbf{k}(\sigma) \cdots D(l, \mathbf{k}(\sigma))\right).$$

We apply Theorem 3 to $\mathbf{k}_\sigma$-segments. Each execution of $\mathbf{k}_\sigma$**-segment LLL** — due to a violated inequality (1) — decreases $\mathbf{D}(\mathbf{k}(\sigma))$ by the factor $\delta^{(\mathbf{k}_\sigma)^2/2}$. $\mathbf{D}(\mathbf{k}(\sigma))$ is product of $n/\mathbf{k}_\sigma$ Gramian determinants. As initially $\mathbf{D}(\mathbf{k}(\sigma)) \leq M_{Sc}^{n/\mathbf{k}_\sigma}$, and upon termination $\mathbf{D}(\mathbf{k}(\sigma)) \geq 1$ we see that

$$\operatorname{decr}(\mathbf{k}(\sigma)) \leq \frac{2n}{(\mathbf{k}_\sigma)^3} \log_{1/\delta} M_{Sc}.$$

In total there are $n/\mathbf{k}_\sigma - 1 + \frac{2n}{(\mathbf{k}_\sigma)^3} \log_{1/\delta} M_{Sc}$ executions of $\mathbf{k}_\sigma$**-segment LLL** each inducing an overhead of $O(\mathbf{k}_\sigma \mathbf{k}_{\sigma+1}^2)$ arithmetic steps. This overhead includes: the orthogonal transform of the $\mathbf{k}_\sigma$-segments B_{l-1}, B_l providing the local R-matrix $R_l \in \mathbf{R}^{2\mathbf{k}_\sigma \times 2\mathbf{k}_\sigma}$ of $[B_{l-1}, B_l]$, the transport of H_l to the transformation matrix $H_{l'}$ of the next higher level and size-reduction of R_l against $R_{l'}$. The total overhead of all $\mathbf{k}_\sigma$**-segment LLL** executions is

$$O(n\mathbf{k}_{\sigma+1}^2 + nk_{\sigma+1}^2 \log_{1/\delta} M_{Sc}).$$

In the particular case $\sigma = s-1$, this overhead is $O(d\mathbf{k}_s^2 + dk_s^2 \log_{1/\delta} M_{Sc})$ as the global transforms are done directly on the basis vectors in $\mathbf{Z}^d$.

In the particular case $\sigma = 0$, the overhead covers a total of $O(n \log_{1/\delta} M_{Sc})$ LLL-exchanges of two consecutive vectors which each uses $O(k_1^2)$ arithmetic steps for a local exchange and a local size-reduction. We see that $\mathbf{k}_s$**-segment LLL** performs $O(n^2 d + d \sum_{\sigma=1}^{s} k_\sigma^2 \log_{1/\delta} M_{Sc})$ arithmetic steps providing the time bound of the theorem.

Conclusion. The time bound of the novel algorithm $\mathbf{k}_s$**-segment LLL** is comparable to that of classical matrix multiplication. The size of the integers occuring in the new reduction algorithm can be bounded by the method of Theorem 4. The algorithm uses integers of bit length $O(s \log_2 M_{Sc})$, where s is the number of levels σ. Each level can add $O(\log_2 M_{Sc})$ bits when transporting the local transformation H_l to the next higher level.

References

[BN00] *D. Bleichenbacher and P.Q. Nguyen*, Noisy Polynomial Interpolation and Noisy Chinese Remaindering, Proc. Eurocrypt'00, LNCS 1807, Springer-Verlag, pp. 53-69, 2000.

[Bo00] *D. Boneh*, Finding Smooth Integers in Small Intervals Using CRT Decoding, Proc. STOC'00, ACM Press, pp. 265-272, 2000.

[Ca00] *J. Cai*, The Complexity of some Lattice Problems, Proc. ANTS'00, LNCS 1838, Springer-Verlag, pp. 1-32, 2000.

[Co97] *D. Coppersmith*, Small Solutions to Polynomial Equations, and Low Exponent RSA Vulnerabilities, *J. Crypt.* **10**, pp. 233-260, 1997.

[K84] *R. Kannan*, Minkowski's Convex Body Theorewm and Integer Programming. Mathematical Operation Research, **12**, pp. 415–440, 1984.

[K01] *H. Koy*, Notes of a Lecture. Frankfurt 2001.

[KS01] *H. Koy and C.P. Schnorr*, LLL-Reduction with Floating Point Orthogonalization. This proceedings CaLC 2001, pp. 81–96.

[LLL82] *A.K. Lenstra, H.W. Lenstra and L. Lovász*, Factoring polynomials with rational coefficients, *Math. Ann.* **261**, pp. 515-534, 1982.

[NS00] *P.Q. Nguyen and J. Stern*, Lattice Reduction in Cryptology, An Update, Proc. ANTS'00, LNCS 1838, Springer-Verlag, pp. 85-112, 2000.

[S87] *C.P. Schnorr*, A hierarchy of polynomial time lattice basis reduction algorithms, *Theoretical Computer Science* **53**, pp. 201-224, 1987.

[S91] *C.P. Schnorr and M. Euchner*, Lattice Basis Reduction and Solving Subset Sum Problems. Proceedings FCT'91, LNCS 591, Springer-Verlag, pp. 68–85, 1991. The complete paper appeared in Mathematical Programming Studies, 66A, **2**, pp. 181–199, 1994.

[S94] *C.P. Schnorr*, Block Reduced Lattice Bases and Successive Minima, *Combinatorics, Probability and Computing*, **3**, pp. 507-522, 1994.

[SH95] *C.P. Schnorr and H. Hörner*, Attacking the Chor-Rivest Cryptosystem by Improved Lattice Reduction. Proceedings Eurocrypt'95, LNCS 921, Springer-Verlag, pp. 1–12, 1995.

[Sc84] *A. Schönhage*, Factorization of univariate integer polynomials by diophantine approximation and improved lattice basis reduction algorithm, *Proc. 11-th Coll. Automata, Languages and Programming, Antwerpen 1984*, LNCS **172**, Springer-Verlag, pp. 436-447, 1984.

Segment LLL-Reduction with Floating Point Orthogonalization

Henrik Koy[1] and Claus Peter Schnorr[2]

[1] Deutsche Bank AG, Frankfurt am Main
henrik.koy@db.com
[2] Fachbereiche Mathematik und Informatik
Universität Frankfurt, PSF 111932
D-60054 Frankfurt am Main, Germany
schnorr@cs.uni-frankfurt.de

Abstract. We associate with an integer lattice basis a *scaled basis* that has orthogonal vectors of nearly equal length. The orthogonal vectors or the QR-factorization of a scaled basis can be accurately computed up to dimension 2^{16} by Householder reflexions in floating point arithmetic (fpa) with 53 precision bits.
We develop a highly practical fpa-variant of the new *segment LLL-reduction* of Koy and Schnorr [KS01]. The LLL-steps are guided in this algorithm by the Gram-Schmidt coefficients of an associated scaled basis. The new reduction algorithm is much faster than previous codes for LLL-reduction and performs well beyond dimension 1000.
Keywords: LLL-reduction, Householder reflexion, floating point arithmetic, stability, scaled basis, segment LLL-reduction, local LLL-reduction.

1 Introduction

Practical algorithms for LLL-reduction compute the orthogonal vectors of a basis in floating point arithmetic (fpa). The corresponding fpa-errors must be small otherwise the size-reduction included in the orthogonalization gets unstable. Up to dimension 250 Gram-Schmidt orthogonalization of an LLL-reduced basis can be done in fpa with some correction steps [SE91]. Householder orthogonalization has better stability, as was shown in [RS96]. In practice it yields up to dimension 350 a sufficient QR-factorization of the basis matrix — but this process gets unstable in dimension 400. [S88] presents a method for correcting the Gram-Schmidt coefficients using Schulz's method of matrix inversion. This method is provably stable but requires $O(n + \log_2 M)$ precision bits, where M bounds the Euclidean length of the basis vectors. It is useless for fpa with 53 precision bits.

In this paper we associate with an integer lattice basis a *scaled basis* that has orthogonal vectors of nearly equal length. The orthogonal vectors or the QR-factorization of a scaled basis can be accurately computed up to dimension 2^{16} by Householder reflexions in fpa with 53 precision bits. We present an algorithm

J.H. Silverman (Ed.): CaLC 2001, LNCS 2146, pp. 81–96, 2001.

HRS that efficiently generates an associated scaled basis. We present a *fpa*-variant of LLL-reduction where LLL-steps are guided by the orthogonal vectors of an associated scaled basis. In particular, size-reduction is done against the scaled basis. The weaker size-reduction does in practice not degrade the quality of the reduced basis.

We present a *fpa*-variant of segment LLL-reduction, a novel concept proposed in [KS01]. The algorithm **segment sLLL** performs scaled segment LLL-reduction, so that all LLL-steps are guided by the orthogonalization of an associated scaled basis. The algorithm **segment sLLL** is a very efficient reduction algorithm. Its efficiency comes from local LLL-reduction of two consecutive segments B_{l-1}, B_l that is done by reducing the local matrix R_l in *fpa*. To make this local LLL-reduction possible in the limits of *fpa* it is necessary to bound the integer transformation matrix of the local LLL-reduction. For this we carefully prepare the local matrix R_l of B_{l-1}, B_l prior to local LLL-reduction.

In practice, **segment sLLL** is much faster than previous codes for LLL-reduction. It performs well beyond dimension 1000 and provides lattice bases that are in practice of similar quality as LLL-reduced bases. For dimension 1000 and basis vectors with 400 bit integer coordinates, the new LLL-reduction takes only about 10 hours on a 800 MHz PC. It is the first code for LLL-reduction that performs well beyond dimension 350.

In this paper we focus on practical aspects related to *fpa*. A rigorous analysis of segment LLL-reduction in the model of integer arithmetic is in the companion paper [KS01]. For general background on *fpa* and Householder transformation see [LH95].

2 Stability Properties of Householder Orthogonalization

For an introduction of LLL-reduced lattice bases and of our notation on lattices we refer to Section 2 of the companion paper [KS01]. For easy reference we recall Definition 1 and Theorem 1 from [KS01]. We let $\delta \in]\frac{1}{4}, 1]$ and $\alpha = 1/(\delta - \frac{1}{4})$.

Definition 1. *An ordered basis $b_1, \ldots, b_n \in \mathbf{Z}^d$ of the lattice L is* LLL-reduced *with $\delta \in]\frac{1}{4}, 1]$ if it has properties* **1.** *and* **2.**:

1. $|\mu_{j,i}| \leq 1/2$ *for* $1 \leq i < j \leq n$,

2. $\delta\, \|\widehat{b}_i\|^2 \;\leq\; \mu_{i+1,i}^2\, \|\widehat{b}_i\|^2 + \|\widehat{b}_{i+1}\|^2$ *for* $i = 1, \ldots, n-1$.

Theorem 1. *A basis $b_1, \ldots, b_n$ of lattice L that is* LLL-*reduced with δ satisfies*:

1. $\|b_i\|^2 \leq \alpha^{n-1}\, \lambda_i^2$ *and* $\|b_1\|^2 \leq \alpha^{i-1}\, \|\widehat{b}_i\|^2$ *for* $i = 1, \ldots, n$,

2. $\|b_1\|^2 \leq \alpha^{\frac{n-1}{2}}\, (\det L)^{\frac{2}{n}}$ *and* $\|\widehat{b}_n\|^2 \geq \alpha^{-\frac{n-1}{2}}\, (\det L)^{\frac{2}{n}}$.

Accuracy of Householder Reflexions. Consider the QR-factorization $B = QR$ of the basis matrix $B = [b_1, \ldots, b_n] \in \mathbf{Z}^{d \times n}$, where $Q \in \mathbf{R}^{d \times d}$ is an orthogonal matrix and $R = [r_{i,j}] = [\mathbf{r}_1, ..., \mathbf{r}_n] \in \mathbf{R}^{d \times n}$ is an upper triangular matrix, $r_{i,j} = 0$

for $i > j$. We have $\mu_{j,i} = r_{i,j}/r_{i,i}$ and $|r_{i,i}| = \|\widehat{b}_i\|$. The vector $\mathbf{r}_l$ is the orthogonal transform of b_l.

Consider the process of orthogonalization of a basis matrix $B = [b_1, ..., b_n]$. In ideal arithmetic we get the QR-factorization $B = QR$ by a sequence of Householder transformations

$$C_1 := B, \; C_{j+1} := Q_j C_j \qquad \text{for } j = 1, \ldots, n,$$

where $Q_j \in \mathbf{R}^{d \times d}$ is an orthogonal matrix — an Householder reflexion — that produces zeros in positions $j+1$ through d of column j of $Q_j C_j$. Thus $C_{j+1} \in \mathbf{R}^{d \times n}$ is upper triangular in the first j columns. Finally,

$$R = C_{n+1} \qquad \text{and} \qquad Q = Q_n \cdots Q_1.$$

In actual computation, however, we use floating point operations

$$\bar{C}_1 = fl(B), \; \bar{C}_{j+1} := fl(\bar{Q}_j \bar{C}_j) \qquad \text{for } j = 1, \ldots, n.$$

We assume the standard *fpa*-model of WILKINSON, see [LH95, page 85] for details. Let $0 < \eta \ll 1$ be the relative precision — each floating point operation induces a normalized relative error bounded in magnitude by η. [1] In this model, it has been shown [LH95, page 87, formula (15.38)] that

$$\|\bar{C}_{j+1} - Q_j \cdots Q_1 B\|_F \leq (6d - 3j + 40) j\eta \, \|B\|_F + O(\eta^2), \tag{1}$$

where $\|A\|_F = (\sum_{i,j} a_{i,j}^2)^{\frac{1}{2}}$ denotes the FROBENIUS NORM of the matrix $A = [a_{i,j}]$. In the following we neglect the low order term $O(\eta^2)$. Thus, in actual computation we get $\bar{R} = [\bar{r}_{i,j}] = \bar{C}_{n+1}$ satisfying for $n \geq 14$,

$$\|\bar{R} - R\|_F \leq 6dn\eta \, \|B\|_F. \tag{2}$$

It follows that the approximate Gram-Schmidt coefficients $\bar{\mu}_{j,i} = \bar{r}_{i,j}/\bar{r}_{i,i}$ satisfy for $i < j$

$$|\bar{\mu}_{j,i} - \mu_{j,i}| \leq 6dj\eta \, \|b_1, \ldots, b_j\|_F / (|r_{i,i}|(1 - \varepsilon)) \tag{3}$$

provided that $|\bar{r}_{i,i} - r_{i,i}| \leq \varepsilon |r_{i,i}|$. Therefore, we get from inequality (2) the following lemma.

Lemma 1. *Inequality* (3) *holds provided that*

$$6dj\eta \, \|b_1, \ldots, b_j\|_F \leq \varepsilon |r_{i,i}|. \tag{4}$$

Instability of Householder QR-Factorization for Dimension 400. Inequalities (1), (2), (3) are rather sharp. To combine Householder transformation and size-reduction we need accurate coefficients $\mu_{j,i}$. Condition (4) characterizes the stability of Householder transformation — stability requires that

$$6dn\eta \, \|B\|_F < \min_i |r_{i,i}|.$$

[1] Standard double length, wired *fpa* has 53 precision bits, $\eta = 2^{-53}$.

On the other hand, Theorem 2 shows that random LLL-reduced bases on the average satisfy $\|b_1\| \approx 1.1^{n-1}\|\widehat{b}_n\|$. For dimension $n = 400$ and $\eta = 2^{-53}$ this yields

$$6dn\eta\,\|b_1\|_F \approx 6dn\eta 1.1^{n-1}\|\widehat{b}_n\| \approx 4 \cdot 10^6\|\widehat{b}_n\|.$$

Inequality (4) is grossly violated. Therefore, Householder transformation of LLL-reduced bases is necessarily unstable in dimension 400 for fpa with 53 precision bits.

Theorem 2. *Consider a random LLL-reduced basis with random coefficients* $\mu_{i+1,i} \in_R [-\frac{1}{2}, \frac{1}{2}]$ *for* $i = 1, ..., n-1$, *and let the inequalities* **2.** *of Definition* 1 *be tight. Then* $\|b_1\| \approx 1.1^{n-1}\|\widehat{b}_n\|$ *holds on the average.*

Proof. While $(\delta - \mu_{i+1,i}^2)\|\widehat{b}_i\|^2 \leq \|\widehat{b}_{i+1}\|^2$ holds for LLL-reduced bases, the converse $(\delta - \varepsilon^2)\|\widehat{b}_i\|^2 \geq \|\widehat{b}_{i+1}\|^2$ holds provided that

$$\delta\|\widehat{b}_i\|^2 = \|\widehat{b}_{i+1}\|^2 + \mu_{i+1,i}^2\|\widehat{b}_i\|^2 \qquad \text{and} \qquad |\mu_{i+1,i}| \leq \varepsilon.$$

The inequality $|\mu_{i+1,i}| \leq \varepsilon$ holds with probability 2ε for random values of $\mu_{i+1,i} \in_R [-\frac{1}{2}, \frac{1}{2}]$. As

$$\int_{-\frac{1}{2}}^{\frac{1}{2}} 2\varepsilon^2 d\varepsilon = \frac{1}{6} \qquad \text{and} \qquad \sqrt{\delta - \frac{1}{6}} \approx 1.1^{-1}$$

holds for $\delta \approx 1$ we see that $\|b_1\| \approx 1.1^{n-1}\|\widehat{b}_n\|$ holds on the average. □

3 The Scaled Basis Matrix

Scaling is a well known method for improving the stability of fpa. We associate with an integer lattice basis $b_1, ..., b_n$ a scaled basis $b_1^s, ..., b_n^s$ that has orthogonal vectors of nearly equal length.

Definition 2. *Let* $b_1, \ldots, b_n$ *be a basis of an integer lattice* L. *We call* $b_1^s, ..., b_n^s \in L$ *an associated* scaled basis *with* scaling factors $2^{e_1}, ..., 2^{e_n}$, *where* $e_1, ..., e_n \in \mathbf{N}$, *if* $b_1^s, ..., b_n^s$ *form a size-reduced basis of a sublattice of* L *satisfying*

$$\|b_1^s\| \leq 2^{e_i}\|\widehat{b}_i\| = \|\widehat{b}_i^s\| < 2\|b_1^s\| \qquad \text{for } i = 1, \ldots, n. \tag{5}$$

We show in Theorem 3 that a given scaled basis yields accurate Gram-Schmidt coefficients by Householder reflexions in fpa. In Section 4 we show how to produce an associated scaled basis efficiently in fpa. In Sections 5 and 7 we use an associated scaled basis to guide LLL-reduction and segment LLL-reduction.

We can easily transform a basis $b_1, ..., b_n$ into an associated scaled basis $b_1^s, ..., b_n^s$ using exact arithmetic — first scale then size-reduce:

1. *scaling.* $e_1 := \lfloor \log_2(\max_j \|\widehat{b}_j\|/\|b_1\|) \rfloor$, $b_1^s := 2^{e_1} b_1$,

for $i = 2, ..., n$ **do** $e_i := \max(0, \lceil \log_2 \|b_1^s\|/\|\widehat{b}_i\| \rceil)$, $b_i^s := 2^{e_i} b_i$

2. *size-reduction.* **for** $j = 2, ..., n$ **for** $i = j-1, ..., 1$ **do** $b_j^s := b_j^s - \left\lceil \frac{\langle b_j^s, \widehat{b}_i^s \rangle}{\|\widehat{b}_i^s\|^2} \right\rfloor b_i^s$.[2]

New Notation. For the remaining of the paper we let the column vector $\mathbf{r}_\nu$ of the R-matrix be related to the scaled vector b_ν^s rather than to the original vector b_ν. *The size of scaled Gram-Schmidt coefficients.* The coefficients $\mu_{\nu,j}^s$ of an arbitrary, unscaled vector

$$b_\nu = \widehat{b}_\nu + \sum_{j=1}^{\nu-1} \mu_{\nu,j}^s \widehat{b}_j^s$$

are uniformly bounded:

Corollary 1.

$$|\mu_{\nu,j}^s| = \frac{|\langle b_\nu, \widehat{b}_j^s \rangle|}{\|\widehat{b}_j^s\|^2} \leq \frac{\|b_\nu\|}{\|\widehat{b}_j^s\|} \overset{(5)}{\leq} 2 \cdot \frac{\|b_\nu\|}{\|b_1^s\|}.$$

In contrast, LLL-reduced bases $b_1, \ldots, b_{\nu-1}$ satisfy

$$b_\nu = \sum_{j=1}^{\nu} \mu_{\nu,j} \widehat{b}_j \qquad \text{with} \qquad |\mu_{\nu,j}|^2 \leq \frac{\|b_\nu\|^2}{\|b_1\|^2} \alpha^{j-1}.$$

The coefficient $\mu_{\nu,\nu-1}$ that enters first into size-reduction of b_ν tends to be very large due to the factor $\alpha^{\nu-1}$. A small relative error of $\mu_{\nu,\nu-1}$ confuses the size-reduction of b_ν.

Corollary 2. *The basis $b_1^s, \ldots, b_n^s$ satisfies*

$$\|b_j^s\| \leq \sqrt{j+3}\, \|b_1^s\| \qquad \text{for } j = 1, \ldots, n.$$

Proof. $\|b_j^s\|^2 = \|\widehat{b}_j^s\|^2 + \sum_{i=1}^{j-1} (\mu_{j,i}^s)^2 \|\widehat{b}_i^s\|^2$

$$\leq \|\widehat{b}_j^s\|^2 + (j-1)/4 \max_{i<j} (\|\widehat{b}_i^s\|)^2$$

$$\overset{(5)}{\leq} 4\|b_1^s\|^2 + (j-1)\, \|b_1^s\|^2 = (j+3)\|\widehat{b}_1^s\|^2.$$ □

Next we study for a given scaled basis the accuracy of the approximate coefficients $\bar{\mu}_{j,i}^s$, computed in *fpa* by Householder reflexions.

Theorem 3. *The approximate $\bar{\mu}_{j,i}^s$ of $b_1^s, \ldots, b_n^s$ satisfy*

$$|\bar{\mu}_{j,i}^s - \mu_{j,i}^s| \leq \varepsilon/(1-\varepsilon) \qquad \text{for } \varepsilon = 6dj^2\eta.$$

[2] Let $\lceil r \rfloor = \lceil r - \frac{1}{2} \rceil$ be the nearest integer to the real number r.

Proof. By scaling and size-reduction we have for $j \neq 2$: [3]

$$\|b_j^s\|^2 \leq \frac{1}{4}\sum_{i=1}^{j-1} \|\widehat{b}_i^s\|^2 + \|\widehat{b}_j^s\|^2 \leq \frac{j+3}{4} \max_{i\leq j}\|\widehat{b}_i^s\|^2 \overset{(5)}{\leq} j \min_{i\leq j}\|\widehat{b}_i^s\|^2.$$

Hence $\|b_1^s,\ldots,b_j^s\|_F = (\sum_{i=1}^j \|b_i^s\|^2)^{1/2} \leq \sqrt{j}\max_{i\leq j}\|b_i^s\| \leq j\,\|\widehat{b}_i^s\|$, and thus

$$\|b_1^s,\ldots,b_j^s\|_F/\|\widehat{b}_i^s\| \leq j \qquad \text{for } i = 1,...,j. \tag{6}$$

Hence, Inequality (4) holds for $\varepsilon = 6dj^2\eta$. By Lemma 1 Inequality (3) holds for that ε and thus

$$\begin{aligned}|\bar{\mu}_{j,i}^s - \mu_{j,i}^s| &\overset{(3)}{\leq} 6dj\eta\,\|b_1^s,\ldots,b_j^s\|_F/(|r_{i,i}|(1-\varepsilon)) \\ &\overset{(6)}{\leq} 6dj^2\eta/(1-\varepsilon) = \varepsilon/(1-\varepsilon).\end{aligned}$$

□

Stability up to Dimension 2^{16}. Consider Theorem 3 in the case $j \leq n = d = 2^{16}$ and $\eta = 2^{-53}$. Then we have $\varepsilon \leq 6n^3\eta \leq 0.19$ and $\varepsilon/(1-\varepsilon) \leq 0.24$. Therefore, Householder reflexions yield up to dimension 2^{16} sufficiently accurate Gram-Schmidt coefficients for the basis $b_1^s,\ldots,b_n^s$.

4 Orthogonalization via Scaling and Size-Reduction

Suppose we are given $b_1^s,\ldots,b_{\nu-1}^s, b_\nu$ and we want to produce a scaled vector b_ν^s. At that point the $\mu_{j,i}^s$ of $b_1^s,\ldots,b_{\nu-1}^s$ are given with high accuracy. We iteratively transform b_ν into b_ν^s using better and better approximations of the $\mu_{\nu,i}^s$. The procedure **HRS** (Householder, Reduction, Scaling) iterates the following steps

1. the first $\nu - 1$ Householder transformations $b_\nu \mapsto \bar{Q}_{\nu-1}\cdots\bar{Q}_1 b_\nu$,

2. size-reduction of b_ν against $b_1^s,\ldots,b_{\nu-1}^s$,

3. scaling of b_ν to b_ν^s.

Steps **1.** and **2.** must be iterated as the size of b_ν is by Inequality (2) crucial for the accuracy of Householder transformation. As the scaling increases the size of b_ν we scale in stages, repeating **1. 2.** after each stage of scaling. Let $Q_j = I_d - 2v_jv_j^\top\|v_j\|^{-2}$, where $I_d \in \mathbf{Z}^{d\times d}$ is the identity matrix and $v_j \in \mathbf{R}^d$ is the Householder vector associated with Q_j. Note that $x \mapsto Q_jx$ reflects x at the hyperplane that is orthogonal to v_j: $Q_jv_j = -v_j$, $Q_ju = u$ for $u \perp v_j$. **HRS** is given for input the Householder vectors $v_1,...,v_{\nu-1} \in \mathbf{R}^d$, the first $\nu - 1$ columns $\bar{\mathbf{r}}_1,...,\bar{\mathbf{r}}_{\nu-1}$ of the matrix $\bar{C}_\nu = \bar{Q}_{\nu-1}\cdots\bar{Q}_1\bar{C}_1$ and the computed scaling exponents $\bar{e}_1,...,\bar{e}_{\nu-1}$. The operations on $b_1^s,...,b_{\nu-1}^s$ are in exact integer arithmetic, the other operations are in *fpa*. Taking *fpa*-errors into account we relax size-reduction of b_ν^s to the relaxed condition $|\mu_{\nu,j}^s| \leq 0.52$.

[3] In the following we neglect the exception $j = 2$.

Suppressing Backward Rescaling. Upon entry of **HRS**(ν), the scaled vectors $b_1^s, ..., b_{\nu-1}^s$ are given while $b_\nu^s, ..., b_n^s$ are unknown. At this stage the scaling factors $2^{\bar{e}_1}, ..., 2^{\bar{e}_{\nu-1}}$ correspond to the subbasis $b_1, ..., b_{\nu-1}$. If $\|\widehat{b}_\nu\| > \|\widehat{b}_1^s\|$ we would need to rescale b_i^s by increasing $\bar{e}_i := \bar{e}_i + \bar{e}$ and $b_i^s := 2^{\bar{e}} b_i^s$ for $\bar{e} := \lfloor \log_2(\|\widehat{b}_\nu\| / \|\widehat{b}_1^s\|) \rfloor$ and $i = 1, ..., \nu - 1$. We suppress this backward rescaling. It is sufficient to store $\bar{e}$ and to do all subsequent size-reductions against $2^{\bar{e}} b_i^s$ rather than against b_i^s, i.e., we replace subsequent reduction steps

$$b := b - \left\lceil \frac{\langle b, \widehat{b}_i^s \rangle}{\langle \widehat{b}_i^s, \widehat{b}_i^s \rangle} \right\rfloor b_i^s \qquad \text{by the steps} \qquad b := b - 2^{\bar{e}} \left\lceil \frac{2^{-\bar{e}} \langle b, \widehat{b}_i^s \rangle}{\langle \widehat{b}_i^s, \widehat{b}_i^s \rangle} \right\rfloor b_i^s.$$

For simplicity, the program **HRS**(ν) does not include the steps required in case that $\|\widehat{b}_\nu\| > \|\widehat{b}_1^s\|$.

HRS(ν) (*Householder Transformation, Reduction and Scaling of* b_ν)
INPUT $b_1^s, \ldots, b_{\nu-1}^s \in \mathbf{Z}^d$, $\bar{\mathbf{r}}_1, ..., \bar{\mathbf{r}}_{\nu-1}$, $v_1, \ldots, v_{\nu-1} \in \mathbf{Q}^d$, $\bar{e}_1, ..., \bar{e}_{\nu-1} \in \mathbf{Z}$
OUTPUT b_ν^s (size-reduced against $b_1^s, ..., b_{\nu-1}^s$), $v_\nu, \bar{\mathbf{r}}_\nu, \bar{e}_\nu$

1. $\bar{e}_\nu := 0$, $b_\nu^s := b_\nu$
2. $\bar{\mathbf{r}}_\nu := \bar{Q}_{\nu-1} \cdots \bar{Q}_1 \cdot b_\nu^s$
3. *size-reduce* b_ν^s *against* $b_1^s, ..., b_{\nu-1}^s$
   for $i = \nu - 1, \ldots, 1$ **do**
   $\bar{\mu}_{\nu,i}^s := \bar{r}_{i,\nu} \;/\; \bar{r}_{i,i}$
   if $|\bar{\mu}_{\nu,i}^s| \geq 0.51$ **then** $b_\nu^s := b_\nu^s - \lceil \bar{\mu}_{\nu,i}^s \rfloor b_i^s$, $\bar{\mathbf{r}}_\nu := \bar{\mathbf{r}}_\nu - \lceil \bar{\mu}_{\nu,i}^s \rfloor \bar{\mathbf{r}}_i$
4. **if** $\exists i : |\bar{\mu}_{\nu,i}^s| \geq 2^{10}$ **then go to 2.**
5. $\tau := (\sum_{i=\nu}^d \bar{r}_{i,\nu}^2)^{\frac{1}{2}}$, $\widetilde{e} := \min(\lceil \log_2(\|b_1^s\| / \tau) \rceil, 30)$
6. **if** $\widetilde{e} > 0$ **then** $\bar{e}_\nu := \bar{e}_\nu + \widetilde{e}$, $b_\nu^s := 2^{\widetilde{e}} b_\nu^s$ **go to 2.**
7. $\bar{\mathbf{r}}_\nu := \bar{Q}_{\nu-1} \cdots \bar{Q}_1 b_\nu^s$, $\sigma := \mathrm{sig}(\bar{r}_{\nu,\nu})$, $\tau := (\sum_{i=\nu}^d \bar{r}_{i,\nu}^2)^{\frac{1}{2}}$
   $v_\nu := (0, \ldots, 0, \bar{r}_{\nu,\nu} + \sigma\tau, \bar{r}_{\nu+1,\nu}, \ldots, \bar{r}_{d,\nu})^\top$
   $\bar{\mathbf{r}}_\nu := (\bar{r}_{1,\nu}, \ldots, \bar{r}_{\nu-1,\nu}, -\,-\sigma\tau, 0, \ldots, 0)^\top$ *end*

How **HRS** *Works.* According to (5) the input vectors $b_1^s, ..., b_{\nu-1}^s$ satisfy

$$|r_{1,1}| \leq |r_{i,i}| < 2|r_{1,1}| \qquad \text{for } i = 1, ..., \nu - 1.$$

Let $6d\nu^2\eta < \frac{1}{2}$ so that Theorem 3 holds for $\varepsilon = \frac{1}{2}$. Then, by (2) and (6) we have that

$$|\bar{r}_{i,i} - r_{i,i}| \leq 6dj^2\eta |r_{i,i}| \qquad \text{for } i = 1, ..., j.$$

Reducing Long $\|b_\nu\|$. By (2) the relative error of $\bar{r}_{i,\nu}/\|b_\nu\|$ in step **2.** is at most $6d\nu\eta$. The subsequent size-reduction reduces the large $\bar{r}_{i,\nu}$ — in the equation $\|b_\nu\|^2 = \sum_{i=1}^d r_{i,\nu}^2$ — to less than $|\bar{r}_{i,i}|$ in absolute value. Steps **2. 3. 4.** decrease the $(\sum_{i=1}^{\nu-1} \bar{r}_{i,\nu}^2)^{\frac{1}{2}}$–part until the inequality

$$\left(\sum_{i=1}^{\nu-1} \bar{r}_{i,\nu}^2\right)^{\frac{1}{2}} \leq \|b_1^s, ..., b_{\nu-1}^s\|_F$$

holds.

Upon entry of step **3.** the coefficients $\mu_{\nu,i}^s$ are uniformly bounded as

$$|\mu_{\nu,j}^s| \leq 2 \cdot \frac{\|b_\nu\|}{\|b_1^s\|},$$

see Corollary 1. Size-reduction of Step **3.** can temporarily increase the coefficients $|\mu_{\nu,j}^s|$ up to a factor $(3/2)^{\nu-j-1}$ in worst case. This could possibly make the size-reduction of step **3.** unstable in worst case. No such instability has been reported in practice, see Figure 1.

Scaling Up Small $\|\widehat{b}_\nu\|$. Let

$$\|b_\nu\| \leq \|b_1^s, ..., b_{\nu-1}^s\|_F \qquad \text{and} |r_{\nu,\nu}| \leq \eta \|b_\nu\|.$$

By (2) and (6) we have after step **2.** that

$$|\bar{r}_{\nu,\nu} - r_{\nu,\nu}| < 6d\nu\eta \, \|b_1^s, ..., b_{\nu-1}^s\|_F \leq 6d\nu^2\eta \|b_1^s\|.$$

This yields a scaling factor $\tilde{e}$ for step **6.** so that $2^{\tilde{e}} \geq (6d\nu^2\eta)^{-1} > 1$. Therefore, steps **2.** to **6.** increase $|\bar{r}_{\nu,\nu}|$ and $\|b_\nu^s\|$ until

$$|\bar{r}_{1,1}| \leq |\bar{r}_{\nu,\nu}| \approx \|b_\nu^s\| < 2|\bar{r}_{1,1}|.$$

Corollary 3. HRS(ν) *produces a scaled vector* b_ν^s *satisfying* $|\mu_{\nu,i}^s| \leq 0.52$ *for* $i = 1, ..., \nu - 1$ *provided that size-reduction of Step* **3.** *remains stable.*

Speeding Up **HRS**. To speed up **HRS** replace the rounded values $\lceil \bar{\mu}_{\nu,i}^s \rfloor$ in step **3.** by single precision integers consisting of the leading bits of $\lceil \bar{\mu}_{\nu,i}^s \rfloor$. This way **HRS** becomes very fast.

The Number of Arithmetic Steps. A matrix-vector multiplication

$$b_\nu \mapsto \bar{Q}_1 b_\nu = b_\nu - \frac{2 v_1 v_1^\top b_\nu}{\|v_1\|^2}$$

requires $2d$ multiplications and one division — as usual we neglect the additions/subtractions. Thus, we get $\bar{\mathbf{r}}_\nu := \bar{Q}_{\nu-1} \cdots \bar{Q}_1 b_\nu$ using $2\nu d$ multiplications/divisions. Size-reduction of step **3.** requires νd multiplications in exact arithmetic. In total, one *round* of steps **2.**, **3.**, and **4.** requires $2\nu d$ multiplications/divisions in fpa and νd exact multiplications of long integers.

The Number of Rounds. In practice each round of steps **2. 3. 4.** either decreases $(\sum_{i=1}^{\nu-1} r_{i,\nu}^2)^{\frac{1}{2}}$ by the factor $6d\nu^2\eta$, or increases $\|\widehat{b}_\nu^s\|$ by the scaling factor

$$2^{\bar{e}_\nu} \geq (6d\nu^2\eta)^{-1} > 1,$$

or does a combination of both, see the explanations for **HRS**. In practice we find that **HRS**(ν) requires $\log_2(\|b_\nu\|/\|\widehat{b}_\nu\|)/|\log_2(6d\nu^2\eta)|$ rounds.

Practical Performance of **HRS.** Consider a random public-key basis of the GGH-cryptosystem [GGH] of dimension 400. Figure 1 shows how **HRS**(400) decreases the coefficients $|\mu_{400,j}|$ of the last basis vector b_{400}. Using 53 bit *fpa* each of five rounds decreases $|\mu_{400,j}|$ by a factor about $10^9 \approx 2^{30}$. After each round all coefficients $|\mu_{400,j}|$ are of nearly equal size because the preceding orthogonal vectors have been scaled to nearly equal length $\|\widehat{b}_j^s\|$. After each of the first 4 rounds however, $|\mu_{\nu,j}^s|$ increases by about a factor 10 as j decreases from 400 to 1. This is due to the temporary increase of $|\mu_{\nu,j}^s|$ by a factor $(3/2)^{\nu-j-1}$ in worst case. Figure 1 shows that this temporary increase has little effect in practice.

Using 106 bit *fpa* — instead of 53 bit *fpa* — the five rounds of **HRS**(400) reduce to two rounds, and the running time of **HRS** reduces accordingly. Software implemented *fpa* with 106 precision bits makes the reduction clearly faster than the standard *fpa* with 53 precision bits.

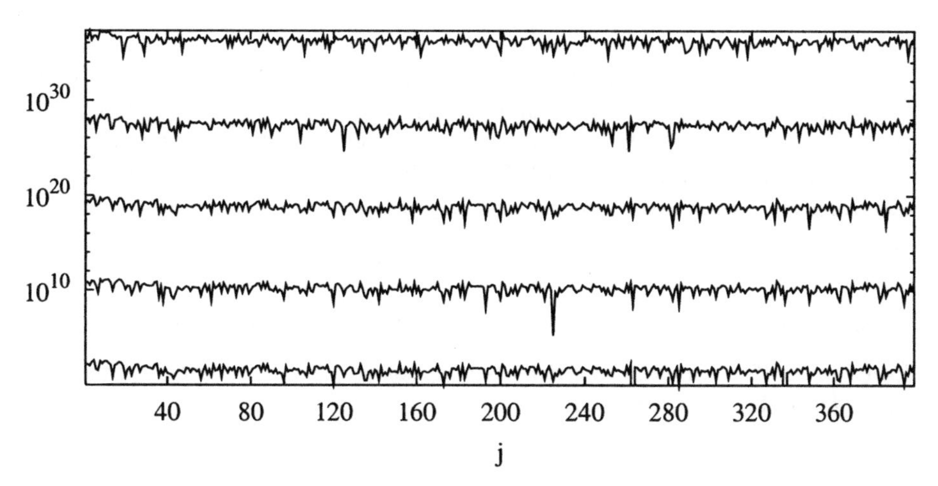

Fig. 1. Displayed are the values $|\mu_{400,j}|$ upon termination of each of five rounds. The $|\mu_{400,j}|$ are measured after step 2 of the next round — after new orthogonalization.

5 Scaled LLL-Reduction

We introduce *scaled* LLL-*reduced* lattice bases, and we present an algorithm **scaled LLL** for scaled LLL-reduction. This algorithm is a useful prelude to the more complicated scaled segment LLL-reduction of Section 7.

Definition 3. *We call a lattice basis* $b_1, \ldots, b_n \in \mathbf{Z}^d$ scaled LLL-reduced *if it has properties* **1.** *and* **2.**

1. *There is an associated scaled basis* $b_1^s, ..., b_n^s$ *so that* $b_1, ..., b_n$ *is size-reduced against* $b_1^s, ..., b_n^s$*, i.e.,*

$$\left| \frac{\langle b_\nu, \widehat{b}_j^s \rangle}{\|\widehat{b}_j^s\|^2} \right| \leq 0.52 \qquad \textit{for} \;\; 1 \leq i < j \leq n,$$

2. $\delta \; \|\widehat{b}_i\|^2 \;\leq\; \alpha \|\widehat{b}_{i+1}\|^2 \qquad$ *for* $i = 1, \ldots, n-1$.

We call a basis with property **1.** *scaled-reduced.* Importantly, the inequalities of Theorem 1 still hold for scaled LLL-reduced lattice bases. The weaker size-reduction does not affect these inequalities. Scaled LLL-reduced bases are in practice as good as LLL-reduced bases.

Next we present an algorithm **scaled LLL** that transforms a lattice basis into a scaled LLL-reduced basis of the same lattice. **Scaled LLL** guides the LLL-steps on the original basis by the orthogonalization of an associated scaled basis. So we keep and update two versions of the basis. LLL-transformations operate on the original basis. Size-reduction is done against the scaled basis. This form of LLL-reduction is quite stable as **HRS** yields an accurate QR-factorization of the scaled basis.

The Procedure **HRS′**. Local LLL-reduction of two basis vectors $b_{\nu-1}, b_\nu$ is guided by the orthogonal vector of a modified scaled vector b_ν^s — with a scaling factor $2^{\bar{e}_\nu}$ that coincides with the scaling factor $2^{\bar{e}_{\nu-1}}$ of $b_{\nu-1}$. We get the modified b_ν^s by the following variant **HRS′** of **HRS**: Set in Step **1.** $\bar{e}_\nu := \bar{e}_{\nu-1}$ and skip Step **6.**. Moreover, we let **HRS′** perform in Step **3.** the same size-reduction steps

$$b_\nu^s := b_\nu^s - \lceil \bar{\mu}_{\nu,\nu-1}^s \rfloor b_{\nu-1}^s \qquad \text{and} \qquad b_\nu := b_\nu - \lceil \bar{\mu}_{\nu,\nu-1}^s \rfloor b_{\nu-1}$$

on b_ν^s against $b_{\nu-1}^s$ and on b_ν against $b_{\nu-1}$. As a consequence we have:

Lemma 2. *The reduction coefficients of* b_ν *against* $b_{\nu-1}$ *and of* b_ν^s *against* $b_{\nu-1}^s$ *coincide in* **HRS′**(ν)*, and thus*

$$\mu_{\nu,\nu-1} = \mu_{\nu,\nu-1}^s.$$

The output vector b_ν *of* **HRS′**(ν) *is size-reduced against* $b_{\nu-1}$*. The coefficients* $r_{i,j}$ *for* $\nu-1 \leq i, j \leq \nu$ *associated with* $b_{\nu-1}, b_\nu$ *are proportional to the coefficients associated with* $b_{\nu-1}^s, b_\nu^s$ *with proportionality factor* $2^{\bar{e}_\nu}$.

Scaled LLL (*Algorithm for scaled* LLL*-reduction.*)

INPUT $b_1, \ldots, b_n \in \mathbf{Z}^d$, δ

OUTPUT $b_1, \ldots, b_n$ scaled LLL-reduced basis

1. $\nu := 1$

2. **while** $\nu \leq n$ **do**

3. **if** $\nu = 1$ **then** $\mathbf{HRS}(1)$, $\nu := 2$

4. $\mathbf{HRS}'(\nu)$

5. **if** $\delta\, \bar{r}^2_{\nu-1,\nu-1} > \bar{r}^2_{\nu-1,\nu} + \bar{r}^2_{\nu,\nu}$

then swap $b_{\nu-1}, b_\nu$, $\nu := \nu - 1$

else $\mathbf{HRS}(\nu)$, $\nu := \nu + 1$ *end*

Scaling Does Not Affect LLL-*Exchanges.* By Lemma 2, the output vector b_ν of $\mathbf{HRS}'(\nu)$ is size-reduced against $b_{\nu-1}$, i.e., $|\mu_{\nu,\nu-1}| \leq 0.52$. Moreover, the decision about swapping $b_{\nu-1}, b_\nu$ in Step **5.** is the same as for the original LLL-algorithm — except for *fpa*-errors.
fpa-errors do not affect the LLL-*exchanges.* We see from $\|b^s_\nu\|^2 = \sum_{i=1}^d r^2_{i,\nu}$ and (2) that the relative error of $\bar{r}_{\nu-1,\nu}/\|b^s_\nu\| \approx \bar{r}_{\nu-1,\nu}/|\bar{r}_{\nu,\nu}|$ and $|r_{\nu,\nu}|/\|b^s_\nu\|$ is at most $6d\nu\eta$. On the other hand, $|\bar{r}_{\nu-1,\nu-1}| \geq \|b^s_1\|$ holds by scaling. Therefore, the decision about swapping $b_{\nu-1}, b_\nu$ in Step **5.** is correct for $6d\nu\eta \ll 1$.

6 Segment LLL-Reduction

We summarize the concept of segment LLL-reduced bases of the companion paper, for full coverage see [KS01].

Segments and Local Coordinates. Let the basis $b_1, \ldots, b_n \in \mathbf{Z}^d$ have dimension $n = k \cdot m$ and the QR-factorization $[b_1, \ldots, b_n] = QR$. We partition the basis-matrix B into m *segments* $B_l = [b_{k(l-1)+1}, \ldots, b_{kl}]$ for $l = 1, ..., m$. Local reduction of two consecutive segments uses the coefficients of the submatrix $R_l := [r_{kl+i,kl+j}]_{-k<i,j\leq k} \in \mathbf{R}^{2k\times 2k}$ of $R \in \mathbf{R}^{d\times n}$, corresponding to two consecutive segments B_{l-1}, B_l. We want to do most of the LLL-exchanges and the corresponding size-reduction in local coordinates of some R_l. Extra global transformations are required after local LLL-reduction. In order to minimize these global costs we introduce *k-segment reduced bases.* We let $D(l) = \|\widehat{b}_{k(l-1)+1}\|^2 \cdots \|\widehat{b}_{kl}\|^2$ denote the *local Gramian determinant* of segment B_l. We have that $D_{kl} = D(1) \cdots D(l)$.

Definition 4. *We call a basis* $b_1, \ldots, b_n \in \mathbf{Z}^d, n = km$, k-segment LLL-reduced *with* $\delta \in]\frac{1}{4}, 1]$ *if it is size-reduced and satisfies for* $\alpha = 1/(\delta - \frac{1}{4})$:

1. $\delta\, \|\widehat{b}_i\|^2 \leq \mu^2_{i+1,i}\|\widehat{b}_i\|^2 + \|\widehat{b}_{i+1}\|^2$ *for* $i \neq 0 \mod k$,

2. $D(l) \leq (\alpha/\delta)^{k^2} D(l+1)$ *for* $l = 1, \ldots, m-1$,

3. $\delta^{k^2}\, \|\widehat{b}_{kl}\|^2 \leq \alpha\|\widehat{b}_{kl+1}\|^2$ *for* $l = 1, \ldots, m-1$.

Theorem 4. [KS01] *Let* $b_1, \ldots, b_n$ *be a basis that is* k*-segment* LLL-*reduced with* δ*. Then we have for* $i = 1, \ldots, n$:

$$\delta^{2k^2+n-1}\, \|b_i\|^2 \leq \alpha^{n-1}\lambda_i^2 \qquad \text{and} \qquad \delta^{k^2+i-1}\, \|b_1\|^2 \leq \alpha^{i-1}\|\widehat{b}_i\|^2,$$

where $\lambda_1 \leq \cdots \leq \lambda_n$ *are the successive minima of the lattice.*

Algorithm for Segment LLL-*Reduction.* The algorithm **segment LLL** transforms a given basis into a k-segment reduced basis using exact integer arithmetic. It iterates local LLL-reduction of two segments $[B_{l-1}, B_l] = [b_{kl-k+1}, ..., b_{kl+k}]$ via:

The Procedure **loc-LLL**(l). Given the orthogonalization of a k-segment reduced basis $b_1, ..., b_{kl-k}$ the procedure **loc-LLL**(l) computes the orthogonalization and size-reduction of the segments B_{l-1}, B_l. In particular it provides the submatrix $R_l \in \mathbf{R}^{2k\times 2k}$ of $R \in \mathbf{R}^{d\times n}$ corresponding to the segments B_{l-1}, B_l. Thereafter it performs a local LLL–reduction of R_l and stores the LLL-transformation in the matrix $H \in \mathbf{Z}^{2k\times 2k}$. Finally, it transforms $[B_{l-1}, B_l]$ into the locally reduced segments $[B_{l-1}, B_l]H$ and size-reduces $[B_{l-1}, B_l]$ globally.

Segment LLL

INPUT $b_1, \ldots, b_n \in \mathbf{Z}^d, k, m, n = km,\ \delta$

OUTPUT $b_1, \ldots, b_n$ k-segment LLL-reduced basis

1. $l := 1$

2. **while** $l \leq m - 1$ **do**

loc-LLL(l)

if $l \neq 1$ **and**

$(\ D(l-1) > (\alpha/\delta)^{k^2} D(l)$ **or** $\delta^{k^2} \|\widehat{b}_{k(l-1)}\|^2 > \alpha\|\widehat{b}_{k(l-1)+1}\|^2\ )$

then $l := l - 1$ **else** $l := l + 1$. *end*

Theorem 5. [KS01] *For* $k = \Theta(m) = \Theta(\sqrt{n})$, **segment LLL** *performs*

$$O(nd \log_{1/\delta} M)$$

arithmetic steps using integers of bit length $O(\log_2 M)$.

7 Scaled Segment LLL-Reduction

We combine the methods of Sections 4 and 6 to a stable algorithm for segment LLL-reduction. Size-reduction is done against an associated scaled basis. Segment LLL-reduction is guided by the orthogonal vectors of the scaled basis.

Definition 5. *We call a basis* $b_1, \ldots, b_n \in \mathbf{Z}^d$ k-segment sLLL-reduced, *if it is k-segment* LLL-*reduced except that* $b_1, ..., b_n$ *is size-reduced in a weaker sense — it is size-reduced against an associated basis* $b_1^s, ..., b_n^s$ *with the properties* **1.** *and* **2.**

1. *The first orthogonal vectors* $\widehat{b}^s_{kl-k+1}$ *of segments are nearly equally long:*

$$\|b_1^s\| \leq 2^{e_{kl}}\|\widehat{b}_{kl-k+1}\| = \|\widehat{b}^s_{kl-k+1}\| < 2\|b_1^s\| \quad \text{for } l = 1, ..., m.$$

2. *There is a* uniform *scaling factor* $2^{e_{kl}}$ *for segment* B_l *so that the coefficients* $r_{kl-k+i,kl-k+j}$, *for* $1 \leq i, j \leq k$ *of the R-matrix corresponding to* $b_1, ..., b_n$ *and those corresponding to* $b_1^s, ..., b_n^s$ *are proportional with proportionality factor* $2^{e_{kl}}$.

The inequalities for k-segment LLL-reduced bases in Theorem 4 also hold for k-segment sLLL-reduced bases. The weaker size-reduction is not important. In practice k-segment sLLL-reduced bases are nearly as good as LLL-reduced bases.

We sketch a procedure **loc-sLLL**(l) — for local scaled LLL-reduction — that replaces **loc-LLL**(l) within **segment LLL**. The algorithm **segment sLLL** iterates local scaled LLL-reduction of two consecutive segments $[B_{l-1}, B_l] = [b_{kl-k+1}, ..., b_{kl+k}]$.

The Procedure **loc-sLLL**(l). Its inputs are a lattice basis, uniform scaling factors $2^{\bar{e}_{k\ell}}$ satisfying $b^s_{k\ell-j} = 2^{\bar{e}_{k\ell}} b_{k\ell-j}$ for $j = 0, ..., k-1$, $\ell = 1, ..., l-1$. The Householder vectors $v_1, ..., v_{k(l-1)}$ of $b_1, ..., b_{k(l-1)}$ and the matrix $R \in \mathbf{R}^{d \times k(l-1)}$ for $[b^s_1, ..., b^s_{k(l-1)}]$ are also given. Local LLL-reduction of B_{l-1}, B_l is done in fpa via the local matrix $R_l \in \mathbf{R}^{2k \times 2k}$.

The procedure **loc-sLLL**(l):

- computes a uniform scaling factor $2^{\bar{e}_{kl}}$ for the two segments $[B_{l-1}, B_l]$,
- computes the local matrix R_l of $2^{\bar{e}_{kl}}[B_{l-1}, B_l]$ via **HRS**$'$,
- performs a local LLL-reduction on $[B_{l-1}, B_l]$ using R_l,
- stores the transformation in the matrix $H \in \mathbf{Z}^{2k \times 2k}$.
- Upon termination it transforms $[B_{l-1}, B_l]$ into $[B_{l-1}, B_l]H$ in exact arithmetic.

Restarting **loc-sLLL**(l). Whenever $\|H\|_\infty$ surpasses the threshold 2^{15} the procedure **loc-sLLL**(l) is restarted with the transformed segments $[B_{l-1}, B_l]H$. This is necessary as the norm $\|H\|_\infty$ directly translates into additional fpa-errors. As the number of restarts is crucial for the running time we carefully prepare R_l as to prevent an early restart. We show below how to predict the size of the matrix $H \in \mathbf{Z}^{2k \times 2k}$ occuring in the subsequent local LLL-reduction of B_{l-1}, B_l. We slash R_l so that $\|H\|_\infty \leq 2^{15}$ holds on the average. The threshold 2^{15} is for wired 53-bit fpa, for 106-bit fpa we increase the threshold accordingly.

Computing the Matrix R_l and $2^{\bar{e}_{kl}}$. First compute via **HRS** the uniform scaling factor $2^{\bar{e}_{kl}}$ and the first vector $\mathbf{r}_{kl-k+1}$ of B_{l-1} so that $\|b^s_1\| \leq 2^{e_{kl}} \|\widehat{b}_{kl-k+1}\| = \|\widehat{b}^s_{kl-k+1}\| < 2\|b^s_1\|$. With the same scaling factor $2^{\bar{e}_{kl}}$ compute $\mathbf{r}_h$ via **HRS**$'$ for $h = kl-k+2, ..., kl+k$. For local LLL-reduction we have to size-reduce b_h against $b_{kl-k+1}, \ldots b_{h-1}$. We generalize Step **3.** of **HRS**$'$ accordingly: perform the same size-reduction steps $b^s_h := b^s_h - \lceil \bar{\mu}^s_{h,j} \rfloor b^s_j$, $b_h := b_h - \lceil \bar{\mu}^s_{h,j} \rfloor b_j$ on b^s_h against b^s_j and on b_h against b_j for $j = kl-k+1, \ldots h-1$. This implies that the local matrix R_l of segments B_{l-1}, B_l corresponding to $b_1, ..., b_n$ and the R_l corresponding to $b^s_1, ..., b^s_n$ are proportional with the factor $2^{\bar{e}_{kl}}$.

The Heuristic for Slashing R_l . We let $r^l_{i,j} = r_{kl-k+i,kl-k+j}$ denote the coefficients of the local matrix R_l. Large values $|r^l_{1,1}/r^l_{j,j}|$ have a double negative effect on the stability of **loc-sLLL**(l). By Lemma 1 the orthogonalization of b_j gets inaccurate. Moreover, $\|H\|_\infty$ will be large due to Lemma 2 [KS01]. A detailed argument shows that $\|H\|_\infty$ is expected to be less than 2^{15} if $\max_j |r^l_{1,1}/r^l_{j,j}| \leq 2^{15}$. This suggests to *slash* R_l so that $\max_j |r^l_{1,1}/r^l_{j,j}| \leq 2^{15}$.

Slashing the Matrix R_l. Slash all values $|r^l_{j,j}|$ — satisfying $|r^l_{1,1}/r^l_{j,j}| < 2^{-15}$ — to $r^l_{j,j} := |r^l_{1,1}|2^{-15}$. Moreover, set $r^l_{h,i} := 0$ for $h = j, ..., 2k$ and $h < i \leq 2k$.

$$R_l := \begin{bmatrix} r^l_{1,1} \cdots & r^l_{1,j-1} & r^l_{1,j} & \cdots & r^l_{1,k} \\ \ddots & \vdots & \vdots & & \vdots \\ 0 & r^l_{j-1,j-1} & r^l_{j-1,j} & \cdots & r^l_{j-1,k} \\ 0 \cdots & 0 & |r^l_{1,1}|2^{-15} & & 0 \\ \vdots & \vdots & & \ddots & \\ 0 \cdots & 0 & 0 & & |r^l_{1,1}|2^{-15} \end{bmatrix}$$

We locally LLL-reduce the slashed matrix R_l, and afterwards we transform $[B_{l-1}, B_l]$ into $[B_{l-1}, B_l]H$. If either $\|H\|_\infty > 2^{15}$ or if the slashing did effectively change R_l we restart the local reduction with the transformed segments $[B_{l-1}, B_l]H$. Note that a restart of **loc-sLLL**(l) also adjusts the uniform scaling factor $2^{\bar{e}_{kl}}$. This method of correction works very well for segment size $k \leq 100$.

Segment sLLL

INPUT $b_1, \ldots, b_n \in \mathbf{Z}^d, k, m, n = km,\ \delta$

OUTPUT $b_1, \ldots, b_n$ k-segment and scaled reduced basis

1. $l := 1$

2. **while** $l \leq m-1$ **do**

  loc-sLLL(l)

  if $l \neq 1$ **and**

   $(\ D(l-1) > (\alpha/\delta)^{k^2} D(l)$ **or** $\delta^{k^2} \|\widehat{b}_{kl}\|^2 > \alpha\|\widehat{b}_{kl+1}\|^2\ )$

  then $l := l-1$ **else** $l := l+1$.   *end*

8 Performance of Segment sLLL

Consider a sample of [GGH]-bases of dimension $n = \nu \cdot 100$ for $\nu = 1, \ldots, 8$. The vectors of the input basis consist of integers with $n/2$ bits. The algorithm uses 106 bit fpa and $\delta = 0.99$ respectively $\delta = 0.999$. All tests have been performed on 350 MHz PC's, the code of **segment sLLL** uses the [NTL] computer algebra library. We present the segment size and the average bit length of the reduced basis vectors in Figure 2.

The running time increases considerably as δ approaches 1. The size of the reduced basis decreases accordingly. Figure 3 shows the time of **segment sLLL** for [GGH]-bases.

Lattice dimension	$\delta = 0.99$ *segment size*	l^{av}	$\delta = 0.999$ *segment size*	l^{av}
100	25	7.9	25	5.2
200	40	13.8	40	9.4
300	45	23.0	45	14.7
400	50	30.5	50	17.7
500	50	34.3	56	22.6
600	60	38.8	60	27.9
700	60	40.9	65	34.3
800	70	46.6	70	37.0

Fig. 2. segment sLLL Reduced Bases. l^{av} = Average Bit Length of the Output Integers.

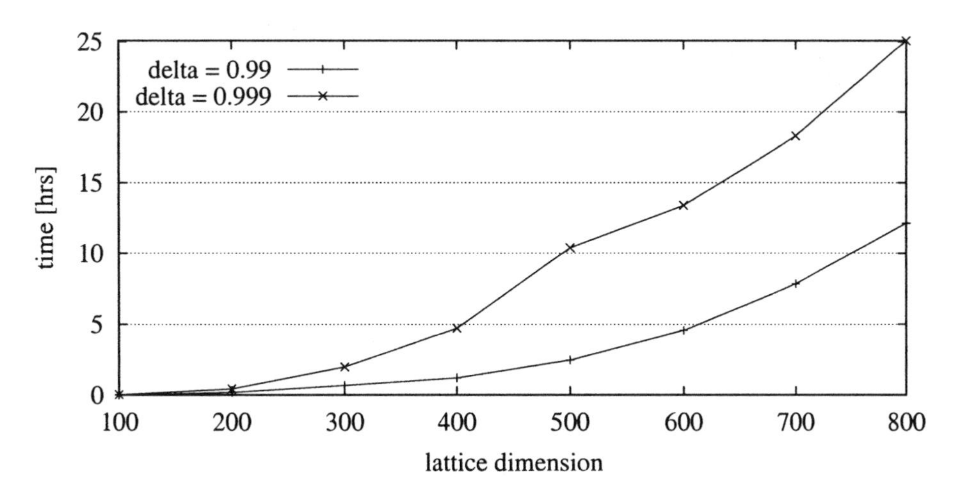

Fig. 3. Running Time of **Segment sLLL** Using $\delta = 0.99$ resp. $\delta = 0.999$.

Acknowledgment

We thank Bartol Filipović for his help in producing and measuring the code of the new reduction algorithm.

References

[GGH] *O. Goldreich, S. Goldwasser, and S. Halevi*, Public-key cryptosystems from lattice reduction problems. Proc. Crypto'97, LNCS 1294, Springer-Verlag, pp. 112–131, 1997.

[KS01] *H. Koy and C.P. Schnorr*, Segment LLL-Reduction of Lattice Bases. Proceedings CaLC 2001, pp. 67–80.

[LLL82] *A.K. Lenstra, H.W. Lenstra, and L. Lovász*, Factoring polynomials with rational coefficients, *Math. Ann.* **261**, pp. 515–534, 1982.

[LH95] *C.L. Lawson and R.J. Hanson*, Solving Least Square Problems, SIAM, Philadelphia, 1995.

[NTL] NTL homepage: `http://www.shoup.net/ntl/`, 2000.

[RS96] *C. Rössner and C.P. Schnorr*, An optimal stable continued fraction algorithm for arbitrary dimension. 5.-th IPCO, LNCS **1084**, pp. 31–43, Springer-Verlag, 1996.

[S87] *C.P. Schnorr*, A hierarchy of polynomial time lattice basis reduction algorithms, *Theoretical Computer Science* **53**, pp. 201-224, 1987.

[S88] *C.P. Schnorr*, A more efficient algorithm for lattice basis reduction, *J. Algorithms* **9**, pp. 47–62, 1988.

[SE91] *C.P. Schnorr and M. Euchner*, Lattice Basis Reduction: Improved Practical Algorithms and Solving Subset Sum Problems, *Proc. Fundamentals of Computation Theory'91*, L. Budach, ed., LNCS **529**, Springer-Verlag, pp. 68-85, 1991. (Complete paper in MATHEMATICAL PROGRAMMING STUDIES **66A**, No 2 , pp. 181–199, 1994.)

[Sc84] *A. Schönhage*, Factorization of univariate integer polynomials by diophantine approximation and improved lattice basis reduction algorithm, *Proc. 11-th Coll. Automata, Languages and Programming, Antwerpen 1984*, LNCS **172**, Springer-Verlag, pp. 436–447, 1984.

The Insecurity of Nyberg–Rueppel and Other DSA-Like Signature Schemes with Partially Known Nonces

Edwin El Mahassni[1], Phong Q. Nguyen[2]*, and Igor E. Shparlinski[3]**

[1] Department of Computing, Macquarie University, NSW 2109, Australia
eelmaha@ics.mq.edu.au
[2] École Normale Supérieure, Département d'Informatique
45 rue d'Ulm, 75005 Paris, France
pnguyen@ens.fr
http://www.di.ens.fr/~pnguyen/
[3] Department of Computing, Macquarie University, NSW 2109, Australia
igor@ics.mq.edu.au
http://www.comp.mq.edu.au/~igor/

Abstract. It has recently been proved by Nguyen and Shparlinski that the *Digital Signature Algorithm* (DSA) is insecure when a few consecutive bits of the random nonces k are known for a reasonably small number of DSA signatures. This result confirmed the efficiency of some heuristic lattice attacks designed and numerically verified by Howgrave-Graham and Smart. Here, we extend the attack to the Nyberg–Rueppel variants of DSA. We use a connection with the *hidden number problem* introduced by Boneh and Venkatesan and new bounds of exponential sums which might be of independent interest.
Keywords: DSA, Closest Vector Problem, Hidden Number Problem, Exponential Sums.

1 Introduction

1.1 The Digital Signature Algorithm (DSA)

Recall the *Digital Signature Algorithm* (DSA) used in the American federal digital signature standard. Let p and $q \geq 3$ be prime numbers with $q|p-1$. As usual $\mathbb{F}_p$ and $\mathbb{F}_q$ denote fields of p and q elements which we assume to be represented by the elements $\{0, \dots, p-1\}$ and $\{0, \dots, q-1\}$ respectively. For a rational γ and integer $m \geq 1$ we denote by $\lfloor \gamma \rfloor_m$ the smallest non-negative residue of γ modulo m. Let $\mathcal{M}$ be the set of messages to be signed and let $h : \mathcal{M} \longrightarrow \mathbb{F}_q$ be an arbitrary hash function. The signer selects a secret key $\alpha \in \mathbb{F}_q^*$ and computes the public key $A = \lfloor g^\alpha \rfloor_p$, where $g \in \mathbb{F}_p$ is a public element of multiplicative order q.

* Part of this work is an output of the "Turbo-signatures" project, supported by the French Ministry of Research.
** Work supported in part by the Australian Research Council.

J.H. Silverman (Ed.): CaLC 2001, LNCS 2146, pp. 97–109, 2001.

For an integer k we define the function

$$r(k) = \left\lfloor \lfloor g^k \rfloor_p \right\rfloor_q. \tag{1}$$

Finally, for a random element $k \in \mathbb{F}_q^*$ called a *nonce* and a *message* $\mu \in \mathcal{M}$ we define the function

$$s(k,\mu) = \left\lfloor k^{-1}\left(h(\mu) + \alpha r(k)\right)\right\rfloor_q \tag{2}$$

and call the pair $(r(k), s(k,\mu))$ the *DSA signature* of the message μ with a nonce k.

1.2 Lattice Attacks on DSA

Howgrave-Graham and Smart [6] noticed that if some most significant bits of k are known for a reasonably small number of signatures then one can apply the LLL lattice reduction algorithm [10] to heuristically recover the secret key α. They also presented numerical results to confirm the efficiency of their attack. Nguyen [14] exhibits a close link between this problem and the *hidden number problem* (HNP) introduced by Boneh and Venkatesan [3, 4], together with improved numerical results. HNP can be stated as follows: recover a number $\alpha \in \mathbb{F}_q$ such that for many known random $t \in \mathbb{F}_q$ a certain number ℓ of the most significant bits of $\lfloor \alpha t \rfloor_q$ are known. Boneh and Venkatesan showed in [3, 4] how to solve HNP in polynomial time when the number of bits was roughly $\log^{1/2} q$ and the distribution of t was uniform, using a reduction to the lattice closest vector problem.

The link between DSA and HNP is the following. Assume that we know the ℓ least significant bits of a nonce $k \in \mathbb{F}_q^*$. That is, we are given an integer a such that $0 \le a \le 2^\ell - 1$ and $k - a = 2^\ell b$ for some integer $b \ge 0$. Given a message μ signed with the nonce k, the signing equation (1) can be rewritten for $s(k,\mu) \ne 0$ as:

$$\alpha r(k) 2^{-\ell} s(k,\mu)^{-1} \equiv \left(a - s(k,\mu)^{-1} h(\mu)\right) 2^{-\ell} + b \pmod{q}.$$

Now define the following two elements

$$t(k,\mu) = \left\lfloor 2^{-\ell} r(k) s(k,\mu)^{-1} \right\rfloor_q$$
$$u(k,\mu) = \left\lfloor 2^{-\ell}\left(a - s(k,\mu)^{-1} h(\mu)\right)\right\rfloor_q$$

and remark that both $t(k,\mu)$ and $u(k,\mu)$ can easily be computed by the attacker from the publicly known information. Recalling that $0 \le b \le q/2^\ell$, we obtain

$$\lfloor \alpha t(k,\mu) - u(k,\mu) \rfloor_q < q/2^\ell.$$

Thus, the ℓ most significant bits of $\lfloor \alpha t(k,\mu) \rfloor_q$ are revealed.

Finally, Nguyen and Shparlinski [15] have proved that the heuristic attack of [6, 14] always succeeds. This has been done by generalizing the lattice-based proof of Boneh-Venkatesan to cases where the distribution of t is not necessarily perfectly uniform, and more importantly, by proving some uniformity results on the distribution of t in the case of DSA, using bounds of *exponential sums*.

1.3 Our Results

In this paper, we extend the results of Nguyen and Shparlinski [15] to several known modifications of DSA. Although our method is similar to that of [15], the exponential sums arising in this paper are different. In particular, we study the five following modifications of the *signing equation* (2) which are outlined in Section 20.4 of [19] (see also Section 11.5.2 in [12] and [18])

$$
\begin{array}{ll}
\text{DSA}_1: & s_1(k,\mu) = \lfloor k^{-1}(r(k) + \alpha h(\mu)) \rfloor_q. \\
\text{DSA}_2: & s_2(k,\mu) = \lfloor \alpha^{-1}(h(\mu) + kr(k)) \rfloor_q; \\
\text{DSA}_3: & s_3(k,\mu) = \lfloor \alpha r(k) + kh(\mu) \rfloor_q; \\
\text{DSA}_4: & s_4(k,\mu) = \lfloor \alpha h(\mu) + kr(k) \rfloor_q; \\
\text{DSA}_5: & s_5(k,\mu) = \lfloor \alpha^{-1}(r(k) + kh(\mu)) \rfloor_q.
\end{array}
$$

The DSA_i signature on a message μ is $(r(k), s_i(k,\mu))$, $i = 1, \ldots, 5$. We remark that for these DSA variants, α, $h(\mu)$ and k have the same meaning as in the original DSA and $r(k)$ is defined by the same equation (1). By alternating signs in these equations, one obtains additional signature schemes which can be studied quite analogously.

We also show that the same method can be applied to yet another modification proposed by Nyberg and Rueppel [18] which provides message recovery, see also Section 11.5.4 in [12] and Section 20.4 of [19]. The initial settings of this scheme are the same except that instead of the hash function h mapping $\mathcal{M}$ to $\mathbb{F}_q$, we are given a function H whose values are elements of $\mathbb{F}_p$. In fact, the main idea of this scheme is to use an easily invertible (and just one-to-one) function $H : \mathcal{M} \longrightarrow \mathbb{F}_p$. Then the NR signature $(r(k,\mu), v(k,\mu)$ is defined by *signing equation*

$$
\text{NR:} \qquad v(k,\mu) = \lfloor \alpha r(k,\mu) + k \rfloor_q,
$$

where

$$
r(k,\mu) = \lfloor H(\mu) g^k \rfloor_p
$$

In this paper we do not discuss advantages and disadvantages of the above schemes (for example, some of them are "inversion"-free and thus the value $k = 0$ can be selected as well) but remark that for the NR signature scheme one derives

$$
H(\mu) \equiv g^{-v(k,\mu)} A^{r(k,\mu)} r(k,\mu) \pmod p
$$

where as before $A = \lfloor g^\alpha \rfloor_p$. So if the function $h(\mu)$ is easy to invert, then the message μ can be recovered from the signature itself.

One can easily verify that if we are given an integer a such that $0 \le a \le 2^\ell - 1$ and $k - a = 2^\ell b$ for some integer $b \ge 0$ then for each of the signature schemes DSA_i, $i = 1, \ldots, 5$ we have,

$$
\lfloor \alpha t_i(k,\mu) - u_i(k,\mu) \rfloor_q < q/2^\ell, \qquad i = 1, \ldots, 5, \tag{3}
$$

where

$$t_1(k,\mu) \equiv \left\lfloor \frac{h(\mu)}{2^l s_1(k,\mu)} \right\rfloor_q, \quad u_1(k,\mu) \equiv \left\lfloor 2^{-l}\left(a - \frac{r(k)}{s_1(k,\mu)}\right) \right\rfloor_q;$$

$$t_2(k,\mu) \equiv \left\lfloor \frac{s_2(k,\mu)}{2^l r(k)} \right\rfloor_q, \quad u_2(k,\mu) \equiv \left\lfloor 2^{-l}\left(\frac{h(\mu)}{r(k)} + a\right) \right\rfloor_q;$$

$$t_3(k,\mu) \equiv \left\lfloor -\frac{r(k)}{2^l h(\mu)} \right\rfloor_q, \quad u_3(k,\mu) \equiv \left\lfloor 2^{-l}\left(a - \frac{s_3(k,\mu)}{h(\mu)}\right) \right\rfloor_q;$$

$$t_4(k,\mu) \equiv \left\lfloor -\frac{h(\mu)}{2^l r(k)} \right\rfloor_q, \quad u_4(k,\mu) \equiv \left\lfloor 2^{-l}\left(a - \frac{s_4(k,\mu)}{r(k)}\right) \right\rfloor_q;$$

$$t_5(k,\mu) \equiv \left\lfloor \frac{s_5(k,\mu)}{2^l h(\mu)} \right\rfloor_q, \quad u_5(k,\mu) \equiv \left\lfloor 2^{-l}\left(\frac{r(k)}{h(\mu)} + a\right) \right\rfloor_q;$$

(for the schemes DSA_3 and DSA_5 we assume that $h : \mathcal{M} \longrightarrow \mathbb{F}_q^*$).

For the signature scheme NR, we have

$$\left\lfloor \alpha 2^{-l} r(k,\mu) - w(k,\mu) \right\rfloor_q < q/2^{\ell}, \tag{4}$$

where

$$w(k,\mu) = \left\lfloor 2^{-l}(v(k,\mu) - a) \right\rfloor_q.$$

It is important to remark that in the above inequalities the "multipliers" $t_i(k,\mu)$, $r(k,\mu)$ and the "approximations" $u_i(k,\mu)$, $w(k,\mu)$, $i = 1,\dots,5$, can be efficiently evaluated from publicly available data.

Thus to obtain analogues of the results of Nguyen and Shparlinski [15], one has to study the distribution of $t_i(k,\mu)$, $i = 1,\dots,5$, and $r(k,\mu)$ for randomly selected nonce $k \in \mathbb{F}_q^*$ and $\mu \in \mathcal{M}$.

2 Preparations

2.1 Preliminaries

Let

$$\mathbf{e}_q(z) = \exp(2\pi i z/q) \qquad \text{and} \qquad \mathbf{e}_p(z) = \exp(2\pi i z/p).$$

We recall a few results about exponential which we use to show that our attacks work. The first one is the well-known *Weil* bound on exponential sums with rational functions which is presented in the form given by C. J. Moreno and O. Moreno [13], see also [11].

Lemma 1. *For any polynomials $g(X), f(X) \in \mathbb{F}_q[X]$ such that the rational function $F(X) = f(X)/g(X)$ is not constant on $\mathbb{F}_q$, the bound*

$$\left| \sum_{\lambda \in \mathbb{F}_q} {}^* \mathbf{e}_q\left(F(\lambda)\right) \right| \le \left(\max\{\deg g, \deg f\} + u - 2\right) q^{1/2} + \delta$$

holds, where $\sum^*$ *means that the summation is taken over all* $\lambda \in \mathbb{F}_q$ *which are not poles of* $F(X)$ *and*

$$(u, \delta) = \begin{cases} (v, 1), & \text{if } \deg f \le \deg g, \\ (v+1, 0), & \text{if } \deg f > \deg g, \end{cases}$$

and v *is the number of distinct zeros of* $g(X)$ *in the algebraic closure of* $\mathbb{F}_q$.

We also use the following well-known statement, see Exercise 11.a in Chapter 3 of [23].

Lemma 2. *For any integers* u *and* $M \ge 1$,

$$\sum_{\lambda=0}^{M-1} \exp(2\pi i z u/M) = \begin{cases} 0, & \text{if } u \not\equiv 0 \pmod M; \\ M, & \text{if } u \equiv 0 \pmod M. \end{cases}$$

We also need a result from [15] about the distribution of $r(k)$. For an integer $\rho \in \mathbb{F}_q$, let $N(\rho)$ be the number of solutions of the equation

$$r(k) = \rho, \qquad k \in \mathbb{F}_q^*.$$

The following estimate has been obtained in [15] (using some bounds for exponential sums from [8]).

Lemma 3. *Let* Q *be a sufficiently large integer. The following statement holds with* $\nu = 1/3$ *for all primes* $p \in [Q, 2Q]$, *and with* $\nu = 0$ *for all primes* $p \in [Q, 2Q]$ *except at most* $Q^{5/6+\varepsilon}$ *of them. For any* $\varepsilon > 0$ *there exists* $\delta > 0$ *such that for any element* $g \in \mathbb{F}_p$ *of multiplicative order* $q \ge p^{\nu+\varepsilon}$ *the bound*

$$N(\rho) = O\left(q^{1-\delta}\right), \qquad \rho \in [0, q-1],$$

holds.

We recall that the *discrepancy* $\mathcal{D}(\Gamma)$ of a sequence $\Gamma = (\gamma_i)_{i=1}^N$ of N elements of the interval $[0,1]$ is defined as

$$\mathcal{D}(\Gamma) = \sup_{\mathcal{I} \subseteq [0,1]} \left| \frac{A(\mathcal{I}, N)}{N} - |\mathcal{I}| \right|,$$

where the supremum is extended over all subintervals $\mathcal{I}$ of $[0,1]$, $|\mathcal{I}|$ is the length of $\mathcal{I}$, and $A(\mathcal{I}, N)$ denotes the number of points γ_n in $\mathcal{I}$ for $0 \le n \le N-1$, see [5,9,17].

Following [15] we say that a finite sequence $\mathcal{T}$ of integers is Δ-*homogeneously distributed modulo* q if for any integer a coprime with q the discrepancy of the sequence $\{\lfloor at \rfloor_q / q\}_{t \in \mathcal{T}}$ is at most Δ.

For an integer x, we denote by $\mathrm{MSB}_{\ell,q}(x)$ any integer u with

$$|x - u| \le q/2^{\ell+1}.$$

Remark that ℓ in this inequality is not necessarily an integer.

The following generalization of Theorem 1 of [3], using Δ-homogeneously distributed multipliers in the hidden number problem, has been obtained in [15].

Lemma 4. *For a prime q, define*

$$\ell = \lceil \log^{1/2} q \rceil + \lceil \log\log q \rceil,$$

and $d = 2\left\lceil \log^{1/2} q \right\rceil$. Let $\mathcal{T}$ be a $2^{-\log^{1/2} q}$-homogeneously distributed modulo q sequence of integer numbers. There exists a deterministic polynomial time algorithm $\mathcal{A}$ such that for any fixed integer α in the interval $[0, q-1]$, given a prime q and $2d$ integers t_i and $u_i = \mathrm{MSB}_{\ell,q}\left(\lfloor \alpha t_i \rfloor_q\right)$ for $i = 1 \dots, d$, its output satisfies for sufficiently large q

$$\Pr_{t_1,\dots,t_d \in \mathcal{T}} \left[\mathcal{A}\left(q, t_1, \dots, t_d; u_1, \dots, u_d\right) = \alpha\right] \geq 1 - 2^{-(\log q)^{1/2} \log\log q}$$

if $t_1, \dots, t_d$ are chosen uniformly and independently at random from the elements of $\mathcal{T}$.

This result and Theorem 1 of [3] rely heavily on the LLL lattice reduction algorithm [10], more precisely on Babai's approximation algorithm [2] for the closest vector problem in a lattice (see [16] for a brief exposition of the proof).

In fact using a combination of the Schnorr modification [20] of the lattice basis reduction with a result of Kannan [7] about reduction of the closest vector problem to the shortest vector problem, one can obtain an analogue of Lemma 4 with a slightly smaller value of ℓ. This result can be improved even further if one uses a more recent result of [1]. All these improvements are of order of some power of $\log\log q$.

2.2 Distribution of Multipliers

Let us denote by $\mathcal{S}_i$ the set of pairs $(k, \mu) \in [1, q-1] \times \mathcal{M}$ for which the corresponding equation for $t_i(k, \mu)$ and $t_i(k, \mu)$ does not contain division by zero. Then

$$|\mathcal{S}_i| = q|\mathcal{M}| \left(1 + O\left(q^{-\delta}\right)\right), \qquad i = 1, \dots, 5, \tag{5}$$

for all p and q satisfying the conditions of Lemma 3.

Our results are based on the bounds of exponential sums

$$\sigma_i(c) = \sum_{(k,\mu) \in \mathcal{S}_i} \mathbf{e}_q\left(c t_i(k, \mu)\right), \qquad i = 1, \dots, 5,$$

and

$$\tau(c) = \sum_{\mu \in \mathcal{M}} \sum_{k \in \mathbb{F}_q^*} \mathbf{e}_p\left(c r(k, \mu)\right).$$

For a hash function $h : \mathcal{M} \to \mathbb{F}_q$ we also denote by W the number of pairs $(\mu_1, \mu_2) \in \mathcal{M}^2$ with $h(\mu_1) = h(\mu_2)$. Thus, $W/|\mathcal{M}|^2$ is a probability of a *collision* and our results for the schemes DSA_i, $i = 1, \dots, 5$, are nontrivial under a reasonable assumption that this probability is of order of magnitude close to $1/q$.

Lemma 5. *Let Q be a sufficiently large integer. The following statement holds with $\nu = 1/3$ for all primes $p \in [Q, 2Q]$, and with $\nu = 0$ for all primes $p \in [Q, 2Q]$ except at most $Q^{5/6+\varepsilon}$ of them. For any $\varepsilon > 0$ there exists $\gamma > 0$ such that for any element $g \in \mathbb{F}_p$ of multiplicative order $q \geq p^{\nu+\varepsilon}$ the bound*

$$\max_{c \in \mathbb{F}_q^*} |\sigma_i(c)| = O(W^{1/2} q^{3/2-\gamma})$$

holds.

Proof. Define $R(\lambda)$ to be the number of $\mu \in M$ with $h(\mu) = \lambda$. Then

$$\sum_{\lambda \in M} R(\lambda)^2 = W. \tag{6}$$

From the definition of $s_i(k, \mu)$ and $t_i(k, \mu)$ we see that

$$\sigma_i = \sum_{\lambda \in \mathbb{F}_q} R(\lambda) \sum_{k \in \mathbb{F}_q^*}{}^{*} \mathbf{e}_q \left(c F_k^{(i)}(\lambda) \right), \qquad i = 1, \ldots, 6,$$

where

$$F_k^{(1)}(\lambda) = \frac{k\lambda}{r(k) + \alpha\lambda}, \qquad F_k^{(2)}(\lambda) = \frac{\lambda - kr(k)}{\alpha r(k)}, \qquad F_k^{(3)}(\lambda) = \frac{-r(k)}{\lambda},$$

$$F_k^{(4)}(\lambda) = \frac{-\lambda}{r(k)}, \qquad F_k^{(5)}(\lambda) = \frac{r(k) + k\lambda}{\alpha\lambda}.$$

Using the Cauchy inequality and the identity (6), we obtain

$$\begin{aligned} |\sigma_i(c)|^2 &\leq W \sum_{\lambda \in \mathbb{F}_q} \left| \sum_{k \in \mathbb{F}_q^*} \mathbf{e}_q \left(c F_{(k,m)}^{(i)}(\lambda) \right) \right|^2 \\ &= W \sum_{k,m \in \mathbb{F}_q^*} \sum_{\lambda \in \mathbb{F}_q}{}^{*} \mathbf{e}_q \left(c \left(F_{k,m}^{(i)}(\lambda) \right) \right), \end{aligned}$$

where $F_{k,m}^{(i)}(\lambda) = F_k^{(i)}(\lambda) - F_m^{(i)}(\lambda)$ and as before $\sum^*$ means that the summation is taken over all $\lambda \in \mathbb{F}_q$ which are not poles of $F_{k,m}^{(i)}$, $i = 1, \ldots, 5$.

For $i = 1, \ldots, 5$ we denote by N_i the number of pairs $(k, m) \in (\mathbb{F}_q^*)^2$ for which the rational function $F_{k,m}^{(i)}(\lambda)$ is constant. For these k and m we estimate the inner sum over λ trivially as q. For other k and m, to estimate the inner sum over λ we use Lemma 1 (for $i = 1, 5$), Lemma 2 (for $i = 2, 4$) and the identity

$$\sum_{\lambda \in \mathbb{F}_q^*} \mathbf{e}_q(a\lambda^{-1}) = \sum_{\lambda \in \mathbb{F}_q^*} \mathbf{e}_q(\lambda) = -1, \qquad a \in \mathbb{F}_q^*,$$

(for $i = 3$). Thus we derive the bound

$$\max_{c \in \mathbb{F}_q^*} |\sigma_i(c)|^2 = O\left(W \left(N_i q + q^2 B_i \right) \right), \qquad i = 1, \ldots, 5, \tag{7}$$

where B_i is the bound on the corresponding exponential sums, that is

$$B_i = \begin{cases} q^{1/2}, & \text{if } i = 1, 5; \\ 0, & \text{if } i = 2, 4; \\ 1, & \text{if } i = 3. \end{cases}$$

Now, we remark that

$$F^{(1)}_{(k,m)}(\lambda) = \frac{k\lambda}{r(k) + \alpha\lambda} - \frac{m\lambda}{r(m) + \alpha\lambda}$$
$$F^{(2)}_{(k,m)}(\lambda) = \frac{\lambda - kr(k)}{\alpha r(k)} - \frac{\lambda - mr(m)}{\alpha r(m)}$$

are constant only when $k = m$, thus $N_1 = N_2 = q - 1$.

For other $i = 3, 4, 5$ the functions

$$F^{(3)}_{(k,m)}(\lambda) = \frac{-r(k)}{\lambda} - \frac{-r(m)}{\lambda},$$
$$F^{(4)}_{(k,m)}(\lambda) = \frac{-\lambda}{r(k)} - \frac{-\lambda}{r(m)},$$
$$F^{(5)}_{(k,m)}(\lambda) = \frac{r(k) + k\lambda}{\alpha\lambda} - \frac{r(m) + m\lambda}{\alpha\lambda}$$

are constant only when $r(k) = r(m)$. Using Lemma 3, we find that $N_i = O(q^{2-\delta})$, $i = 3, 4, 5$. Without loss of generality we may assume that $\delta < 1/4$. Substituting the bounds on $B_i \le q^{1/2}$ and $N_i = O(q^{2-\delta})$, $i = 1, \ldots, 5$, in the inequality (7), we obtain the desired result. □

For a function $H : \mathcal{M} \to \mathbb{F}_p$ we also denote by U the number of pairs $(\mu_1, \mu_2) \in \mathcal{M}^2$ with $H(\mu_1) = H(\mu_2)$. As we have mentioned, in practical applications of the NR scheme H is a one to one function thus $U = |\mathcal{M}|$ in this case (although our results remain nontrivial for U of large order).

The following statement is a variant of Exercise 14.a, of Chapter 6 of [23] and can be proved quite similarly.

Lemma 6. *The bound*

$$\max_{c \in \mathbb{F}_p^*} |\tau(c)| \le U^{1/2} p^{1/2} q^{1/2}$$

holds.

Proof. Define $Q(\lambda)$ to be the number of $\mu \in \mathcal{M}$ with $H(\mu) = \lambda$. Then as in the proof of Lemma 5 we have

$$\tau(c) = \sum_{\lambda \in \mathbb{F}_p} Q(\lambda) \sum_{k \in \mathbb{F}_q} \mathbf{e}_p\left(c\lambda g^k\right), \qquad b \in \mathbb{F}_p.$$

From the Cauchy inequality we derive

$$|\tau(c)|^2 \le \sum_{\lambda \in \mathbb{F}_p} Q(\lambda)^2 \sum_{\lambda \in \mathbb{F}_p} \left| \sum_{k=1}^{q-1} \mathbf{e}_p\left(c\lambda g^k\right) \right|^2$$
$$= U \sum_{k,m \in \mathbb{F}_q} \sum_{\lambda \in \mathbb{F}_q} \mathbf{e}_p\left(c\lambda(g^k - g^m)\right).$$

Using Lemma 2, we see that if $\gcd(c,p) = 1$ the sum over λ is equal p if $g^k \equiv g^m \pmod{q}$, that is, if $k = m$, and vanishes otherwise and the result follows. □

Lemma 7. *Let Q be a sufficiently large integer. The following statement holds with $\nu = 1/3$ for all primes $p \in [Q, 2Q]$, and with $\nu = 0$ for all primes $p \in [Q, 2Q]$ except at most $Q^{5/6+\varepsilon}$ of them. For any $\varepsilon > 0$, there exists $\gamma > 0$ such that for any element $g \in \mathbb{F}_p$ of multiplicative order $q \ge p^{\nu+\varepsilon}$ the sequences $t_i(k,\mu)$, $(k,\mu) \in \mathcal{S}_i$, are $2^{-\log^{1/2} q}$ homogeneously distributed modulo q for every $i = 1, \ldots, 5$, provided that*

$$W \le \frac{|\mathcal{M}|^2}{q^{1-\gamma}}.$$

Proof. Let us fix an integer a coprime with q. According to a general discrepancy bound, given by ([17], Corollary 3.11) for the discrepancy $D_i(a)$ of the set

$$\left\{ \frac{\lfloor a t_i(k,\mu)_q \rfloor}{q} : (k,\mu) \in \mathcal{S}_i \right\}, \qquad i = 1, \ldots, 5,$$

we have

$$\max_{a \in \mathbb{F}_q^*} D_i(a) \le \frac{\log q}{|\mathcal{S}_i|} \max_{c \in \mathbb{F}_q^*} |\sigma_i(c)|.$$

From Lemma 5 and (5) we obtain

$$\max_{a \in \mathbb{F}_q^*} D_i(a) = O(W^{1/2} q^{1/2-\gamma} |\mathcal{M}|^{-1} \log q)$$
$$= O(q^{-\gamma/2} \log q) = O(2^{-\log^{1/2} q}),$$

provided that q is sufficiently large, $i = 1, \ldots, 5$. □

Similarly, from Lemma 6 we have a similar result for the NR scheme.

Lemma 8. *For any element $g \in \mathbb{F}_p$ of multiplicative order q the sequence $2^{-l} r(k,\mu)$, $(k,\mu) \in \mathbb{F}_q \times \mathcal{M}$, are $2^{-\log^{1/2} q}$ homogeneously distributed modulo q, provided that*

$$U \le \frac{|\mathcal{M}|^2 q^{3-\gamma}}{p^3}.$$

for some $\gamma > 0$.

Proof. Let a be an integer with $\gcd(a,q) = 1$.

Let $\mathcal{I} \in [0,1]$ be an interval. Then the condition

$$\frac{\lfloor a2^{-l}r(k,\mu)\rfloor_q}{q} \in \mathcal{I} \tag{8}$$

is equivalent to the congruence

$$a2^{-l}r(k,\mu) \equiv q\vartheta + \rho \pmod p, \tag{9}$$

with some integer ρ such that $\rho/q \in \mathcal{I}$ and some integer ϑ such that $0 \le q\vartheta+\rho \le p-1$.

Let $L = (p-1)/q$ and let the integer I be defined is the length of the largest interval of the form

$$\left[\frac{J+1}{q}, \frac{J+I}{q}\right] \in \mathcal{I}$$

with integer I. Thus $I = q|\mathcal{I}| + O(1)$.

For each fixed $\vartheta \in [0, L-1]$ as before, from Lemma 2 we conclude that the number of pairs $(k,\mu) \in \mathbb{F}_q \times \mathcal{M}$ satisfying (9) with $\rho/q \in \mathcal{I}$ is

$$\frac{1}{p}\sum_{c\in\mathbb{F}_p}\sum_{\mu\in\mathcal{M}}\sum_{k\in\mathbb{F}_q^*}\sum_{\rho=J+1}^{J+I} \mathbf{e}_p\left(c\left(2^{-l}r(k,\mu) - q\vartheta - \rho\right)\right)$$

$$= \frac{|\mathcal{M}|qI}{p} + O\left(\frac{1}{p}\sum_{c\in\mathbb{F}_p^*}|\tau(c2^{-l})|\left|\sum_{\rho=J+1}^{J+I}\mathbf{e}_p(c\rho)\right|\right).$$

From Lemma 6 and the estimate

$$\sum_{c\in\mathbb{F}_p^*}\left|\sum_{\rho=J+1}^{J+I}\mathbf{e}_p(c\rho)\right| = O(p\log p),$$

see Exercise 11.c of Chapter 3 of [23], we conclude that there are

$$\frac{|\mathcal{M}|qI}{p} + O\left(U^{1/2}p^{1/2}q^{1/2}\right) = \frac{|\mathcal{M}|q^2|\mathcal{I}|}{p} + O\left(U^{1/2}p^{1/2}q^{1/2} + |\mathcal{M}|qp^{-1}\right)$$

such pairs $(k,\mu) \in \mathbb{F}_q \times \mathcal{M}$.

For $\vartheta = L$ the only possible value of ρ is $\rho = 0$. Obviously for each $\mu \in \mathcal{M}$ there is at most one value of k which satisfies the congruence

$$a2^{-l}r(k,\mu) \equiv p-1 \pmod p,$$

thus it holds for at most $|\mathcal{M}|$ pairs $(k,\mu) \in \mathbb{F}_q \times \mathcal{M}$. Putting everything together we see that (8) holds for

$$\frac{|\mathcal{M}|q^2|\mathcal{I}|}{p}L + O\left(LU^{1/2}p^{1/2}q^{1/2} + L|\mathcal{M}|qp^{-1} + |\mathcal{M}|\right)$$

$$= |\mathcal{M}|q|\mathcal{I}| + O\left(U^{1/2}p^{3/2}q^{-1/2} + |\mathcal{M}|\right)$$

pairs $(k,\mu) \in \mathbb{F}_q \times \mathcal{M}$. One verifies that

$$U^{1/2}p^{3/2}q^{-1/2} \le |\mathcal{M}|q^{1-\gamma/2}$$

under the condition of the theorem and the result follows. □

3 Lattice Attacks on DSA-Like Algorithms

3.1 Main Results

For $i = 1, \dots, 5$ and an integer ℓ we define the oracle $\mathcal{O}_{i,\ell}^{DSA}$ which, for any given DSA signature $(r(k), s_i(k,\mu))$, $(k,\mu) \in \mathcal{S}_i$, returns the ℓ least significant bits of k. Combining the inequality (3), Lemma 4 and Lemma 7, we obtain:

Theorem 1. *Let Q be a sufficiently large integer. The following statement holds with $\nu = 1/3$ for all primes $p \in [Q, 2Q]$, and with $\nu = 0$ for all primes $p \in [Q, 2Q]$ except at most $Q^{5/6+\varepsilon}$ of them. For $i = 1, \dots, 5$ and any $\varepsilon > 0$ there exists $\gamma > 0$ such that for any element $g \in \mathbb{F}_p$ of multiplicative order $q \ge p^{\nu+\varepsilon}$, and any hash function h with*

$$W \le \frac{|\mathcal{M}|^2}{q^{1-\gamma}},$$

given an oracle $\mathcal{O}_{i,\ell}^{DSA}$ with $\ell = \left\lceil \log^{1/2} q \right\rceil + \lceil \log\log q \rceil$, there exists a probabilistic polynomial time algorithm to recover the signer's DSA$_i$ secret key α, from $O\left(\log^{1/2} q\right)$ signatures $(r(k), s_i(k,\mu))$ with $k \in [0, q-1]$ and $\mu \in \mathcal{M}$ selected independently and uniformly at random. The probability of success is at least $1 - 2^{-(\log q)^{1/2}\log\log q}$.

For an integer ℓ we define the oracle $\mathcal{O}_{\ell}^{NR}$ which, for any given DSA signature $(r(k,\mu), v(k,\mu))$, $k \in \mathbb{F}_q$, $\mu \in \mathcal{M}$, returns the ℓ least significant bits of k. Combining the inequality (4), Lemma 4 and Lemma 8, we obtain:

Theorem 2. *For any element $g \in \mathbb{F}_p$ of multiplicative order $q \le p^{1-\varepsilon}$, any fixed $\gamma > 0$ and any function H with*

$$U \le \frac{|\mathcal{M}|^2 q^{3-\gamma}}{p^3}$$

given an oracle $\mathcal{O}_{\ell}^{NR}$ with $\ell = \left\lceil \log^{1/2} q \right\rceil + \lceil \log\log q \rceil$, there exists a probabilistic polynomial time algorithm to recover the signer's NR secret key α, from $O\left(\log^{1/2} q\right)$ signatures $(r(k,\mu), v(k,\mu))$ with $k \in [0, q-1]$ and $\mu \in \mathcal{M}$ selected independently and uniformly at random. The probability of success is at least $1 - 2^{-(\log q)^{1/2}\log\log q}$.

As noticed previously, it is reasonable to expect that W is close to $|\mathcal{M}|^2/q$ so that the corresponding condition of Theorem 1 is almost always satisfied.

Furthermore, for the function H in the NR signature, in the most interesting case we have $U = |\mathcal{M}|$ thus the corresponding inequality of Theorem 2 takes the form $|\mathcal{M}| \geq p^3 q^{-3+\gamma}$. In particular, if the message set $\mathcal{M}$ is "dense" (that is, of order p) then Theorem 2 applies for $q \geq p^{2/3+\gamma}$. On the other hand, in almost all practical applications q is much smaller than this bound and it would be very interesting to extend Theorem 2 to smaller values of q. We remark that studying the NR scheme has turned out to be harder than studying the classical DSA scheme in [15] and its modifications DSA$_i$, $i = 1, \ldots, 5$, in this paper. The reason is that for the NR scheme the multiplier $\lfloor r(k, \mu) \rfloor_q$ is not a product of two distinct quantities taken modulo q (although $r(k, \mu)$ is such a product taken modulo p this property is lost after reducing modulo q). Accordingly the technique of estimation of double exponential sums which is used in [15] and in this paper for DSA$_i$, $i = 1, \ldots, 5$ cannot be applied to the studying $r(k, \mu)$.

Similar results can be obtained for other modifications of the DSA which are outlined in [12, 18, 19].

3.2 Experimental Results

We experimented the attack with the NTL library [22]. The running time is less than half an hour for a number of signatures d less than a hundred, on a 500 MHz DEC Alpha. We used a 160-bit prime q, and a 512-bit prime q. For each choice of parameters size, we ran the attack several times on newly generated parameters (including the prime q and the multipliers of the DSA–HNP).

The results are exactly the same as those obtained in [15]. The proof of Lemma 4 relies on the ability to approximate the closest vector problem in a lattice. Due to the well-known experimental fact that lattice reduction algorithms behave better than theoretically expected, it is in practice possible to obtain much better bounds than those of Lemma 4. In [15], it is shown that $\ell = 2$ instead of roughly $\ell \approx \log^{1/2} q$ is sufficient for the attack to work, if ideal lattice reduction is assumed.

Using Schnorr's improved lattice reduction [21] to solve the closest vector problem in a lattice, we were always able to solve the DSA-like problems with $\ell = 3$ and $d = 100$ (on more than 50 trials).

We always failed with $\ell = 2$ and $d = 150$, perhaps because current lattice basis reduction algorithms are more suited to the Euclidean norm than the infinity norm.

Finally, as in [15] we remark that one of the possible ways to obtain several most significant bits of the nonce k is to use *timing* or *power* attacks and select signatures corresponding to small values of k, thus to values whose most significant bits are zeros.

References

1. M. Ajtai, R. Kumar and D. Sivakumar, *A sieve algorithm for the shortest lattice vector problem,* Proc. 33rd ACM Symp. on Theory of Comput., Crete, Greece, July 6-8, 2001 601–610.
2. L. Babai, *On Lovász lattice reduction and the nearest lattice point problem,* Combinatorica, **6** (1986), 1–13.
3. D. Boneh and R. Venkatesan, *Hardness of computing the most significant bits of secret keys in Diffie–Hellman and related schemes,* Lect. Notes in Comp. Sci., Springer-Verlag, Berlin, **1109** (1996), 129–142.
4. D. Boneh and R. Venkatesan, *Rounding in lattices and its cryptographic applications,* Proc. 8-rd Annual ACM-SIAM Symp. on Discr. Algorithms, ACM, NY, 1997, 675–681.
5. M. Drmota and R. Tichy, *Sequences, discrepancies and applications,* Springer-Verlag, Berlin, 1997.
6. N. A. Howgrave-Graham and N. P. Smart, *Lattice attacks on digital signature schemes,* Designs, Codes and Cryptography, (to appear).
7. R. Kannan, *Algorithmic geometry of numbers,* Annual Review of Comp. Sci., **2** (1987), 231–267.
8. S. V. Konyagin and I. E. Shparlinski, *Character sums with exponential functions and their applications,* Cambridge Univ. Press, Cambridge, 1999.
9. R. Kuipers and H. Niederreiter, *Uniform distribution of sequences,* Wiley-Interscience, NY, 1974.
10. A. K. Lenstra, H. W. Lenstra and L. Lovász, *Factoring polynomials with rational coefficients,* Mathematische Annalen, **261** (1982), 515–534.
11. R. Lidl and H. Niederreiter, *Finite fields,* Cambridge University Press, Cambridge, 1997.
12. A. J. Menezes, P. C. van Oorschot and S. A. Vanstone, *Handbook of Applied Cryptography,* CRC Press, Boca Raton, FL, 1996.
13. C. J. Moreno and O. Moreno, *Exponential sums and Goppa codes, I,* Proc. Amer. Math. Soc., **111** (1991), 523–531.
14. P. Q. Nguyen, *The dark side of the hidden number problem: Lattice attacks on DSA,* Proc. Workshop on Cryptography and Computational Number Theory, Singapore 1999, Birkhäuser, 2001, 321–330.
15. P. Q. Nguyen and I. E. Shparlinski, *The insecurity of the Digital Signature Algorithm with partially known nonces,* J. of Cryptology, to appear.
16. P. Q. Nguyen and J. Stern, *The hardness of the hidden subset sum problem and its cryptographic implications,* Lect. Notes in Comp. Sci., Springer-Verlag, Berlin, **1666** (1999), 31–46.
17. H. Niederreiter, *Random number generation and quasi–Monte Carlo methods,* SIAM, Philadelphia, 1992.
18. K. Nyberg and R. A. Rueppel, *Message recovery for signature schemes based on the discrete logarithm problem,* J. Cryptology, **8** (1995), 27–37.
19. B. Schneier, *Applied cryptography,* J. Wiley, NY, 1996.
20. C. P. Schnorr, *A hierarchy of polynomial time basis reduction algorithms,* Theor. Comp. Sci., **53** (1987), 201–224.
21. C. P. Schnorr and M. Euchner, *Lattice basis reduction: improved practical algorithms and solving subset sum problems,* Math. Programming, **66** (1994), 181–199.
22. V. Shoup, *Number Theory C++ Library (NTL),* Available at `http://www.shoup.net/ntl/`.
23. I. M. Vinogradov, *Elements of number theory,* Dover Publ., New York, 1954.

Dimension Reduction Methods for Convolution Modular Lattices

Alexander May and Joseph H. Silverman

[1] Department of Mathematics and Computer Science
University of Paderborn, 33095 Paderborn, Germany
alexx@uni-paderborn.de
[2] NTRU Cryptosystems, Inc., 5 Burlington Woods, Burlington, MA 01803 USA
and Mathematics Department, Brown University
Providence, RI 02912 USA
jhs@ntru.com, jhs@math.brown.edu

Abstract. We describe a dimension reduction method for convolution modular lattices. Its effectiveness and implications for parallel and distributed computing are analyzed.
Keywords: Lattice reduction, cryptography, convolution modular lattice.

Introduction

The theory of lattices provides a tempting source of hard problems for cryptography, because the linearity of lattice operations offers speed advantages over group based cryptosystems. Two proposed cryptosystems based on (approximate) shortest and closest vector problems are GGH [4] and NTRU [6]. Unfortunately (for cryptographers), current lattice reduction methods such as LLL and its variants [11–13, 18, 19] are quite effective at finding short vectors in lattices of moderate dimension. This renders the GGH cryptosystem impractical [17], since its public keys consist of a full basis for the lattice, so the key size is proportional to the square of the lattice dimension.

The NTRU public key cryptosystem [6] and the related NSS signature scheme [7, 8] overcome this difficulty by using a class of lattices in which an entire basis can be specified by a single vector. These *convolution modular lattices* are defined using convolution products and reduction modulo q, and the resulting cryptosystems have key size that is proportional to $n \log q$ for a lattice of dimension n.

Convolution modular lattices have a cyclic structure that makes them nice to use in cryptography, but that same structure is a potential liability, since it offers an additional avenue for solving the underlying hard lattice problem. In this note we explore a method, originally proposed by the first author [14, 15], that exploits the cyclic structure in order to decrease the dimension of the underlying lattice. Since the solution time is roughly proportional to the dimension of the lattice, this offers the possibility of significant savings; and indeed in low

J.H. Silverman (Ed.): CaLC 2001, LNCS 2146, pp. 110–125, 2001.

dimensions, the method is fairly effective. However, in higher dimensions such as those used in commercial implementations of NTRU [9], the speedup does not significantly affect basic security estimates. We also find that although the dimension reduction method can be parallelized, the effect is not linear in the number of processors.

1 Convolution Products and Convolution Modular Lattices

The *convolution product* of two vectors

$$\mathbf{u} = [u_0, u_1, \ldots, u_{N-1}] \in \mathbb{Z}^N \qquad \text{and} \qquad \mathbf{v} = [v_0, v_1, \ldots, v_{N-1}] \in \mathbb{Z}^N$$

is the vector $\mathbf{w} = [w_0, w_1, \ldots, w_{N-1}] = \mathbf{u} * \mathbf{v}$ whose k^{th} coordinate is given by the formula

$$\begin{aligned} w_k &= u_0 v_k + u_1 v_{k-1} + \cdots + u_k v_0 + u_{k+1} v_{N-1} + \cdots + u_{N-1} v_{k+1} \\ &= \sum_{i+j \equiv k \pmod N} u_i v_j . \end{aligned}$$

Convolution product makes the lattice $\mathbb{Z}^N$ into a ring. In a similar manner, if we start with vectors whose coordinates are in the finite field $\mathbb{F}_q$, then convolution product makes the vector space $\mathbb{F}_q^N$ into a ring, and reduction modulo q gives a natural ring homomorphism $\mathbb{Z}^N \to \mathbb{F}_q^N$. In this note we will study certain sublattices of $\mathbb{Z}^N$ that are defined by convolution congruences.

Remark 1. An alternative description of the convolution ring $\mathbb{Z}^N$ is the polynomial quotient ring $\mathbb{Z}[X]/(X^N - 1)$. Under this isomorphism, the vector $\mathbf{v}$ is identified with the polynomial $v_0 + v_1 X + v_2 X^2 + \cdots + v_{N-1} X^{N-1}$. We will mainly use the vector description, but it is sometimes useful to deal with polynomials, especially when discussing the mod q multiplicative inverse of a vector.

Let $q \geq 1$ be an integer, which we will call the *modulus*, and let $\mathbf{c} \in \mathbb{Z}^N$ be a fixed vector. The collection of pairs of vectors $[\mathbf{u}, \mathbf{v}] \in \mathbb{Z}^{2N}$ satisfying the congruence

$$\mathbf{c} * \mathbf{u} \equiv \mathbf{v} \pmod q \tag{1}$$

forms a lattice in $\mathbb{Z}^{2N}$. We denote this lattice by $L(\mathbf{c}, q)$ and call it the *Convolution Modular Lattice* (CML) associated to $\mathbf{c}$ and q. It is clear that $L(\mathbf{c}, q)$ is a lattice of dimension $2N$, since directly from (1) we see that $L(\mathbf{c}, q)$ is a subgroup of $\mathbb{Z}^{2N}$ and satisfies

$$q\mathbb{Z}^{2N} \subset L(\mathbf{c}, q) \subset \mathbb{Z}^{2N}.$$

Using the formula for the convolution product, the lattice $L(\mathbf{c}, q)$ may be described very explicitly as the lattice spanned by the rows of the following matrix:

$$L = L(\mathbf{c}, q) = \text{RowSpan} \left(\begin{array}{cccc|cccc} 1 & 0 & \cdots & 0 & c_0 & c_1 & \cdots & c_{N-1} \\ 0 & 1 & \cdots & 0 & c_{N-1} & c_0 & \cdots & c_{N-2} \\ \vdots & \vdots & \ddots & \vdots & \vdots & \vdots & \ddots & \vdots \\ 0 & 0 & \cdots & 1 & c_1 & c_2 & \cdots & c_0 \\ \hline 0 & 0 & \cdots & 0 & q & 0 & \cdots & 0 \\ 0 & 0 & \cdots & 0 & 0 & q & \cdots & 0 \\ \vdots & \vdots & \ddots & \vdots & \vdots & \vdots & \ddots & \vdots \\ 0 & 0 & \cdots & 0 & 0 & 0 & \cdots & q \end{array} \right)$$

Remark 2. The N-by-N upper righthand block of the matrix for $L(\mathbf{c}, q)$ is the circulant matrix of the sequence $[c_0, c_1, \ldots, c_{N-1}]$. For this reason an alternative name for $L(\mathbf{c}, q)$ might be *Circulant Modular Lattice.*

Remark 3. In typical cryptographic applications such as NTRU and NSS [6–8], convolution modular lattices are constructed to contain a particular short vector $[\mathbf{a}, \mathbf{b}]$. For example, $\mathbf{a}$ and $\mathbf{b}$ might be chosen from the set of binary (or trinary) vectors with a specified number of ones (and negative ones). The corresponding CML is given by the solutions $[\mathbf{u}, \mathbf{v}]$ to the congruence

$$\mathbf{b} * \mathbf{u} \equiv \mathbf{a} * \mathbf{v} \pmod{q}.$$

Equivalently, it is the lattice $L(\mathbf{c}, q)$ with

$$\mathbf{c} \equiv \mathbf{a}^{-1} * \mathbf{b} \pmod{q}.$$

The public information is the generating vector $\mathbf{c}$, and the secret information used as a trapdoor is the short vector $[\mathbf{a}, \mathbf{b}]$.

2 The Shortest Vector Problem and the Closest Vector Problem in Convolution Modular Lattices

Two classical and much-studied problems in the theory of lattices are the shortest vector problem (SVP) and the closest vector problem (CVP). These are, respectively, the problem of finding a shortest nonzero vector in L and the problem of finding a vector in L that is closest to a given target vector $\mathbf{t}$ not lying in L. An overview of the theoretical study of these problems may be found in [3]. The practical problem of solving SVP and CVP has also attracted considerable attention. The LLL algorithm [13] and its improvements and variants (e.g., [11, 12, 18, 19]) provide a practical method for finding moderately short and moderately close vectors, but the computation of shortest and closest vectors remains very difficult from both a theoretical [1, 16] and a practical perspective [2].

The standard notation for the length of a shortest nonzero vector in a lattice L is

$$\lambda_1 = \lambda_1(L) = \min \{ \|\mathbf{v}\| : \mathbf{v} \in L, \ \mathbf{v} \neq \mathbf{0} \}.$$

If $\mathbf{t}$ is any vector, we use a similar notation

$$\lambda_1(\mathbf{t}) = \lambda_1(\mathbf{t}, L) = \min\{\|\mathbf{t} - \mathbf{v}\| : \mathbf{v} \in L\}$$

to denote the distance from $\mathbf{t}$ to a closest vector in L.

In terms of theoretical complexity theory, CVP may be a little harder to solve than SVP (but see [5]). However, for most problems there appears to be very little difference in practice. More precisely, if $(L, \mathbf{t})$ is an n-dimensional CVP to be solved, then one forms the $(n+1)$-dimensional lattice

$$L' = \{(\mathbf{v}, 0) : \mathbf{v} \in L\} + \{k(\mathbf{t}, c) : k \in \mathbb{Z}\} \subset \mathbb{R}^{n+1}.$$

For an appropriately chosen value of c, a shortest nonzero vector in L' will often have the form $(\mathbf{u}, c)$, and then it is likely that the vector $\mathbf{t} - \mathbf{u}$ (which is in L) will be a closest vector to $\mathbf{t}$. See, e.g., [4, 17] for further details.

Example 1. Typical convolution modular lattices used in cryptography will contain short (probably shortest) vectors $\mathbf{w} = [\mathbf{u}, \mathbf{v}]$ that are binary or trinary of known form. For ease of exposition, we will consider the case of binary vectors, the trinary case being similar. For example, if we let

$$\mathcal{B}_N(d) = \{\text{binary vectors with } d \text{ ones and } N - d \text{ zeros}\},$$

then an adversary may know that $L(\mathbf{c}, q)$ contains a vector $\mathbf{w} = [\mathbf{u}, \mathbf{v}]$ satisfying $\mathbf{u}, \mathbf{v} \in \mathcal{B}_N(d)$. His objective is to find this vector (or one of its rotations, see Section 3).

The most straightforward approach is to solve SVP. The vector $\mathbf{w}$ has length $\|\mathbf{w}\| = \sqrt{2d}$, so if $\mathbf{w}$ is a smallest vector than solving SVP will find $\mathbf{w}$. However, it will generally be easier to solve CVP with target vector

$$\mathbf{t} = \left(\frac{d}{N}, \frac{d}{N}, \frac{d}{N}, \ldots, \frac{d}{N}\right),$$

since the distance from $\mathbf{t}$ to $\mathbf{w}$ satisfies

$$\|\mathbf{t} - \mathbf{w}\| = \sqrt{\frac{2d(N-d)}{N}} < \sqrt{2d} = \|\mathbf{w}\|.$$

More generally, the two halves of the shortest vector may have unequal sizes, say $\mathbf{u} \in \mathcal{B}_N(d_1)$ and $\mathbf{v} \in \mathcal{B}_N(d_2)$. In this case the lattice should be balanced before formulating the appropriate CVP. For ease of exposition, we will restrict attention in this note to the case that $\|\mathbf{u}\| = \|\mathbf{v}\|$, but see Section 5 for a brief discussion of the balancing process.

Extensive experiments with convolution modular lattices using NTL's implementation [10] of current lattice reduction methods have yielded the following result.

Heuristic 1. *Let $\mathcal{L} = \{(L, \mathbf{t})\}$ be a collection of pairs consisting of a convolution modular lattice L and a CVP target vector $\mathbf{t}$, and suppose that the collection has the following properties:*

1. *The ratio N/q is (approximately) constant for the lattices in $\mathcal{L}$.*
2. *The ratio $\lambda_1(\mathbf{t})/\sqrt{q}$ is (approximately) constant for the lattices in $\mathcal{L}$.*
3. *$\lambda_1(\mathbf{t})$ is significantly smaller than $\sqrt{Nq/\pi e}$, say by at least a factor of 2.*

Then there are positive constants α and β so that the time required to solve the CVP for every $(L, \mathbf{t})$ in $\mathcal{L}$ is given by

$$\log_{10}(\text{Time}) \approx \alpha N + \beta.$$

In other words, the time to solve CVP is exponential in the dimension of the lattice.

Remark 4. The key conditions in Heuristic 1 are conditions (1) and (2) requiring that N/q and $\lambda_1(\mathbf{t})/\sqrt{q}$ be held approximately constant. Condition (3) is less important, it simply says that the lattice contains a vector that is significantly closer to $\mathbf{t}$ than would have been predicted by the Gaussian heuristic. More precisely, the Gaussian heuristic suggests that a (random) lattice L of dimension n and discriminant D will generally have few vectors that are significantly closer than the Gauss constant

$$\gamma = \gamma(L) = D^{1/n}\sqrt{n/2\pi e}$$

to a given vector $\mathbf{t}$. See [6] or [9, Annex A.3.5] for further details of this observation.

Remark 5. We can rephrase Heuristic 1 more intrinsically without direct reference to convolution modular lattices, although we do not know whether it holds for more general lattices. The Gaussian heuristic says that in a "random" lattice L, SVP and CVP have solutions of size approximately $\gamma(L)$. Then the heuristic says that in a collection of (CML) lattices and target vectors satisfying

$$\frac{\gamma(L)}{\dim(L)} \approx c_1 \qquad \text{and} \qquad \lambda_1(\mathbf{t}) \approx c_2 \frac{\gamma(L)}{\sqrt{\dim(L)}},$$

the solution time is approximately exponential in the dimension of L.

Example 2. A series of experiments on convolution modular lattice CVP's (with $67 \le N \le 82$) satisfying

$$\frac{N}{q} \approx 1.96 \qquad \text{and} \qquad \frac{\lambda_1(\mathbf{t})}{\sqrt{q}} \approx 0.896$$

yielded a solution time

$$\log_{10}(\text{Time}) \approx 0.083N - 7.5,$$

where the time is measured in MIPS-years. (A MIPS-year is an approximate amount of computation that a machine capable of performing one million arithmetic instructions per second would perform in one year, about $3 \cdot 10^{13}$ arithmetic instructions.) These values correspond to taking target binary vectors (as described in Example 1) in which each half of the target vector has approximately $0.287N$ ones and $0.713N$ zeros. Extrapolating to the higher value $N = 251$ gives a search time of approximately $2.1 \cdot 10^{13}$ MIPS-years for target vector halves having 72 ones and 179 zeros.

3 Rotation Invariance of Convolution Modular Lattices

For any vector $\mathbf{u} = [u_0, u_1, \ldots, u_{N-1}]$, define the *(right) rotation of u* by

$$\rho(\mathbf{u}) = [u_{N-1}, u_0, u_1, \ldots, u_{N-2}].$$

(In terms of the polynomial ring $\mathbb{Z}[X]/(X^N - 1)$ described in Remark 1, right rotation is simply multiplication by X.) Applying ρ several times gives a k-fold rotation, which we denote by

$$\mathbf{u}^{(k)} = \rho^k(\mathbf{u}) = [u_{N-k}, u_{N-k+1}, \ldots, u_{N-k-1}].$$

An easy computation using the definition of the convolution product shows that $\rho(\mathbf{u} * \mathbf{v}) = \mathbf{u} * \rho(\mathbf{v})$, and repeated application of this rule gives

$$\rho^k(\mathbf{u} * \mathbf{v}) = \mathbf{u} * \rho^k(\mathbf{v}) \quad \text{for all } k \in \mathbb{Z}.$$

(Negative powers of ρ are defined as left rotations.)

Let $L(\mathbf{c}, q)$ be a CML. Applying ρ^k to the defining congruence (1) for $L(\mathbf{c}, q)$ yields

$$\mathbf{c} * \rho^k(\mathbf{u}) \equiv \rho^k(\mathbf{v}) \pmod{q},$$

so we see that if $[\mathbf{u}, \mathbf{v}]$ is in $L(\mathbf{c}, q)$, then every rotation $[\mathbf{u}^{(k)}, \mathbf{v}^{(k)}]$ is also in $L(\mathbf{c}, q)$.

Note that the rotations of a vector have the same length as the original vector. In particular, the SVP for a CML will have up to N different solutions corresponding to the N rotations of any one shortest vector. Similarly, if the target vector $\mathbf{t} \in \mathbb{R}^n$ for a CML CVP is rotation invariant (i.e., has all of its coordinates the same), then the CVP will have up to N different solutions, since if $\mathbf{w} \in L$ solves the CVP for $\mathbf{t}$, then all of the rotations $\rho^k(\mathbf{w}) \in L$ will also solve the CVP for $\mathbf{t}$. See Example 1 for a typical situation where this occurs.

4 The Pattern Method for Convolution Modular Lattices

The convolution lattices typically used in cryptography [6–8] are constructed to contain short binary (or trinary) vectors, which means in particular they tend to contain short vectors in which many of the coordinates are equal to zero. More precisely, they are constructed by choosing two short vectors $\mathbf{a}$ and $\mathbf{b}$ and taking the lattice L defined by

$$\mathbf{b} * \mathbf{u} \equiv \mathbf{a} * \mathbf{v} \pmod{q}.$$

(See Remark 3. The lattice L is the convolution modular lattice $L(\mathbf{c}, q)$ with $\mathbf{c} = \mathbf{a}^{-1} * \mathbf{b} \pmod{q}$.)

In this situation, the first author [14, 15] suggested looking for vectors $[\mathbf{u}, \mathbf{v}]$ in L such that the first r coordinates of $\mathbf{u}$ (or $\mathbf{v}$) are all zero. The hope, of course, is that the generating vector $[\mathbf{a}, \mathbf{b}] \in L$ has this property. If it does, then this method helps to efficiently find it. More precisely, one multiplies the first r columns of the matrix for L by a large constant θ, which has the effect of biasing lattice reduction against nonzero coordinates in those columns.

It is further noted in [14, 15] that the method is more effective for convolution modular lattices than for general lattices because of the rotation invariance inherent in a convolution modular lattice. Thus for a CML it is not necessary that $\mathbf{a}$ (or $\mathbf{b}$) have its first r coordinates zero, it suffices that the vector contains a string of r consecutive zeros somewhere in its list of coordinates. This is because $L(\mathbf{c}, q)$ also contains all of the rotations $[\mathbf{a}^{(k)}, \mathbf{b}^{(k)}]$ of each of its vectors, so if $\mathbf{a}$ (or $\mathbf{b}$) contains a run of r consecutive zeros, then one of the rotations of $[\mathbf{a}, \mathbf{b}]$ will have the zero run in the correct position.

More generally, as noted by the authors and others, rather than choosing a run of r consecutive coordinates, one can choose a random pattern of r coordinates and hope that L contains a short vector with the chosen pattern of zeros. And just as before, since the CML is rotation invariant, it suffices that the generating vectors have some rotation containing the preselected pattern of zeros. Even more generally, one might choose a pattern consisting of both zero and nonzero entries. We will call this the *CML Pattern Method*. For simplicity, we will concentrate in this note on patterns of zeros, but see Remark 8 for some brief comments on the more general situation.

From a cryptographic viewpoint, it is clearly better to choose essentially random patterns. Thus if $L(\mathbf{c}, q) = L(\mathbf{a}^{-1} * \mathbf{b}, q)$ is being used for cryptography, then the person who creates the lattice by choosing $[\mathbf{a}, \mathbf{b}]$ can easily thwart any particular pattern (e.g., long runs of zeros) by discarding $\mathbf{a}$ and $\mathbf{b}$ if they contain the chosen pattern.

There is a second, less obvious, reason why it is disadvantageous for an attacker to choose a pattern consisting of consecutive zero. The attacker "wins" if the pattern of coordinates that he chooses corresponds to zero coordinates in any one of the shifts $\mathbf{a}^{(k)}$ of the unknown vector $\mathbf{a}$. Now it may happen that there are actually two different shifts $\mathbf{a}^{(k_1)}$ and $\mathbf{a}^{(k_2)}$ that win. This means that $L(\mathbf{c}, q)$ contains two different target vectors. In practice having two target vectors does not seem to help lattice reduction very much. (Indeed, underlying May's original idea is taking the lattice $L(\mathbf{c}, q)$, which has N shortest vectors, and breaking the symmetry until there is only one shortest vector.) It turns out that multiple winners are more likely to occur if the attacker chooses consecutive coordinates than if he chooses random coordinates. Since the attacker's goal is simply to choose a single winning set of coordinates, he does best if most winners are only single winners. Thus by choosing random coordinates, he will spread the winning entries more widely.

We illustrate this last point with a small example. We take vectors of length $N = 13$ containing $d = 4$ non-zero entries. There are 715 such vectors. We consider various patterns and for each vector $\mathbf{a}$ we count how many times that pattern appears in some cyclic rotation of the coordinates of $\mathbf{a}$. For example, if we take the pattern $\langle 1, 2, 3, 4 \rangle$, then we are counting how many times there are 4 consecutive zeros in $\mathbf{a}$, while if we take $\langle 1, 3, 5, 8 \rangle$, we are counting how many times the vector **a** contains consecutive coordinates of the form 0#0#0##0. (Note that that the coordinates of **a** wrap, so for example the vector $\mathbf{a} = (0,1,0,1,1,1,0)$ contains the pattern $\langle 1,2,4\rangle$.) We write m_k for the number of vectors in which the given pattern appears k times. In particular, m_0 is the number of vectors which contain no copies of the pattern. Thus in terms of guessing patterns, it is best to have m_0 as small as possible. Table 1 gives the results of this experiment.

Table 1. Multiplicity of Patterns — $(N,d) = (13,4)$

Pattern	m_0	m_1	m_2	m_3	m_4	m_5
$\langle 1,2,3,4,5\rangle$	260	195	130	78	39	13
$\langle 1,2,6,10,12\rangle$	130	299	247	39	0	0
$\langle 1,2,4,7,11\rangle$	117	338	208	52	0	0
$\langle 1,2,3,6,10\rangle$	117	338	208	52	0	0
$\langle 1,2,5,8,10\rangle$	78	377	247	13	0	0
$\langle 1,3,5,8,9\rangle$	78	377	247	13	0	0

Thus for vectors of length 13 containing 4 nonzero entries, the probability of containing the pattern $\langle 1,2,3,4,5\rangle$, that is, the probability of containing five consecutive zeros, is $(715-260)/715 = 64\%$. This is reasonably high, but notice that the probability of containing the pattern $\langle 1,2,4,7,11\rangle$ is 84%, while the probability of containing the pattern $\langle 1,2,5,8,10\rangle$ is 89%. It seems likely that the pattern $\langle 1,2,\dots,r\rangle$ always gives the lowest probability. It would be interesting to try to prove this.

In conclusion, it appears to be in the attacker's best interest either to choose random coordinates or to do a further analysis and determine patterns which have particularly high probability of being successful. However, the disadvantage of choosing a particular pattern is that if the lattice creator knows which pattern(s) the attacker will use, he can always ensure that the shortest vectors in his lattices do not contain that pattern.

Remark 6. It appears that there are some patterns that discourage multiple winners, just as a pattern of consecutive zeros appears to encourage multiple winner. More precisely, it seems that there are some patterns J with the property that if $\mathbf{a}^{(k)}$ does not contain J for some given k, then it is more likely that other rotations will contain J. It seems to be a difficult combinatorial problem to even quantify this precisely, and from a practical perspective, using randomly selected patterns appears to give reasonable performance. We formulate some natural questions and conjectures which we believe are of interest.

- *Conjecture.* Among patterns of length r, the pattern $J = [1,2,\dots,r]$ of consecutive zeros gives the fewest distinct winners.
- *Question.* What is the average number of distinct winners as J runs over all patterns of length r?

- *Question.* What pattern (or patterns) J gives the maximum number of distinct winners?

5 Balancing the Target in a Convolution Modular Lattice

Let $L(\mathbf{c}, q)$ be a CML that is constructed by setting $\mathbf{c} \equiv \mathbf{a}^{-1} * \mathbf{b} \pmod{q}$ with short vectors $\mathbf{a}$ and $\mathbf{b}$ as described in Remark 3. If $\mathbf{a}$ and $\mathbf{b}$ have different lengths, it may be easier to find them by using a balanced CML.

Given $\mathbf{c}$, q and a real number λ, the associated *λ-balanced CML* is the lattice generated by the rows of the following matrix:

$$L(\mathbf{c}, q, \lambda) = \text{RowSpan} \left(\begin{array}{cccc|cccc} \lambda & 0 & \cdots & 0 & c_0 & c_1 & \cdots & c_{N-1} \\ 0 & \lambda & \cdots & 0 & c_{N-1} & c_0 & \cdots & c_{N-2} \\ \vdots & \vdots & \ddots & \vdots & \vdots & \vdots & \ddots & \vdots \\ 0 & 0 & \cdots & \lambda & c_1 & c_2 & \cdots & c_0 \\ \hline 0 & 0 & \cdots & 0 & q & 0 & \cdots & 0 \\ 0 & 0 & \cdots & 0 & 0 & q & \cdots & 0 \\ \vdots & \vdots & \ddots & \vdots & \vdots & \vdots & \ddots & \vdots \\ 0 & 0 & \cdots & 0 & 0 & 0 & \cdots & q \end{array} \right)$$

The real number λ is called the *balancing constant*. It is selected to make lattice reduction algorithms work more efficiently.

More precisely, lattice reduction works best if we minimize the ratio between the length of the shortest vector and the length of the next shortest (independent) vector. [For a CML, $[\mathbf{u}', \mathbf{v}']$ is independent of $[\mathbf{u}, \mathbf{v}]$ if it does not lie in the subspace generated by $[\mathbf{u}, \mathbf{v}]$ and all of its rotations.] The shortest vector in $L(\mathbf{c}, q, \lambda)$ is probably the vector $[\lambda \mathbf{a}, \mathbf{b}]$, while the Gaussian heuristic predicts that the next shortest independent vector has length approximately

$$\sqrt{\frac{\dim L(\mathbf{c}, q, \lambda)}{2\pi e}} \cdot \text{Disc}(L(\mathbf{c}, q, \lambda))^{1/\dim L(\mathbf{c}, q, \lambda)} = \sqrt{\frac{Nq\lambda}{\pi e}}.$$

Thus we want to choose λ to minimize the ratio

$$\sqrt{\frac{\lambda^2 ||\mathbf{a}||^2 + ||\mathbf{b}||^2}{Nq\lambda/\pi e}}.$$

The optimal value is $\lambda = ||\mathbf{b}||/||\mathbf{a}||$. Note that this equalizes the lengths of the two halves of the vector $[\lambda \mathbf{a}, \mathbf{b}]$. Further, the fundamental ratio that influences the difficulty of lattice reduction is equal to

$$\frac{\text{Length of shortest vector}}{\text{Length of next independent vector}} \approx \sqrt{\frac{\lambda ||\mathbf{a}||^2 + ||\mathbf{b}||^2}{Nq\lambda/\pi e}} = \sqrt{\frac{2\pi e ||\mathbf{b}||^2}{Nq}}. \quad (2)$$

Remark 7. In practice, one generally approximates the optimal value of λ with a rational number $\alpha/\beta \approx \lambda$ satisfying $\gcd(\alpha, q) = \gcd(\beta, q) = 1$. The balanced CML for the balancing constant α/β is the set of solutions $[\mathbf{u}, \mathbf{v}] \in \mathbb{Z}^N$ to the congruence

$$\mathbf{c} * \beta\mathbf{u} \equiv \alpha\mathbf{v} \pmod{q}.$$

6 Probability of CML Pattern Success

The CML Pattern Method will not work unless the lattice contains a vector with the selected pattern. It is thus important to calculate the probability that a random vector will contain a given pattern. For example, among all binary vectors of dimension 251 containing 72 ones, what is the probability that a randomly selected vector will contain a particular pattern of 17 zeros? As we have seen in Section 4, the exact probability will depend on the chosen pattern, but by making a certain independence assumption we can obtain a reasonable approximate formula. This estimate is given in the next result.

Theorem 1. *Fix positive integers N, d, r and a set of indices $J = \{j_1, j_2, \dots, j_r\}$ with $0 \le j_1 < j_2 < \dots < j_r < N$. Let*

$$\mathcal{B}_N(d) = \{\textit{binary vectors of dimension } N \textit{ with exactly } d \textit{ ones}\}$$

(a)

$$\operatorname*{Prob}_{\mathbf{a}\in\mathcal{B}_N(d)}(a_{j_1} = a_{j_2} = \dots = a_{j_r} = 0) = \frac{\binom{N-r}{d}}{\binom{N}{d}} = \prod_{i=0}^{d-1}\left(1 - \frac{r}{N-i}\right).$$

(b)

$$\operatorname*{Prob}_{\mathbf{a}\in\mathcal{B}_N(d)}\begin{pmatrix} a_{j_1+k} = a_{j_2+k} = \dots = a_{j_r+k} = 0 \\ \textit{for some } 0 \le k < N \end{pmatrix} \approx 1 - \left(1 - \prod_{i=0}^{d-1}\left(1 - \frac{r}{N-i}\right)\right)^N$$

(The indices on $\mathbf{a} = [a_0, a_1, \dots, a_{N-1}]$ are taken modulo N.)

Proof. (a) In order to create a vector $\mathbf{a}$, we choose d coordinates of $\mathbf{a}$ equal to 1, and the others will be 0. There are $\binom{N}{d}$ ways to do this. If we specify a particular set of r coordinates that are required to be 0, then there are only $\binom{N-r}{d}$ possible choices for $\mathbf{a}$. This proves the first formula for the probability of success, and the second formula follows from a little bit of algebra.
(b) For any particular k, the probability of success is given by the formula in (a). If we assume that the success probabilities for different values of k are independent, then we can use the formula

$$\begin{aligned}\text{Prob(success for some } k) &= 1 - \text{Prob(failure for every } k) \\ &= 1 - \prod_{0\le k<N} \text{Prob(failure for } k) \\ &= 1 - \prod_{0\le k<N} (1 - \text{Prob(success for } k)).\end{aligned}$$

Substituting in the formulas from (a) completes the proof of (b), assuming that the different k events are independent. (See Section 4 for a discussion on the extent to which the different values of k give independent events. The precise amount of independence depends on the particular set of indices J.) □

In order to test the accuracy, on average, of Theorem 1, we performed experiments to compute the probability that an r-pattern wins in $\mathcal{B}_N(d)$. For each r we fixed a random vector $\mathbf{a} \in \mathcal{B}_N(d)$ and choose 10,000 random patterns of length r. The results, given in Table 2, show an excellent match between theory and experiment.

Table 2. Probability that an r-Pattern Wins in $\mathcal{B}_N(d)$.

$(N,d) = (167,30)$	r	15	20	25	30	35	40	45
	Prob-Theory	0.999	0.913	0.533	0.204	0.063	0.017	0.004
	Prob-Exp	1.000	0.918	0.534	0.209	0.059	0.017	0.004

$(N,d) = (251,72)$	r	5	10	15	20	25	30	35
	Prob-Theory	1.000	1.000	0.734	0.189	0.031	0.005	0.001
	Prob-Exp	1.000	1.000	0.731	0.188	0.032	0.004	0.001

$(N,d) = (347,64)$	r	20	25	30	35	40	45	50
	Prob-Theory	0.995	0.823	0.432	0.166	0.056	0.017	0.005
	Prob-Exp	0.995	0.834	0.426	0.167	0.059	0.017	0.005

Remark 8. As noted in Section 4, it may be advantageous to use patterns that contain nonzero entries, as well as zero entries. For example, suppose that $\mathbf{a}$ is a binary vector of dimension N with d ones and $N-d$ zeros, and suppose that an attacker guesses a pattern of length r. The numbers N, d, and r are fixed, but we are free to specify r_1 ones and r_0 zeros, where $r_1 + r_0 = r$. There are two issues to consider. First, it may be easier to guess a mixed pattern of ones and zeros. Second, the length of the associated CVP may be smaller using a mixed pattern. We briefly consider these two issues.

First consider the probability of guessing a correct pattern. If we take indices $J = \{j_1, \dots, j_{r_0}\}$ for the zeros and other indices $L = \{\ell_1, \dots, \ell_{r_1}\}$ for the ones, then the probability in Theorem 1(a) becomes

$$\operatorname*{Prob}_{\mathbf{a}\in\mathcal{B}_N(d)} \left(\begin{matrix} a_{j_1} = a_{j_2} = \cdots = a_{j_{r_0}} = 0 \text{ and} \\ a_{\ell_1} = a_{\ell_2} = \cdots = a_{\ell_{r_1}} = 1 \end{matrix} \right) = \frac{\binom{N-r}{d-r_1}}{\binom{N}{d}}.$$

The attacker wants to maximize this probability. In order to maximize the probability of guessing correctly, he chooses a pattern of length r such that the difference of the number of zeros and the number of ones in the remaining $N-r$ vector entries is minimal. Thus, the optimal choice is

$$r_1 = \frac{r-(N-2d)}{2} \qquad \text{and} \qquad r_0 = \frac{r+(N-2d)}{2}, \tag{3}$$

but there is also the obvious restriction that r_0 and r_1 must be nonnegative. In particular, if $d \approx N/2$, then he should take $r_1 \approx r_2 \approx r/2$. On the other hand, if d and r are small, more precisely if $r+2d \leq N$, then the best that the attacker can do is choose $r_1 = 0$ and $r_0 = r$, i.e., he should choose a pattern consisting entirely of zeros.

Next consider the CVP that arises after a pattern has been correctly guessed. The fundamental length that appears in the associated CVP (see Example 1) is proportional to

$$\sqrt{\frac{(d-r_1)(N-d-r_0)}{N-r}},$$

so the attackers wants to choose r_0 and r_1 (subject to $r_0+r_1=r$) to minimize this quantity. But the choice made in (3) maximizes this function. If the attacker wants to make the CVP easiest, then he chooses a pattern of length r such that the difference between the number of zeros and the number of ones in the remaining $N-r$ vector entries is maximal. For example, if he can guess all zeros or all ones in his pattern, he obtains length zero in the associated CVP.

We will concentrate in this note on maximizing the probability of guessing correct patterns. It is an open question how to choose an optimal pattern that takes also the vector length in the associated CVP into account.

7 The Net Advantage of the CML Pattern Method

The effect of the CML Pattern Method is to reduce the dimension of the search lattice. More precisely, a correct r-fold pattern has the effect of reducing the effective dimension from $2N$ to $2N-2r$. This can be quite significant, since as explained in Section 2, the search time is roughly exponential in N.

However, one does not know, a priori, that the given lattice has a vector with the given pattern. So balanced against the gain from the reduction in dimension is the loss from choosing a pattern that has no chance of working. Thus the effective gain in efficiency is the probability of guessing a correct pattern multiplied by the reduction in search time.

Proposition 1. *Let $\mathcal{L} = \{(L,\mathbf{t})\}$ be a collection of pairs consisting of a convolution modular lattice L and a CVP target vector $\mathbf{t}$. Suppose that there are constants α and β so that the time required to solve the CVP for each $(L,\mathbf{t}) \in \mathcal{L}$ is given by*

$$\log_{10}(\text{Time}) \approx \alpha N + \beta.$$

Then the gain in applying the CML Pattern Method with a random set of indices $J = \langle j_1, j_2, \ldots, j_r \rangle$ *is approximately*

$$\text{Gain} = \begin{pmatrix} \textit{Probability that the lattice} \\ \textit{contains the pattern } J \textit{ of zeros} \end{pmatrix} \cdot \frac{\begin{pmatrix} \textit{Time to solve CVP} \\ \textit{in the original lattice} \end{pmatrix}}{\begin{pmatrix} \textit{Time to solve CVP in} \\ \textit{the } J\textit{-eliminated lattice} \end{pmatrix}}$$

$$\approx \left(1 - \left(1 - \prod_{i=0}^{d-1}\left(1 - \frac{r}{N-i}\right)\right)^N\right) 10^{\alpha r} \tag{4}$$

The optimal value of r and approximate corresponding gain may be determined from the formula (4) once an experimental value of α is determined.

Example 3. We use formula (4) to estimate the approximate gain in speed when the CML Pattern Method is applied to a lattice of dimension $N = 251$ with short vectors having $d = 72$. We use the value $\alpha = 0.083$ from Example 2. The results are given in Table 3. We see that the optimal choice of $r = 15$ gives a speed gain of approximately 13.

Table 3. Speed Gain for $(N, d, \alpha) = (251, 72, 0.083)$.

r	1	2	3	4	5	6	7	8	9	10
Gain	1.21	1.47	1.77	2.15	2.60	3.15	3.81	4.61	5.58	6.76

r	11	12	13	14	15	16	17	18	19	20
Gain	8.16	9.71	11.22	12.36	12.90	12.79	12.14	11.12	9.92	8.66

8 Parallelization of the CML Pattern Method

The CML Pattern Method allows a partial parallelization of the CVP in convolution modular lattices. The basic idea is to have a large number of separate processors each choose a possible pattern and attempt to solve CVP using that pattern. The method is successful if any one machine guesses a pattern that appears in the solution vector.

In a typical situation, one has a certain number of machines available and one wants to choose a value for r (the pattern length) in order to minimize the expected running time. An important observation is that for a fixed value of r, the gain from adding more and more machines becomes less and less. This is due to the fact that it does not help the running time if more than one machine guesses a valid pattern.

We now quantify these remarks. Let

$$T(r) = \text{time to run if we guess a valid pattern of length r},$$

$$P(r) = \text{probability of guessing a valid pattern of length r}.$$

Using one machine, the running time is more-or-less $T(r)/P(r)$. Suppose we instead use two machines. If $P(r)$ is small, the running time is approximately $T(r)/2P(r)$, so we gain a factor of 2. But if $P(r)$ is large, say 50%, then we save much less (on average), because there is a good chance that both machines will guess right the first time, which doesn't help.

So suppose that we have K machines, each of which tries to guess a pattern of length r. Then the formula for the expected running time is given by the following computation.

$$\begin{aligned}\text{TotalTime}(K,r) &= \begin{pmatrix}\text{total expected running time using } K \text{ machines}\\ \text{and guessing patterns of length } r\end{pmatrix}\\ &= \frac{T(r)}{\text{probability that at least one machine guesses correctly}}\\ &= \frac{T(r)}{1-\text{probability that all machines guess wrong}}\\ &= \frac{T(r)}{1-(1-P(r))^K}\end{aligned}$$

Notice that if $P(r)$ is small (compared to K), then we obtain

$$\text{TotalTime}(K,r) \approx \frac{T(r)}{K\cdot P(r)},$$

which agrees with our intuition that using K machines gives a K-fold speedup. And as K gets very large with r fixed, we find that

$$\lim_{K\to\infty} \text{TotalTime}(K,r) = T(r),$$

which also makes sense, since if there are a huge number of machines, then at least one of them will almost certainly guess a valid pattern the first time.

Hence given access to K machines, the optimal strategy is to choose r so as the minimize the running time function

$$\text{TotalTime}(K,r) = \frac{T(r)}{1-(1-P(r))^K}. \tag{5}$$

Letting r_K denote this optimal value of r, we see that as a function of the number of machines K, the expected running time is given by the function $\text{TotalTime}(K,r_K)$, a highly nonlinear function of K.

In order to analyze parallelization more deeply, we thus need to know the functions $T(r)$ and $P(r)$. The formula for $P(r)$ is given in Theorem 1(b). Further, as noted in Section 2, for certain classes of lattices it appears that the running time is exponential in the dimension of the lattice. Since use of a pattern of length r has the effect of reducing the dimension by $2r$ (equivalently, reducing the value of N by r), it is reasonable to take $T(r)$ to have the form

$$T(r) = 10^{\alpha(N-r)+\beta}$$

for certain constants α and β that are independent of r. However, we note that this neglects the fact that the length of the associated CVP problem is also reduced (see Remark 8), so the following computation is only a rough estimate.

With this caveat, we consider the rough approximation

$$\mathrm{TotalTime}(K,r) = \frac{T(r)}{1-(1-P(r))^K} \approx \frac{10^{\alpha(N-r)+\beta}}{1-\left(1-\prod_{i=0}^{d-1}\left(1-\frac{r}{N-i}\right)\right)^{KN}}$$

and ask for the value r_K of r that minimizes this function of K, treating α, β, d, and N as constants. Then the gain obtained by using K processors is approximately

$$\mathrm{Gain}(K) = \frac{\mathrm{TotalTime}(1,r_1)}{\mathrm{TotalTime}(K,r_K)}.$$

It seems quite difficult to obtain a closed form for this expression. We have estimated it numerically for the parameters $(N,d,\alpha,\beta) = (251,72,0.083,-7.5)$ and list the results in Table 4. (The value of β does not affect the gain.)

Table 4. Gain Using K Processors—$(N,d,\alpha,\beta) = (251,72,0.083,-7.5)$.

K	r_K	$P(r_K)$	$K \cdot P(r_K)$	TotalTime(r_K)	Gain(K)
1	15	0.734	0.73	$1.6698 \cdot 10^{12}$	1.000
2	17	0.471	0.94	$1.1607 \cdot 10^{12}$	1.439
3	18	0.357	1.07	$9.4145 \cdot 10^{11}$	1.774
5	20	0.189	0.95	$7.2489 \cdot 10^{11}$	2.303
10	21	0.135	1.35	$5.0929 \cdot 10^{11}$	3.279
20	23	0.066	1.32	$3.5717 \cdot 10^{11}$	4.675
50	26	0.021	1.07	$2.2631 \cdot 10^{11}$	7.378
75	27	0.015	1.10	$1.8491 \cdot 10^{11}$	9.030
100	27	0.015	1.46	$1.6044 \cdot 10^{11}$	10.408
150	28	0.010	1.49	$1.3151 \cdot 10^{11}$	12.697
200	29	0.007	1.35	$1.1376 \cdot 10^{11}$	14.678
300	30	0.005	1.37	$9.3382 \cdot 10^{10}$	17.881

The numbers in Table 4 are puzzling at first, since as noted above, if $P(r)$ is small, then we expect the gain to be approximately K. However, this will only be true if

$$1-(1-P(r))^K \approx KP(r), \tag{6}$$

so in particular $P(r)$ must be considerably smaller than $1/K$. But in Table 4, the optimal choice of $P(r)$ is approximately equal to $1/K$, so the approximation (6) is not at all accurate.

References

1. M. Ajtai, *The shortest vector problem in ℓ_2 is NP-hard for randomized reductions*, Proc. 30th ACM Symposium on the Theory of Computing, pages 10-19, 1998
2. M. Ajtai, R. Kumar, D. Sivakumar, *A sieve algorithm for the shortest lattice vector problem*, Proc. 33rd ACM Symposium on Theory of Computing, 2001 (to appear)
3. J.W.S. Cassels, *An Introduction to the Geometry of Numbers*, Die Grundlehren Der Mathematischen Wissenschaften, Springer-Verlag, 1959.
4. O. Goldreich, S. Goldwasser, S. Halvei, *Public-key cryptography from lattice reduction problems*, CRYPTO'97, Lect. Notes in Computer Science 1294, Springer-Verlag, 1997, 112-131.
5. O. Goldreich, D. Micciancio, S. Safra and J.P. Seifert, *Approximating shortest lattice vectors is not harder than approximating closest vectors*, Information Processing Letters, vol. 71, pp. 55-61, 1999.
6. J. Hoffstein, J. Pipher, J.H. Silverman, *NTRU: A new high speed public key cryptosystem*, in Algorithmic Number Theory (ANTS III), Portland, OR, June 1998, Lecture Notes in Computer Science 1423 (J.P. Buhler, ed.), Springer-Verlag, Berlin, 1998, 267–288.
7. J. Hoffstein, J. Pipher, J.H. Silverman, *NSS: An NTRU Lattice-Based Signature Scheme*, Advances in Cryptology—Eurocrypt 2001, Lecture Notes in Computer Science, Springer-Verlag, 2001.
8. J. Hoffstein, J. Pipher, J.H. Silverman, *The NTRU Signature Scheme: Theory and Practice*, preprint, June 2001.
9. IEEE P1363.1, Standard Specification for Public-Key Cryptographic Techniques Based on Hard Problems over Lattices, Draft 2, 2001.
10. Number Theory Library, Victor Shoup, `http://www.cs.wisc.edu/~shoup/ntl`
11. H. Koy, C.-P. Schnorr, *Segment LLL-Reduction of Lattice Bases*, Cryptography and Lattice Conference (CaLC 2001), Lecture Notes in Computer Science, Springer-Verlag, this volume.
12. H. Koy, C.-P. Schnorr, *Segment LLL-Reduction with Floating Point Orthogonalization*, Cryptography and Lattice Conference (CaLC 2001), Lecture Notes in Computer Science, Springer-Verlag, pp. 81-96.
13. A.K. Lenstra, H.W. Lenstra Jr., L. Lovśz, *Factoring polynomials with rational coefficients*, Mathematische Ann. 261 (1982), 513-534.
14. A. May, *Auf Polynomgleichungen basierende Public-Key-Kryptosysteme*, Johann Wolfgange Goethe-Universitat, Frankfurt am Main, Fachbereich Informatik. (Masters Thesis in Computer Science, 4 June, 1999; Thesis advisor C.P. Schnorr).
15. A. May, *Cryptanalysis of NTRU-107*, preprint, April 1999 (unpublished).
16. D. Micciancio, *The Shortest Vector in a Lattice is Hard to Approximate within Some Constant*, Proc. 39th IEEE Symposium on Foundations of Computer Science, pages 92-98, 1998
17. P. Nguyen, *Cryptanalysis of the Goldreich-Goldwasser-Halevi Cryptosystem*, Advances in Cryptology - Proceedings of CRYPTO'99, M. Wiener (ed.), Lecture Notes in Computer Science, Springer-Verlag, 1999.
18. C.P. Schnorr, M. Euchner, *Lattice basis reduction: improved practical algorithms and solving subset sum problems*, Math. Programming 66 (1994), no. 2, Ser. A, 181-199.
19. C.P. Schnorr, *A hierarchy of polynomial time lattice basis reduction algorithms*, Theoretical Computer Science 53, pages 201-224, 1987

Improving Lattice Based Cryptosystems Using the Hermite Normal Form

Daniele Micciancio*

Department of Computer Science and Engineering
University of California, San Diego
9500 Gilman Drive, La Jolla, CA 92093, USA
daniele@cs.ucsd.edu

Abstract. We describe a simple technique that can be used to substantially reduce the key and ciphertext size of various lattice based cryptosystems and trapdoor functions of the kind proposed by Goldreich, Goldwasser and Halevi (GGH). The improvement is significant both from the theoretical and practical point of view, reducing the size of both key and ciphertext by a factor n equal to the dimension of the lattice (i.e., several hundreds for typical values of the security parameter.) The efficiency improvement is obtained without decreasing the security of the functions: we formally prove that the new functions are at least as secure as the original ones, and possibly even better as the adversary gets less information in a strong information theoretical sense. The increased efficiency of the new cryptosystems allows the use of bigger values for the security parameter, making the functions secure against the best cryptanalytic attacks, while keeping the size of the key even below the smallest key size for which lattice cryptosystems were ever conjectured to be hard to break.
Keywords: Lattices, trapdoor functions, public-key encryption.

1 Introduction

Recent results on the complexity of lattices [1] have drawn considerable attention to lattice problems as potential candidates to design cryptographic primitives, and encryption schemes in particular. The two most notable proposals are the Ajtai-Dwork cryptosystem (AD, [2]) and the Goldreich-Goldwasser-Halevi cryptosystem (GGH, [14]). Other lattice based cryptosystems designed along the lines of [2, 14] were subsequently proposed by Fischlin and Seifert [10] and Cai and Cusick [6]. While the AD cryptosystem is mainly of theoretical interest, the GGH cryptosystem was suggested as a practical alternative to number theory based schemes currently in use (e.g., the RSA cryptosystem [28]).

Two other related cryptosystems are McEliece's [21] and NTRU [17]. Neither of them is a lattice based cryptosystem in the strict meaning of the term, as they use ideas from other areas of mathematics (polynomial ring and finite

* Research supported in part by NSF Career Award CCR-0093029.

J.H. Silverman (Ed.): CaLC 2001, LNCS 2146, pp. 126–145, 2001.

field arithmetic respectively). However, the security of NTRU is related to the hardness of certain lattice problems (see Section 7 for further discussion), and it definitely deserves a mention as the most practical of the above proposals, yielding keys of size $O(n \log n)$ instead of the $\Omega(n^2)$ key size of McEliece, GGH and related schemes. (In fact, we'll see that GGH keys can be as big as $O(n^3 \log n)$.) McEliece's scheme is based on the hardness of coding theoretic problems, rather than lattices, but it bears much resemblance to the GGH cryptosystem, and we will further discuss this scheme in Section 7, after introducing our new lattice based trapdoor function.

As is the case for RSA, no proof of security for the GGH cryptosystem (or any of the aforementioned schemes with the only exception of Ajtai and Dwork's theoretical proposal) is currently known, and its conjectured security is based on the empirical evidence that certain lattice problems are hard. In the attempt of stimulating further cryptanalytic efforts against their system and determining appropriate key size, the authors of GGH published five numerical challenges [13] corresponding to increasing values of the security parameter $n = 200, 250, 300, 350, 400$, resulting in public key sizes ranging from 330 KB to over 2 MB. Despite the big key size even for the smallest dimension (330 KB), this cryptosystem was still competitive with number theoretic cryptosystems because the encryption time is essentially linear in the size of the key, while modular exponentiation typically requires $O(n^3)$ operations.

At the time the GGH cryptosystem was presented at Crypto'97, challenges in dimension $n = 150$ were known to be breakable [31] using lattice reduction techniques. Still these techniques did not seem to apply to lattices in higher dimension $n > 200$. At Crypto'99 Nguyen [24] showed how to exploit a weakness specific to the way the GGH challenges were chosen and break the first four of the five GGH challenges. As the only unbroken challenge had key size over 2 MB, the practical value of the GGH cryptosystem (and variants) seemed quite questionable and [24] concluded that "unless major improvements are found, lattice-based cryptography cannot provide a serious alternative to existing public-key encryption algorithms." In this paper we present one such improvement.

Although quite simple, the techniques described in this paper can be used to significantly reduce the key size of GGH-like cryptosystems (e.g., those presented in [14, 10]). In particular, we show how to build lattice based trapdoor functions with public keys of size below 330 KB, and still secure against the best known lattice attacks. The improvement gets even better as the dimension of the lattice grows: our techniques can be used to asymptotically reduce the key size from $O(n^3 \log n)$ to $O(n^2 \log n)$. The improvement in the encryption time is analogous, being roughly proportional to the size of the public key, and the size of the ciphertext is also significantly reduced going from $O(n^2 \log n)$ to $O(n \log n)$.

Surprisingly, we can achieve these efficiency improvements without decreasing the security of the scheme: any attack to the new scheme can be provably transformed into an (at least) equally effective attack against the original GGH (and variant) schemes. The increased efficiency allows one to use greater values of the security parameter than considered in [14] and [24], while maintaining the scheme reasonably practical. In particular, our techniques give encryption schemes that resist the best known lattice attacks and still have public keys even smaller than the weakest of the GGH challenges, bringing the feasibility of lattice based cryptography back into discussion.

The rest of the paper is organized as follows. In Section 2 we recall some basic definitions and properties of lattices. In Section 3 we describe the original GGH scheme and discuss its weaknesses. In Section 4 we define a new cryptographic function that significantly improves the GGH scheme both in terms of security and efficiency. The new scheme is analyzed in Section 5. To be precise, the GGH scheme, as well as the new one we are proposing in this paper, should be considered as *trapdoor functions* instead of ready-to-use *encryption schemes.* The reason is explained in Section 6, where we also discuss how to transform these functions into encryption schemes. In Section 6 we also describe some cryptanalytic experiments that we performed to validate our theoretical results. The McEliece and NTRU cryptosystems are discussed in Section 7. Section 8 concludes with final remarks and open problems.

2 Preliminaries

Let $B = \{\boldsymbol{b_1}, \ldots, \boldsymbol{b_n}\}$ be a set of n linearly independent vectors in $\mathbb{R}^m$. The *lattice* generated by B is the set $\mathcal{L}(B) = \{\sum_i x_i \boldsymbol{b_i} \mid x_i \in \mathbb{Z}\}$ of all *integer* linear combinations of the vectors in B. The set B is called *basis* and it is usually identified with the matrix $\boldsymbol{B} = [\boldsymbol{b_1}, \ldots, \boldsymbol{b_n}] \in \mathbb{R}^{m \times n}$ having the vectors $\boldsymbol{b_i}$ as columns. In matrix notation $\mathcal{L}(\boldsymbol{B}) = \{\boldsymbol{Bx} \mid \boldsymbol{x} \in \mathbb{Z}^n\}$. The lattice $\mathcal{L}(\boldsymbol{B})$ is *full rank* if $n = m$, i.e. if $\boldsymbol{B}$ spans the entire vector space $\mathbb{R}^m$ over the reals. For simplicity, in the rest of this paper we will consider only full rank lattices. Moreover, for computational purposes, we will assume the basis vectors $\boldsymbol{b_i} \in \mathbb{Z}^n$ have integer entries. Then, a basis is just a non-singular integer matrix $\boldsymbol{B} \in \mathbb{Z}^{n \times n}$.

The basis of a lattice is not unique. A particularly convenient basis for some applications is the *Hermite Normal Form* (HNF). A basis $\boldsymbol{B}$ is in HNF if it is upper triangular, all elements on the diagonal are strictly positive, and any other element $b_{i,j}$ satisfies $0 \leq b_{i,j} < b_{i,i}$. It is easy to see that every integer lattice $\mathcal{L}(\boldsymbol{B})$ has a unique basis in Hermite Normal Form, denoted HNF($\boldsymbol{B}$). Moreover, given any basis $\boldsymbol{B}$ for the lattice, HNF($\boldsymbol{B}$) can be efficiently computed (see [23] for a recent algorithm and a survey of previous results). Notice that HNF($\boldsymbol{B}$) does not depend on the particular basis $\boldsymbol{B}$ with started from, and it is uniquely defined by the lattice $\mathcal{L}(\boldsymbol{B})$ generated by $\boldsymbol{B}$.

For every basis $\boldsymbol{B}$, the *orthogonalized basis* $\boldsymbol{B}^* = [\boldsymbol{b}_1^*, \ldots, \boldsymbol{b}_n^*]$ is defined by the usual Gram-Schmidt orthogonalization process:

$$\boldsymbol{b}_1^* = \boldsymbol{b}_1$$

$$\boldsymbol{b}_i^* = \boldsymbol{b}_i - \sum_{j<i} \frac{\langle \boldsymbol{b}_i, \boldsymbol{b}_j^* \rangle}{\langle \boldsymbol{b}_j^*, \boldsymbol{b}_j^* \rangle} \boldsymbol{b}_j^* \quad .$$

Notice that $\boldsymbol{B}^*$ is a basis for the vector space $\mathbb{R}^n$, but is not in general a lattice basis for $\mathcal{L}(\boldsymbol{B})$. If $\boldsymbol{B}$ is in HNF, then $\boldsymbol{B}^*$ is simply the diagonal matrix with $b_{1,1}, \ldots, b_{n,n}$ on the diagonal. The determinant of a lattice $\mathcal{L}(\boldsymbol{B})$ is the absolute value of the determinant of the matrix $\boldsymbol{B}$. The determinant is a lattice invariant, i.e. it does not depend on the particular choice of the basis $\boldsymbol{B}$. If $\boldsymbol{B}$ is in HNF, then the determinant $\det(\boldsymbol{B})$ is just the product of the elements on the diagonal $\prod_{i=1}^n b_{i,i}$. An important property of integer lattices is that if two vectors $\boldsymbol{v}$ and $\boldsymbol{w}$ are congruent modulo $\det(\boldsymbol{B})$ (i.e. $\det(\boldsymbol{B})$ divides $v_i - w_i$ for all i), then $\boldsymbol{v} \in \mathcal{L}(\boldsymbol{B})$ if and only if $\boldsymbol{w} \in \mathcal{L}(\boldsymbol{B})$. In other words, the lattice repeats identically if translated by multiples of $\det(L)$ along the direction of any of the main axes.

The distance between two vectors $\boldsymbol{v}$ and $\boldsymbol{w}$ is defined by

$$\operatorname{dist}(\boldsymbol{v}, \boldsymbol{w}) = \|\boldsymbol{v} - \boldsymbol{w}\| = \sqrt{\sum_i (v_i - w_i)^2} \ .$$

The distance function is extended to sets of vectors as usual

$$\operatorname{dist}(S_1, S_2) = \inf\{\|\boldsymbol{v} - \boldsymbol{w}\| : \boldsymbol{v} \in S_1, \boldsymbol{w} \in S_2\} \ .$$

In particular the distance of a vector from a lattice is given by $\operatorname{dist}(\boldsymbol{v}, \mathcal{L}(B)) = \min\{\|\boldsymbol{v} - \boldsymbol{w}\| : \boldsymbol{w} \in \mathcal{L}(B)\}$. In the *closest vector problem* (CVP), one is given a basis $\boldsymbol{B}$ and a target vector $\boldsymbol{v}$ (usually not in the lattice) and must find the lattice vector in $\mathcal{L}(\boldsymbol{B})$ closest to $\boldsymbol{v}$. CVP was proved NP-hard in [36], and it remains hard even if the lattice basis can be arbitrarily preprocessed [22], or one allows for approximate solutions with approximation factor $2^{\lg^{1-\epsilon} n}$ [3, 9]. To date, the best polynomial time algorithm to approximate CVP achieves only a worst case approximation factor which is almost exponential in the dimension of the lattice [19, 4, 29].

A closely related problem is the shortest vector problem (SVP): given a lattice $L = \mathcal{L}(\boldsymbol{B})$, find the length $\lambda(L)$ of the shortest non-zero vector in $\mathcal{L}(\boldsymbol{B})$. By linearity, $\lambda(\boldsymbol{B})$ equals the minimum distance between any two lattice points $\min\{\|\boldsymbol{v} - \boldsymbol{w}\| : \boldsymbol{v}, \boldsymbol{w} \in \mathcal{L}(\boldsymbol{B}), \boldsymbol{v} \neq \boldsymbol{w}\}$. It is easy to see that for any vector $\boldsymbol{v}$ (not necessarily in the lattice) there exists at most one lattice point within distance $\lambda/2$ from $\boldsymbol{v}$. One can easily show that the length of the shortest vector in a lattice $L = \mathcal{L}(\boldsymbol{B})$ satisfies $\lambda(L) \geq \min_i \|\boldsymbol{b}_i^*\|$. Moreover, given a vector $\boldsymbol{v}$ within distance $\rho = \frac{1}{2}\min_i \|\boldsymbol{b}_i^*\|$ from the lattice, the (unique) lattice vector within distance ρ from $\boldsymbol{v}$ can be efficiently computed from $\boldsymbol{B}$ and $\boldsymbol{v}$ using Babai's *nearest plane* algorithm [4]. (See also [18].)

3 The GGH Encryption Scheme

The GGH encryption scheme [14] works essentially as follows. The private and public keys of the scheme are two bases $\boldsymbol{B}, \boldsymbol{R}$ of the same lattice $L = \mathcal{L}(\boldsymbol{B}) = \mathcal{L}(\boldsymbol{R})$. The private key $\boldsymbol{R}$ is an exceptionally good basis. In particular, $\boldsymbol{R}$ is chosen in such a way that the quantity $\rho = \frac{1}{2}\min \|\boldsymbol{r}_i^*\|$ is relatively large, so that all errors of length less than ρ can be efficiently corrected using $\boldsymbol{R}$. However, given the public basis this same task should be computationally hard. In particular, $\frac{1}{2}\min\|\boldsymbol{b}_i^*\|$ is much smaller than ρ.

The two bases are used to define a trapdoor function that takes as input an integer vector $\boldsymbol{x}$ and an error vector $\boldsymbol{r}$ of length at most ρ, and returns the vector $\boldsymbol{c} = \boldsymbol{B}\boldsymbol{x} + \boldsymbol{r}$, i.e. the lattice vector with public coefficients $\boldsymbol{x}$ perturbed by a small additive error $\boldsymbol{r}$. Notice that $\boldsymbol{B}\boldsymbol{x}$ can be recovered from $\boldsymbol{c}$ using the private basis $\boldsymbol{R}$. Once $\boldsymbol{B}\boldsymbol{x}$ is recovered, one can easily compute $\boldsymbol{x}$ and $\boldsymbol{r}$ using simple linear algebra, therefore inverting the function.

A message m is "encrypted" by first encoding m in the input $(\boldsymbol{x}, \boldsymbol{r})$, and then applying the trapdoor function to $(\boldsymbol{x}, \boldsymbol{r})$. Two encoding method are considered in [14]:

1. In the first method, the message m is encoded in the error vector $\boldsymbol{r}$, and $\boldsymbol{x}$ is chosen at random
2. In the second method, the message m is encoded in the coefficients $\boldsymbol{x}$, and $\boldsymbol{r}$ is chosen at random.

For concreteness, in the rest of the paper we will assume the first encoding method, but most of the techniques we describe can be easily adapted to the second method as well.

In order to fully specify the trapdoor function the following questions must be answered:

1. How is the private basis $\boldsymbol{R}$ chosen?
2. How is the public basis $\boldsymbol{B}$ obtained from $\boldsymbol{R}$?
3. How is the random vector $\boldsymbol{x}$ chosen?
4. How is the error vector $\boldsymbol{r}$ chosen?

The authors of [14] suggest[1] to take $\boldsymbol{R} = \sqrt{n} \cdot \boldsymbol{I} + \boldsymbol{Q}$, where $\boldsymbol{I}$ is the identity matrix, and $\boldsymbol{Q}$ is a random perturbation matrix with entries in $\{-4, \ldots, +4\}$. Then obtain the public basis $\boldsymbol{B}$ applying a sufficiently long sequence of random elementary column operations to $\boldsymbol{R}$. Then the message m is "encrypted" by encoding it in an error vector $\boldsymbol{r}$ with entries $r_i = \pm 3$, and adding it to a lattice vector $\boldsymbol{B}\boldsymbol{x}$ chosen at random from a sufficiently large region of space.

Notice that in order to avoid attacks based on exhaustive search, the sequence of operations applied to $\boldsymbol{R}$ to obtain the public basis, and the region of space from which the lattice vector $\boldsymbol{B}\boldsymbol{x}$ is chosen must be sufficiently large. Since the lattice repeats identically if translated by $\det(L)$ along any of the main axes, we can always assume that the entries of $\boldsymbol{B}$ and $\boldsymbol{x}$ are reduced modulo $\det(L)$ without decreasing the security of the scheme. We can use this observation to estimate the proper size of the public key $\boldsymbol{B}$ and the ciphertext $\boldsymbol{c} = \boldsymbol{B}\boldsymbol{x} + \boldsymbol{r}$ as $O(n^2 \cdot \lg(\det(L)))$ and $O(n \cdot \lg(\det(L)))$. The determinant $\det(L)$ can also be estimated to be $2^{O(n \lg n)}$ applying Hadamard inequality to the private basis,

[1] These are the choices used to generate the challenges in [13] and considered in cryptanalytic attacks [24]. In fact a broader range of possible choices is considered in the paper [14].

resulting in $O(n^3 \lg n)$ and $O(n^2 \lg n)$ estimates for the key and ciphertext size. At this point one can point out how the particular way questions 1-4 were answered in [13] introduced some serious weakness in the system. In fact:

1. The only property of $\boldsymbol{R}$ required for the decryption algorithm to work is that the orthogonalized vectors $\boldsymbol{r}_i^*$ are sufficiently long. Choosing them to be close to the main axes $\sqrt{n}\cdot\boldsymbol{e}_i$ seems a quite peculiar restriction that might make the scheme much weaker. Other work in the design of lattice based encryption schemes [35, 10] suggests that rotations might play a fundamental role in making these schemes secure.
2. The size of the GGH challenges published at [13] are about an order of magnitude smaller than what we estimated to be the proper size. This suggests that the public key of the GGH challenges had not been "randomized" enough, making them particularly easy to break.
3. The same arguments apply to the choice of the lattice vector $\boldsymbol{Bx}$.
4. Choosing error vectors whose entries have all the same absolute value also introduces serious weaknesses, as shown in [24].

Now it shouldn't be a surprise that the numerical challenges in [13] have been broken, but it should also be clear that known attacks say very little about the general methodology used by the GGH cryptosystem. In particular, Nguyen attack [24] relies, in an essential way, on the property that all the entries in the error vector have the same absolute value and is based on the following observation: if all entries of $\boldsymbol{r}$ have absolute value σ, then $\boldsymbol{c} + \boldsymbol{s} \equiv \boldsymbol{Bx} \pmod{2\sigma}$, where $\boldsymbol{s}$ is the vector $[\sigma, \ldots, \sigma]^T$. This allows to find $\boldsymbol{x}$ modulo 2σ and reduce the problem of finding a lattice vector within distance $\|\boldsymbol{r}\|$ from $\boldsymbol{c}$ to the problem of finding a lattice vector within smaller distance $\|\boldsymbol{r}\|/(2\sigma)$ from $(\boldsymbol{c} - \boldsymbol{B}(\boldsymbol{x} \bmod 2\sigma))/(2\sigma)$. (In the GGH challenges $\sigma = 3$, reducing the length of the error by a factor 6.) As already noted in [24], removing the restriction that all entries in the error vector have the same modulus, immediately fixes the problem, and the main question is whether the security parameter in the GGH scheme can be made sufficiently large to achieve security, and maintaining the scheme practical at the same time.

4 An Optimal GGH-Like Trapdoor Function

We first describe a general scheme to define GGH-like trapdoor functions, of which GGH is a special case. Then we show how to instantiate the scheme in an essentially optimal way, defining a specific trapdoor function that is both more secure and efficient than any other function from the scheme. In particular, the new trapdoor function is considerably more efficient than the original GGH.

Fix a probability distribution on the set of private bases $\boldsymbol{R}$ and let ρ be a correction radius such that using $\boldsymbol{R}$ one can correct any error of length less than ρ. (e.g. $\rho = \frac{1}{2}\min_i \|\boldsymbol{r}_i^*\|$.) We define a family of functions $f_{\beta,\gamma}$ parametrized by two algorithms β and γ as follows.

1. Let β be a (possibly randomized) function that on input a matrix $\boldsymbol{R}$ outputs another basis $\boldsymbol{B}$ for the same lattice that will be used as a public key.
2. γ is a (possibly randomized) function that on input the public basis $\boldsymbol{B}$ and an error vector $\boldsymbol{r}$, outputs the coefficients $\boldsymbol{x}$ of a lattice point $\boldsymbol{Bx}$ to be added to the error vector.
3. Let $f_{\beta,\gamma}$ be the (possibly randomized) function with domain the set of vectors shorter than ρ defined as follows:

$$f_{\beta,\gamma}(\boldsymbol{r}) = \boldsymbol{Bx} + \boldsymbol{r}$$

where $\boldsymbol{B} = \beta(\boldsymbol{R})$ and $\boldsymbol{x} = \gamma(\boldsymbol{B}, \boldsymbol{r})$.

Notice that if the input $\boldsymbol{r}$ has length less then ρ, then the basis $\boldsymbol{R}$ can be used to find the lattice point $\boldsymbol{Bx}$ closest to $f_{\beta,\gamma}(\boldsymbol{r})$ and recover $\boldsymbol{r}$.

Therefore, for any fixed β and γ, the probability distribution on $\boldsymbol{R}$ defines a family of trapdoor functions $f_{\beta,\gamma}$ with public key $(\boldsymbol{B}, \rho)$ and trapdoor information $\boldsymbol{R}$. The GGH trapdoor function can be defined as a special case of this scheme when $\beta(\boldsymbol{R})$ applies a sequence of elementary random operations to $\boldsymbol{R}$, and $\gamma(\boldsymbol{B}, \boldsymbol{r})$ outputs an integer vector $\boldsymbol{x}$ chosen at random from a sufficiently large region of space.

We are now ready to define different β and γ, that greatly increase both the security and the efficiency of the scheme. The idea is to replace the random choice of the public basis $\boldsymbol{B}$ and vector $\boldsymbol{x}$ with deterministic choices that can be formally proved to be optimal from the security point of view. In the following subsections we first give some definitions that will be useful in the sequel, then show how to compute the public basis, and finally define the new trapdoor function.

4.1 Reducing Vectors Modulo a Basis

Every lattice $L = \mathcal{L}(\boldsymbol{B})$ induces an equivalence relation over $\mathbb{Z}^n$ defined as follows: $\boldsymbol{v} \equiv_L \boldsymbol{w}$ if and only if $\boldsymbol{v} - \boldsymbol{w} \in L$. It is easy to see that for every point $\boldsymbol{v} \in \mathbb{Z}^n$, there exists a unique point $\boldsymbol{w}$ in the *orthogonalized parallelepiped* $\mathcal{P}(\boldsymbol{B}^*) = \{\sum_i x_i \boldsymbol{b}_i^* \mid 0 \leq x_i < 1\}$ such that $\boldsymbol{w} \equiv_L \boldsymbol{v}$. Vector $\boldsymbol{w}$ can be easily computed from $\boldsymbol{v}$ and $\boldsymbol{B}$ as follows. For all $i = n, \ldots, 1$ (in decreasing order), let $\alpha_i = \frac{\langle \boldsymbol{v}, \boldsymbol{b}_i^* \rangle}{\langle \boldsymbol{b}_i^*, \boldsymbol{b}_i^* \rangle}$ the component of $\boldsymbol{v}$ along $\boldsymbol{b}_i^*$ and subtract $\lfloor \alpha \rfloor \boldsymbol{b}_i$ from $\boldsymbol{v}$. Let $\boldsymbol{w}$ the final result. Since $\boldsymbol{w}$ and $\boldsymbol{v}$ differ only by integer multiples of basis vectors, we have $\boldsymbol{w} \equiv_L \boldsymbol{v}$. Moreover, it is easy to check that $\frac{\langle \boldsymbol{w}, \boldsymbol{b}_i^* \rangle}{\langle \boldsymbol{b}_i^*, \boldsymbol{b}_i^* \rangle} \in [0, 1)$ for all i, and therefore $\boldsymbol{w} \in \mathcal{P}(\boldsymbol{B}^*)$. The unique element of $\mathcal{P}(\boldsymbol{B}^*)$ congruent to $\boldsymbol{v}$ modulo L is denoted $\boldsymbol{v} \bmod \boldsymbol{B}$. Notice that although the equivalence relation $\boldsymbol{v} \equiv_L \boldsymbol{w}$ does not depend on the particular choice of the basis $\boldsymbol{B}$ for the lattice $L = \mathcal{L}(\boldsymbol{B})$, the definition of the reduced vector ($\boldsymbol{v} \bmod \boldsymbol{B}$) is basis dependent.

Pictorially, we can think the vector space $\mathbb{R}^n$ as partitioned into parallelepipeds $\{\mathcal{P}(\boldsymbol{B}^*) + \boldsymbol{z} \mid \boldsymbol{z} \in \mathcal{L}(\boldsymbol{B})\}$. Then, the reduced vector $\boldsymbol{v} \bmod \boldsymbol{B}$ is the relative position of $\boldsymbol{v}$ in the parallelepiped $\mathcal{P}(\boldsymbol{B}^*) + \boldsymbol{z}$ it belongs to. Notice that if $\boldsymbol{v}$ is an integer vector, then also $\boldsymbol{v} \bmod \boldsymbol{B}$ is an integer vector. Therefore the function $\boldsymbol{x} \mapsto (\boldsymbol{x} \bmod \boldsymbol{B})$ defines a function from $\mathbb{Z}^n$ to $\mathcal{P}(\boldsymbol{B}^*) \cap \mathbb{Z}^n$.

If $\boldsymbol{B}$ is in Hermite Normal Form, then $\boldsymbol{w} = \boldsymbol{v} \bmod \boldsymbol{B}$ is an integer vector satisfying $0 \leq w_i < b_{i,i}$. In particular $\boldsymbol{w}$ can be represented using roughly $\sum_{i=1}^{n} \lg b_{i,i} = \lg(\det(L))$ bits. This representation is essentially optimal because $\equiv_L$ induces exactly $\det(L)$ congruence classes on $\mathbb{Z}^n$.

4.2 Choosing the Public Basis

Let's consider the choice of the public basis. The private basis $\boldsymbol{R}$ we start from is an exceptionally good basis that allows to solve the closest vector problem in the lattice $\mathcal{L}(\boldsymbol{R})$ and consequently decrypt messages. We would like to transform it into another basis for the same lattice L that gives the least possible amount of information about $\boldsymbol{R}$. Instead of computing $\boldsymbol{B}$ by applying a complex random transformation to $\boldsymbol{R}$, we set the public key of the new cryptosystem to be the Hermite Normal Form $\boldsymbol{B} = \mathrm{HNF}(\boldsymbol{R})$ of $\boldsymbol{R}$. Since the $\mathrm{HNF}(\boldsymbol{R})$ depends solely on the lattice $\mathcal{L}(\boldsymbol{R})$ generated by $\boldsymbol{R}$ (and not on the the particular basis $\boldsymbol{R}$ we used to compute it), the new public key gives *no* information about the private key $\boldsymbol{R}$, other then the lattice L it generates. More formally, one can prove that any information about $\boldsymbol{R}$ that can be efficiently computed from $\boldsymbol{B} = \mathrm{HNF}(\boldsymbol{R})$ can also be efficiently computed starting from any other (possibly random) basis $\boldsymbol{B}'$. This is because if $\boldsymbol{B}'$ and $\boldsymbol{R}$ generate the same lattice $L = \mathcal{L}(\boldsymbol{B}') = \mathcal{L}(\boldsymbol{R})$ then $\boldsymbol{B} = \mathrm{HNF}(\boldsymbol{B}') = \mathrm{HNF}(\boldsymbol{R})$ and $\boldsymbol{B}$ can be efficiently computed from $\boldsymbol{B}'$.

4.3 Adding a "Random" Lattice Point

Let's now look at how to simulate the addition of a "random" lattice vector $\boldsymbol{Bx}$ to the error vector $\boldsymbol{r}$. Ideally, we would like $\boldsymbol{Bx}$ to be a uniformly chosen vector from L. Unfortunately this is neither a computationally feasible nor a mathematically well defined operation. However, we notice that we can achieve exactly the same result by mapping the error vector $\boldsymbol{r}$ to its equivalence class $[\boldsymbol{r}]_L$ modulo $\equiv_L$. An efficient way to do this is to use the reduced vector $\boldsymbol{r} \bmod \boldsymbol{B}$ as a representative for this class. So, instead of adding to $\boldsymbol{r}$ a random lattice point $\boldsymbol{Bx}$, we reduce $\boldsymbol{r}$ modulo the public basis $\boldsymbol{B}$ to obtain the ciphertext $\boldsymbol{c} \in \mathcal{P}(\boldsymbol{B}^*)$.

The new trapdoor function is then defined as follows:

$$f(\boldsymbol{r}) = \boldsymbol{r} \bmod \boldsymbol{B}$$

where $\boldsymbol{B} = \mathrm{HNF}(\boldsymbol{R})$ is the Hermite Normal Form of the trapdoor information $\boldsymbol{R}$. The triangular form of $\boldsymbol{B}$ also makes the trapdoor function (i.e., the reduction modulo $\boldsymbol{B}$) extremely simple. Given $\boldsymbol{r}$, the reduced vector $\boldsymbol{r} \bmod \boldsymbol{B}$ can be easily determined as follow. Compute the integer vector $\boldsymbol{x}$ one coordinate at a time (starting from x_n) using the formula

$$x_i = \left\lfloor \frac{r_i - \sum_{j>i} b_{i,j} x_j}{b_{i,i}} \right\rfloor .$$

The output of the trapdoor function is $\boldsymbol{c} = \boldsymbol{r} - \boldsymbol{B}\boldsymbol{x} \equiv \boldsymbol{r} \bmod \boldsymbol{B}$. The reader can easily check that for every i, $0 \leq c_i < b_{i,i}$, i.e., the result is the unique point in the parallelepiped $\mathcal{P}(\boldsymbol{B}^*) = \{\boldsymbol{w} \mid 0 \leq w_i < b_{i,i}\}$ which is congruent to $\boldsymbol{r}$ modulo $\mathcal{L}(B)$.

4.4 The New Trapdoor Function

We now put all pieces together and define the new trapdoor function. Let $\boldsymbol{R}$ be a private basis chosen in such a way that $\rho = \frac{1}{2}\min_i \|\boldsymbol{r}_i^*\|$ is relatively big (see Fig. 1). The public basis $\boldsymbol{B}$ is the Hermite Normal Form of $\boldsymbol{R}$ (see Fig. 2). One can see that the public basis $\boldsymbol{B}$ and the corresponding orthogonalized parallelepiped $\mathcal{P}(\boldsymbol{B}^*)$ are very skewed. The public basis $\boldsymbol{B}$ defines a function with domain the set of vectors of length at most ρ (the shaded circle in Fig. 2). The result of applying the function to vector $\boldsymbol{r}$ is the point $(\boldsymbol{r} \bmod \boldsymbol{B})$ in the parallelepiped $\mathcal{P}(\boldsymbol{B}^*)$ congruent to $\boldsymbol{r}$ modulo the lattice. Notice that even if we always start from a vector $\boldsymbol{r}$ close to the origin, the result of performing the reduction operation is a point of $\mathcal{P}(\boldsymbol{B}^*)$ possibly closest to some other lattice point (see black regions in Fig. 2 and the corresponding closest lattice points). Notice that recovering the input vector $\boldsymbol{r}$ from $f(\boldsymbol{r})$ involves finding the lattice point closest to $f(\boldsymbol{r})$, which is conjectured to be infeasible using only the public basis $\boldsymbol{B}$. However the lattice vector closest to $\boldsymbol{r} \bmod \boldsymbol{B}$ can be computed using the private basis $\boldsymbol{R}$ as discussed in Section 2 because $\operatorname{dist}(f(\boldsymbol{r}), L) = \operatorname{dist}(\boldsymbol{r}, L) \leq \rho$. Fig. 3 shows the orthogonalized parallelepipeds $\mathcal{P}(\boldsymbol{R}^*)$ centered at every lattice point. Notice that the lattice point closest to $f(\boldsymbol{r})$ (i.e., any point of the black regions in the picture) is just the center of the parallelepiped $\mathcal{P}(\boldsymbol{R}^*)$ containing $f(\boldsymbol{r})$, which can be found using the private basis $\boldsymbol{R}$.

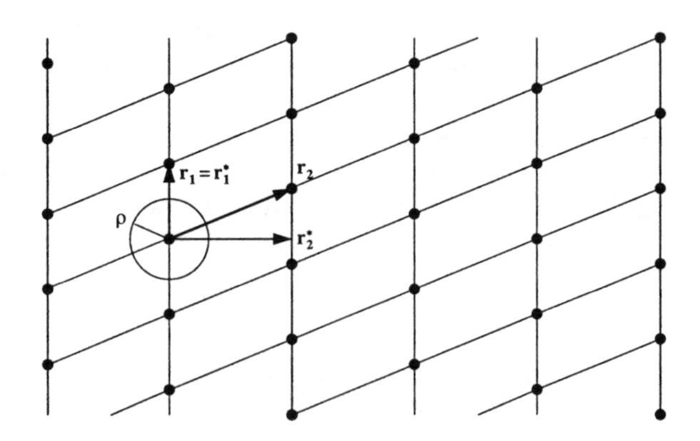

Fig. 1. A good lattice basis and the corresponding correction radius.

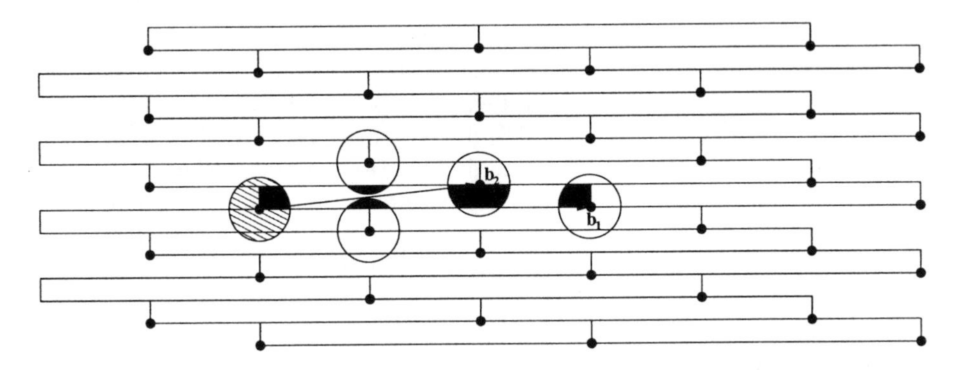

Fig. 2. HNF basis and corresponding orthogonalized parallelepiped.

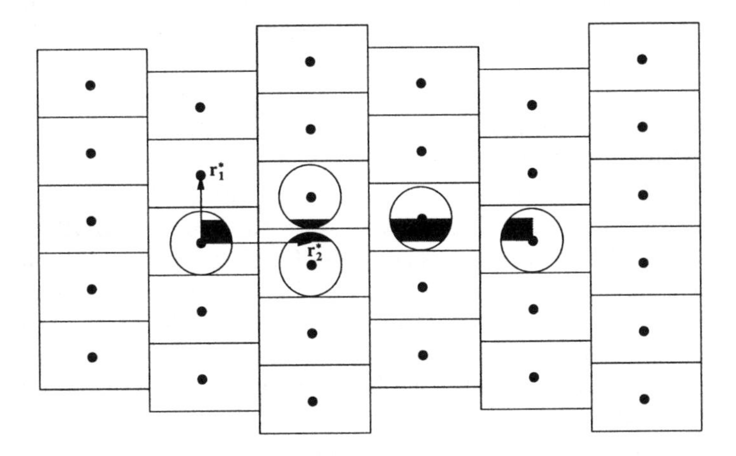

Fig. 3. Correcting small errors using the private basis.

5 Analysis

In this section we discuss the security and performance of the new scheme.

5.1 Security

We want to prove that the new trapdoor function $f(\boldsymbol{r})$ is at least as secure as the original GGH function. We actually prove that $f(\boldsymbol{r})$ is at least as secure as *any* GGH-like function $f_{\beta,\gamma}$ as defined in the previous section.

Theorem 1. *For any (efficiently computable) functions β, γ, and for any (efficient) algorithm that on input $f(\boldsymbol{r})$ finds some partial information about $\boldsymbol{r}$ with non-trivial probability, there exists an efficient algorithm that on input $f_{\beta,\gamma}(\boldsymbol{r})$ finds the same partial information with the same success probability.*

Proof. The proof is a simple reduction argument. Assume A is an algorithm that breaks f. We show how to attack $f_{\beta,\gamma}$ using A as a subroutine. We are given public basis $\boldsymbol{B} = \beta(\boldsymbol{R})$ and a ciphertext $\boldsymbol{c} = \boldsymbol{B} \cdot \gamma + \boldsymbol{r}$. The task is to find (some partial information about) $\boldsymbol{r}$. We first compute $\boldsymbol{B}' = \text{HNF}(\boldsymbol{B})$ and $\boldsymbol{c}' = \boldsymbol{c} \bmod \boldsymbol{B}'$. Notice that $\boldsymbol{B}' = \text{HNF}(\boldsymbol{R})$ and $\boldsymbol{c}' = (\boldsymbol{B}\gamma + \boldsymbol{r}) \bmod \boldsymbol{B}' = \boldsymbol{r} \bmod \boldsymbol{B}'$. Therefore $\boldsymbol{B}'$ and $\boldsymbol{c}'$ have the right distribution for algorithm A. Running A on $\boldsymbol{B}'$ and $\boldsymbol{c}'$ we will recover (the partial information about) $\boldsymbol{r}$ with the same success probability as A. □

5.2 Space Efficiency

We now analyze the size of the keys and the ciphertext of the new encryption algorithm. We assume that the private key satisfies $|r_{i,j}| < poly(n)$. Therefore the size of the private key can be bounded by $O(n^2 \lg n)$. Using the Hadamard inequality we can also bound the size of the determinant by $O(n \lg n)$, and using the bounds proved in Section 3, we get that also the public basis is $O(n^2 \lg n)$ and the ciphertext has size $O(n \lg n)$.

Estimates of the key and ciphertext sizes for the GGH and the modified scheme are shown in Fig. 1. The estimates are based on the GGH challenges published at [13]. Notice that the size of the GGH challenges is much smaller than it should have been to assure adequate randomization. This discrepancy might be explained noticing the the authors of GGH applied the LLL reduction algorithm to the public basis to somehow reduce their size. Nevertheless, the new scheme results in keys and ciphertexts more than an order of magnitude smaller than GGH. We remark that the sizes relative to the modified scheme are only upper bounds obtained using the Hadamard inequality to estimate the determinant of the lattice, and the actual sizes of the keys and ciphertexts of the modified cryptosystem can be even smaller than shown in the table. On the other hand, public-key and ciphertext size might be bigger than the estimate shown in the table if the secret key is generated differently than the GGH challenges.

Table 1. Comparison of the key and ciphertext sizes in the GGH scheme and the modified scheme. All sizes are in kilobytes (KB).

	Basis Size		Ciphertext	
dimension	GGH	New scheme	GGH	New scheme
200	330	32	2	0.16
250	620	50	3	0.20
300	990	75	4	0.25
350	1630	100	5	0.30
400	2370	140	6	0.35

5.3 Running Time

The experiments described in the next section have been performed using a highly unoptimized prototype implementation, so we do not have meaningful experimental data on the running time of the new lattice scheme. However a few remarks regarding the running time are due.

Encryption time for GGH-like schemes is roughly proportional to the size of the public key. So, if the original GGH cryptosystem was competitive with RSA (see [14]), we should expect the new scheme to outperform RSA in terms of encryption speed because of the reduced key size.

Key Generation was one of the major problems in GGH, requiring the application of LLL on a high dimensional matrix with very large entries. Here, key generation essentially consists of a Hermite Normal Form computation, a much simpler task than lattice reduction. Moreover, recent progress on the design of HNF algorithms [23] might lead to efficient key generation procedures.

The most critical part of lattice based encryption schemes at this point is probably *decryption*. In this paper, and in our experiments we have used Babai's nearest plane algorithm [4], as we believe this is a quite natural choice. Although polynomial time, this algorithm is very slow compared to the linear time encryption procedure, and decrypting lattices in dimension 400 (using the private basis) can take several minutes with a straightforward implementation. In fact, [14] suggested to use the simpler (but less accurate) rounding off algorithm (also from [4]) instead of nearest plane. It should be noted that the nearest plane algorithm can be made considerably faster if the orthogonalized basis needed by the nearest plane algorithm is precomputed and stored as part of the secret key. Of course, this would increase the size of the secret key, but a relatively poor approximation of the orthogonalized basis should be enough to achieve reasonable correction radius. Decryption methods based on probabilistic rounding procedures as described in [18] are also an interesting alternative to be explored, and much work is still to be done. However, since the decryption procedure is strongly related to the choice of the secret basis, it is probably not worth focusing on the optimization of the decryption algorithm until we get more confidence on what's a good way to generate the private basis for the lattice. (See Section 6 for further discussion about the choice of the private key.)

6 Discussion and Experiments

We defined a trapdoor function f that is at least as hard to break as the GGH "encryption" scheme. Notice that although the original GGH function was a randomized one, the new function is deterministic. Since any semantically secure encryption scheme must be probabilistic [16], the new function f cannot be a secure encryption scheme in the sense of [16]. It is now clear that also GGH (which is less secure than our new function) cannot be a semantically secure encryption scheme, although it is randomized. The situation is similar to other popular "encryption" functions, like RSA: also RSA is a deterministic function, and therefore cannot be a semantically secure encryption scheme if directly applied to the message. In fact, encrypting with RSA usually involves padding the message with some random bits, or applying some other randomized procedure.

In fact, standard techniques can be used to turn trapdoor functions into semantically secure encryption schemes. In the seminal paper [16] Goldwasser and Micali showed that if h is a hard-core predicate for a trapdoor function f, then $E(b,r) = (f(r), h(r) \oplus b)$ is a semantically secure encryption function for one bit messages $b \in \{0,1\}$. Hard-core predicates from any one-way (or trapdoor) function can be easily constructed following [37, 15], giving a first theoretical construction of a semantically secure cryptosystem. The construction is easily generalized to hard-core functions. Namely, if h is a hard-core function for f, then $E(m,r) = (f(r), h(r) \oplus m)$ is a semantically secure encryption scheme for messages of length equal to the output size of h. Practical instantiations of this scheme can be obtained in the random oracle model [5], simply observing that random oracles are hard-core functions for any one-way function. See [5] also for simple constructions (still in the random oracle model) achieving security against chosen ciphertext attacks [27]. The constructions in [5] have also been recently extended in [26, 11, 12] to allow for probabilistic trapdoor functions, but these extensions are not required because our trapdoor function is deterministic.

A problem that needs further investigation is how to choose the private basis $\boldsymbol{R}$. In the GGH cryptosystem $\boldsymbol{R}$ was chosen as the sum of a multiple of the identity matrix $\sqrt{n}\boldsymbol{I}$ plus a perturbation matrix $\boldsymbol{Q}$ with small entries $q_{i,j} \in \{-4, \ldots, +4\}$. It is not clear why one should prefer this probability distribution to other distributions, and in fact we think that disclosing the approximate direction of the vectors $\boldsymbol{b}_i$ might actually weaken the system. At this point of our research, we believe it is better to leave the set of matrices from which $\boldsymbol{R}$ is chosen as big as possible. A possible way to choose $\boldsymbol{R}$ is[2] to choose each entry at random in the interval $\{-n, \ldots, n\}$. It turns out that random matrices are pretty good on average [8] and running the LLL algorithm [19] on them rapidly yields matrices $\boldsymbol{R}$ with relatively large correction radius $\rho = \min_i \|\boldsymbol{r}_i^*\| = O(n)$. On the other hand, if one applies the LLL basis reduction algorithm to the public basis $\boldsymbol{B} = \mathrm{HNF}(\boldsymbol{R})$, although this time LLL takes a much longer time to terminate, the correction radius obtained is much smaller even in relatively low dimension. In Fig. 4 we show some preliminary experimental results obtained running the LLL algorithm on random matrices, and their Hermite normal forms. We observe that random matrices give a correction radius approximately equal to $n/2$, while running the LLL algorithm on the Hermite Normal Forms of the same matrices results in a correction radius that approaches zero as the dimension of the lattice grows.

These preliminary data clearly show that applying the HNF algorithm reduces the effectiveness of lattice reduction. Still, the plot in Fig. 4 is not clear evidence of the increased security of the scheme for various reasons: first of all better result can be achieved using more sophisticated basis reduction algorithms

[2] Similar distributions on $\boldsymbol{R}$ were already considered in [14], thought not used in the construction of the challenges [13].

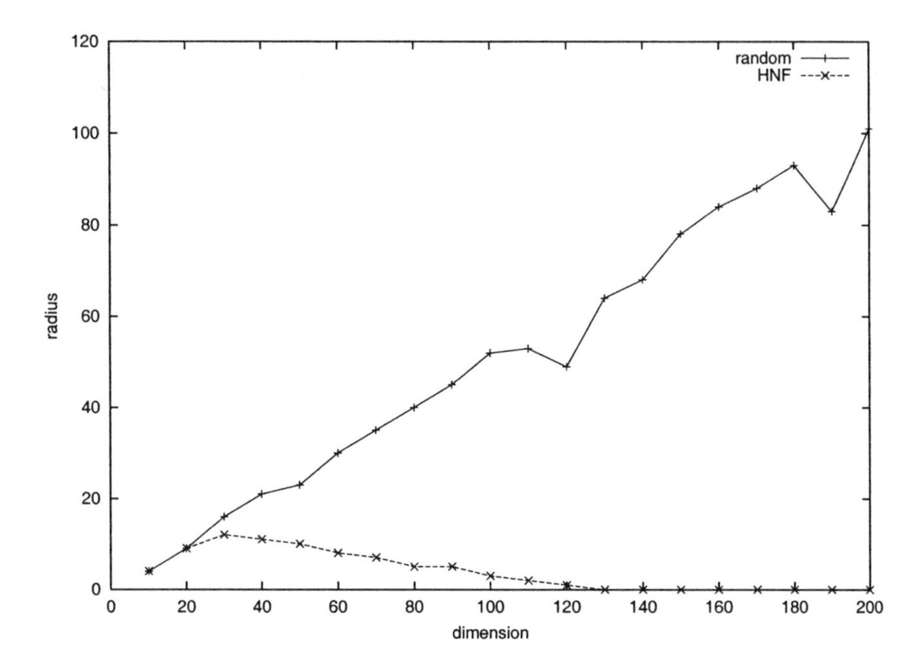

Fig. 4. Correction radius obtained applying the LLL algorithm to random matrices and their Hermite Normal Forms.

(e.g., [29, 30, 32]) than LLL. Moreover, the correction radius is only a worst case measure of the quality of a basis. It is still possible that the HNF basis allows to recover from most small errors.

To support our claim that the modified scheme is secure against the strongest lattice attacks we performed the following experiment. We generated a private basis $\boldsymbol{R}$ in dimension 400×400 by choosing each entry at random in the interval $\{-400, \ldots, +400\}$. After running the fast LLL reduction algorithm, the correction radius of the random matrix was $\rho = 340$. The public basis $\boldsymbol{B} = \text{HNF}(R)$ had size 260 KB. Notice that although the size of $\boldsymbol{B}$ is even smaller than the first GGH challenge (which was about 330 KB), our theoretical analysis suggests that the new function should still be secure because the underlying lattice has much higher dimension.

We then generated random error messages $\boldsymbol{r}$ choosing each entry at random in the interval $\{-28, \ldots, +28\}$ and computed the ciphertexts $\boldsymbol{c} = \boldsymbol{r} \bmod \boldsymbol{B}$. The size of $\boldsymbol{c}$ was around 800 B. The error vector had length 320 (i.e. less than the correction radius) and could be correctly recovered from $\boldsymbol{c}$ using $\boldsymbol{R}$. We then tried to recover $\boldsymbol{r}$ using the public basis $\boldsymbol{B}$.

Following [24], we first applied a strong basis reduction algorithm (Block Korkine-Zolotarev reduction [29] with block size 20, as implemented in Victor Shoup's Number Theory Library [33]) to the public basis $\boldsymbol{B}$ to obtain a reduced basis $\boldsymbol{G}$. The computation took over 10 days on a 700 MHz Pentium III workstation. Still the correction radius was only 6.77. Finally, we applied the "embedding technique" (see [24]) to recover $\boldsymbol{r}$ from $\boldsymbol{c}$ and $\boldsymbol{G}$ using the Block KZ reduction algorithm with block size 60 and pruning factor 14 [32]. The clear-text could not be recovered, and even after several days of computation the best lattice vector found by the attack was 500,000 away from the target, over three orders of magnitude worse than the optimal solution $\|\boldsymbol{r}\| = 320$.

Previous cryptanalytic results [24] warn to be cautious about the security of the cryptosystem, and dimension $n = 400$ should be considered only borderline secure. In fact we suggest the dimension should be at least $n \geq 500$, but only after careful cryptanalysis the minimum dimension for which the system is secure in practice can be determined.

7 Other Cryptosystems

In this section we discuss the McEliece and NTRU cryptosystems, and compare them to the general scheme presented in this paper.

7.1 Comparison with the McEliece Cryptosystem

In 1978, McEliece [21] suggested a "cryptosystem" based on the hardness of coding problems that in retrospect is very similar to the GGH cryptosystem. The proposal is to use as a secret key the generating matrix $\boldsymbol{G}$ of a Goppa code, together with a random permutation matrix $\boldsymbol{P}$ and a random invertible matrix $\boldsymbol{S}$ (over GF(2)). The public key is given by the product $\boldsymbol{G}' = \boldsymbol{PGS}$. Then the trapdoor function is defined by $f(\boldsymbol{x},\boldsymbol{r}) = \boldsymbol{G}'\boldsymbol{x} + \boldsymbol{r}$ (all arithmetic performed over GF(2)), where $\boldsymbol{x}$ is a random binary vector, and $\boldsymbol{r}$ is chosen at random among the binary vectors with small Hamming weight. Let $\boldsymbol{c} = \boldsymbol{G}'\boldsymbol{x} + \boldsymbol{r}$ be the output of the trapdoor function. Using the secret key, we can first compute the permuted target $\boldsymbol{P}^{-1}\boldsymbol{c} = \boldsymbol{GS} + \boldsymbol{P}^{-1}\boldsymbol{r}$. Then we can correct from the small error $\boldsymbol{P}^{-1}\boldsymbol{r}$ using the decoding algorithm for Goppa codes, retrieving codeword $\boldsymbol{G}(\boldsymbol{Sx})$. Finally, we can compute $\boldsymbol{x}$ from $\boldsymbol{Sx}$ solving a system of linear equations over GF(2). Notice that this is essentially the same of the GGH cryptosystem with the message m encoded in the coefficients of the lattice vector. A variant of McEliece cryptosystem roughly corresponding to encoding the message in the error vector has also been proposed [25] and the two are known to be equivalent [20]. Interestingly, essentially the same techniques presented in this paper for lattices, can also be used for codes (e.g., they can be applied to [21, 25] and many of their variants). Here instead of the Hermite normal form, we use systematic form for the public code $\boldsymbol{G}'$. If $[\boldsymbol{I}|\boldsymbol{H}]^T$ is the systematic generating matrix, then the trapdoor function is given by $f(\boldsymbol{x},\boldsymbol{y}) = \boldsymbol{y} - \boldsymbol{H}^T\boldsymbol{x}$, where $\boldsymbol{x}$ and $\boldsymbol{y}$ have total weight less than the correction radius of the code. However, the advantage of this transformation for codes is not as good as in the lattice case, giving only a constant (typically a factor 2) improvement over the original McEliece scheme.

A detailed comparison of lattice and coding based cryptosystems is beyond the scope of this paper. However, an apparent advantage of coding based schemes is a potentially smaller public key: since the matrix defining the code has $\{0, 1\}$ entries, the size of the public key is only $O(n^2)$, as opposed to $O(n^2 \log n)$. Of course, this kind of comparisons is not necessarily meaningful if we first do not achieve a better understanding of the relation between the hardness of lattice and coding problems. For example, if decoding binary linear codes in dimension n is easier than breaking lattices in the same dimension, then for a fair comparison of the two schemes one should use different values of the security parameter. In fact, the original McEliece scheme [21] was already proposing codes with block-length $n = 1024$, and recent cryptanalytic work [7] shows that even this large dimension might be insufficient. Interestingly, the nearest codeword problem for binary (or ternary) codes can be efficiently reduced to the closest vector problem over the integers: in order to find the codeword $\boldsymbol{Cx}$ closest to $\boldsymbol{y}$, look for a lattice vector in $[\boldsymbol{C}|2\boldsymbol{I}]$ closest to $\boldsymbol{y}$. This suggests that lattice problems might be harder than coding problems, at least for the binary case. For larger alphabets, the relation between codes and lattices is less clear. However, if large alphabets are used, then the public key size for code based cryptosystems would increase. For example, if Reed-Solomon codes were used, then the alphabet size would be n, giving keys of total size $O(n^2 \log n)$, matching the asymptotic key size of lattice based cryptosystems. (Here Reed-Solomon codes are just an hypothetical example, as these codes are known not to be secure [34].) We hope that our work will stimulate further investigations on the relation between lattice and coding problems.

7.2 The NTRU Cryptosystem

NTRU is a cryptosystem based on polynomial ring arithmetic proposed by Hoffstein, Pipher and Silverman in [17]. The system works essentially as follows. Let n, p, q be system parameters where n is a security parameter (say $n = 200$), p is a small prime (say, $p = 3$) and q is a relatively large prime (typically $q = \Omega(n)$). The secret key is a pair of degree $n - 1$ polynomials $a(X), b(X) \in \mathbb{Z}[X]$ with small coefficients, such that $a(X)$ is invertible modulo $(X^n - 1, pq)$. The public key is given by $c(X) = p \cdot g(X)/f(X) \bmod (X^n - 1, q)$. The encryption function takes as input two polynomials $m(X), r(X) \in \mathbb{Z}[X]/(X^N - 1)$ with small coefficients (the first interpreted as a message, and the second as a randomizer), and outputs $c(X) \cdot r(X) + m(X) \bmod (X^n - 1, q)$. (For the decription procedure, as well as a more detailed description of the system, see the original article [17].) As for the GGH and the McEliece cryptosystem, NTRU does not provide semantic security, so it is better described as a deterministic trapdoor function, instead of a full fledged probabilistic cryptosystem. Interestingly, this function can be formulated in terms of lattices as follows. Consider the lattice generated by the $2n \times 2n$ matrix

$$\left[\begin{array}{c|c} c \cdot \boldsymbol{I} & \boldsymbol{C} \\ \hline \boldsymbol{0} & \boldsymbol{I} \end{array}\right]$$

where $\boldsymbol{C}$ is the $n \times n$ matrix whose rows are given by all cyclic permutations of the coefficients $[c_0, c_1, \ldots, c_{n-1}]$ of polynomial $c(X) = \sum_{i=0}^{n} c_i X^i$, i.e., $\boldsymbol{C}_{i,j} =$

$c_{(j-i \bmod n)}$. Notice that this public basis is in Hermite normal form. Moreover, it is easy to see that the output $c(X)r(X) + m(X)$ of the trapdoor function is exactly the result of reducing vector $[m_{n-1}, \ldots, m_0, -r_{n-1}, \ldots, -r_0]^T$ modulo the public lattice basis. So, when viewed as a lattice based trapdoor function, NTRU has the same high level structure of the functions described in this paper: the public key is an HNF lattice basis, and the function is computed reducing small "error" vector modulo the public basis. What sets NTRU apart from all other lattice based functions is the use of a class of lattices with special structure: convolutional modular lattices. While the security offered by this class of lattices is still largely to be investigated, the performance advantage is clear: as the HNF basis can be represented by only n ($\log n$)-bit numbers, the public key is much smaller than those obtained using general lattices which require ($n^2 \log n$) bits.

8 Conclusion

We presented a new trapdoor function based on the hardness of lattice problems. The trapdoor function can be transformed into a full fledged encryption algorithm using standard techniques [16, 5]. The new function can be formally proved to be at least as secure as any other function from a general scheme to design lattice based trapdoor functions that includes the GGH trapdoor function [14] and the tensor based trapdoor function [10] as special cases. Moreover, the new function substantially improves previous proposals from the efficiency point of view: for the same level of security the new function reduces both the time and space requirements by a factor $O(n)$. The improved efficiency allows to use bigger values of the security parameter while maintaining the scheme reasonably practical. One last advantage of the new scheme is simplicity. While previous schemes computed the public key and function values using a substantial amount of randomness, in the new function these operations are substituted by simple deterministic procedures. This is important both from the theoretical and practical point of view, because it makes the algorithms easier to implement and also easier to analyze.

At this point the main question about lattice based cryptography is how to choose the private key, i.e., finding families of easily decodable lattices for which decoding becomes infeasible when the lattice is presented in Hermite normal form. We believe that in order to be really competitive with RSA, key sizes even smaller than those achieved in this paper would be desirable. However, the public key size cannot be further reduced unless one considers classes of lattices with special structure. (A simple counting argument shows that the number of lattices in a certain dimension is exponential in the bit-size representation of their Hermite normal forms, so the HNF representation is essentially optimal if one considers arbitrary lattices.) The search for easily decodable lattices for which decoding is hard when the lattice is presented in Hermite normal form becomes particularly interesting if one could find special classes of hard lattices that have small HNF representation. We observed that one such family of lattices is given by the NTRU trapdoor function. Again, the main question is security. Are the convolutional modular lattices used by NTRU really hard to decode? We hope that our work will stimulate further research on the computational complexity of decoding this and other classes of lattices.

Acknowledgements

The author wishes to thank Shai Halevi, Shafi Goldwasser and Salil Vadhan for many stimulating conversations on lattice based cryptography. Thanks also to the anonymous referees for their useful comments.

References

1. M. Ajtai. Generating hard instances of lattice problems (extended abstract). In *Proceedings of the Twenty-Eighth Annual ACM Symposium on the Theory of Computing*, pages 99–108, Philadelphia, Pennsylvania, 22–24 May 1996.
2. M. Ajtai and C. Dwork. A public-key cryptosystem with worst-case/average-case equivalence. In *Proceedings of the Twenty-Ninth Annual ACM Symposium on Theory of Computing*, pages 284–293, El Paso, Texas, 4–6 May 1997.
3. S. Arora, L. Babai, J. Stern, and E.Z. Sweedyk. The hardness of approximate optima in lattices, codes, and systems of linear equations. *J. Comput. Syst. Sci.*, 54(2):317–331, Apr. 1997. Preliminary version in FOCS'93.
4. L. Babai. On Lovasz' lattice reduction and the nearest lattice point problem. *Combinatorica*, 6(1):1–13, 1986.
5. M. Bellare and P. Rogaway. Random oracles are practical: A paradigm for designing efficient protocols. In *Proceedings of the first ACM Conference on Computer and Communications Security*. ACM, Nov. 1993.
6. J.-Y. Cai and T.W. Cusick. A lattice-based public-key cryptosystem. *Information and Computation*, 151(1–2):17–31, May–June 1999.
7. A. Canteaut and N. Sendrier. Cryptanalysis of the original McEliece cryptosystem. In K. Ohta and D. Pei, editors, *Advances in Cryptology — Proceedings of Asiacrypt'98*, volume 1514 of *Lecture Notes in Computer Science*, pages 187–199, Beijing, China, 1998.
8. H. Daude and B. Vallée. An upper bound on the average number of iterations of the LLL algorithm. *Theoretical Computer Science*, 123(1):95–115, Jan. 1994.
9. I. Dinur, G. Kindler, and S. Safra. Approximating CVP to within almost-polynomial factors is NP-hard. In *39th Annual Symposium on Foundations of Computer Science*, Palo Alto, California, 7–10 Nov. 1998. IEEE.
10. R. Fischlin and J.-P. Seifert. Tensor-based trapdoors for CVP and their application to public key cryptography. In *7th IMA International Conference "Cryptography and Coding"*, volume 1746 of *Lecture Notes in Computer Science*, pages 244–257. Springer-Verlag, 1999.
11. E. Fujisaki and T. Okamoto. Secure integration of asymmetric and symmetric encryption schemes. In M. Wiener, editor, *Advances in Cryptology—CRYPTO'99*, volume 1666 of *Lecture Notes in Computer Science*, pages 537–554, University of California, Santa Barbara, Aug. 1999. IACR, Springer-Verlag.
12. E. Fujisaki and T. Okamoto. How to enhance the security of public-key encryption at minimum cost. *IEICE Transaction of Fundamentals of electronic Communications and Computer Science*, E38-A(1):24–32, Jan. 2000.

13. O. Goldreich, S. Goldwasser, and S. Halevi. The GGH cryptosystem, challenge page. `http://theory.lcs.mit.edu/~cis/lattice/challenge.html`.
14. O. Goldreich, S. Goldwasser, and S. Halevi. Public-key cryptosystems from lattice reduction problems. In B. S. Kaliski Jr., editor, *Advances in Cryptology—CRYPTO'97*, volume 1294 of *Lecture Notes in Computer Science*, pages 112–131. Springer-Verlag, 17–21 Aug. 1997.
15. O. Goldreich and L. Levin. A hard predicate for all one-way functions. In *Proceedings of the 21st Annual Symposium on Theory of Computing (STOC)*. ACM, 1989.
16. S. Goldwasser and S. Micali. Probabilistic encryption. *Journal of Computer and System Sience*, 28(2):270–299, 1984. Preliminary version in STOC'82.
17. J. Hoffstein, J. Pipher, and J.H. Silverman. NTRU: A ring based public key cryptosystem. In J. Buhler, editor, *Algorithmic Number Theory (ANTS III)*, volume 1423 of *Lecture Notes in Computer Science*, pages 267–288, Portland, OR, 1998. Springer.
18. P. Klein. Finding the closest lattice vector when it's unusually close. In *Proceedings of the 11th Symposium on Discrete Algorithms*, San Francisco, California, Jan. 2000. SIAM.
19. A.K. Lenstra, H.W. Lenstra, Jr., and L. Lovász. Factoring polynomials with rational coefficients. *Mathematische Annalen*, 261:513–534, 1982.
20. Y.X. Li, R.H. Deng, and X.M. Wang. On the equivalence of McEliece's and Niederreiter's public-key cryptosystems. *IEEE Transactions on Information Theory*, 40(1):271–273, Jan. 1994.
21. R.J. McEliece. A public-key cryptosystem based on algebraic coding theory. DSN Progress Report 42-44, Jet Propulsion Laboratory, Pasadena, 1978.
22. D. Micciancio. The hardness of the closest vector problem with preprocessing. *IEEE Transactions on Information Theory*, 2001. To Appear.
23. D. Micciancio and B. Warinschi. A linear space algorithm for computing the Hermite Normal Form. In B. Mourrain, editor, *International Symposium on Symbolic and Algebraic Computation*. ACM 2001. To Appear.
24. P. Nguyen. Cryptanalysis of the Goldreich-Goldwasser-Halevi cryptosystem from Crypto'97. In M. Wiener, editor, *Advances in Cryptology—CRYPTO'99*, volume 1666 of *Lecture Notes in Computer Science*. Springer-Verlag, Aug. 1999.
25. H. Niederreiter. Knapsack-type cryptosystems and algebraic coding theory. *Problems of Control and Information Theory*, 15(2):159–166, 1986.
26. T. Okamoto and D. Pointcheval. React: Rapid enhanced-security asymmetric cryptosystem transform. In D. Naccache, editor, *Proceedings of the Cryptographers' Track of the RSA Conference '2001 (RSA 2001)*, Lecture Notes in Computer Science, San Francisco, California, USA, 8–12Apr. 2001. Springer-Verlag.
27. C. Rackoff and D. R. Simon. Non-interactive zero-knowledge proof of knowledge and chosen ciphertext attack. In J. Feigenbaum, editor, *Advances in Cryptology: Proceedings of Crypto'91*, volume 576 of *Lecture Notes in Computer Science*, University of California, Santa Barbara, Aug. 1991. IACR, Springer-Verlag.
28. R. L. Rivest, A. Shamir, and L. Adleman. A method for obtaining digital signatures and public-key cryptosystems. *Communications of the ACM*, 21:120–126, 1978.
29. C.-P. Schnorr. A hierarchy of polynomial time lattice basis reduction algorithms. *Theoretical Computer Science*, 53(2–3):201–224, 1987.
30. C.-P. Schnorr and M. Euchner. Lattice basis reduction: Improved practical algorithms and solving subset sum problems. In L. Budach, editor, *Proceedings of Fundamentals of Computation Theory*, volume 529 of *LNCS*, pages 68–85. Springer-Verlag, 1991.

31. C.-P. Schnorr, M. Fischlin, H. Koy, and A. May. Lattice attacks on GGH cryptosystem. Rump session of Crypto'97, 1997.
32. C.-P. Schnorr and H. H. Hörner. Attacking the Chor–Rivest cryptosystem by improved lattice reduction. In L. C. Guillou and J.-J. Quisquater, editors, *Advances in Cryptology—EUROCRYPT'95*, volume 921 of *Lecture Notes in Computer Science*, pages 1–12. Springer-Verlag, 21–25 May 1995.
33. V. Shoup. NTL: A library for doing number theory. Available on-line at URL `http://www.shoup.net/ntl/index.html`.
34. V. Sidelnikov and S. Shestakov. On cryptosystems based on generalized Reed-Solomon codes. *Diskretnaya Math*, 4(3):57–63, 1992. In Russian.
35. N. J. A. Sloane. Encryption by random rotations. In *Workshop on Cryptography Burg Feuerstein 1982*, volume 149 of *Lecture Notes in Computer Science*, pages 71–129, 1983.
36. P. van Emde Boas. Another NP-complete problem and the complexity of computing short vectors in a lattice. Technical Report 81-04, Mathematische Instituut, Universiry of Amsterdam, 1981. Available on-line at URL `http://turing.wins.uva.nl/~peter/`.
37. A. Yao. Theory and applications of trapdoor functions. In *Proceedings of the 23rd IEEE Symposium on Foundations of Computer Science (FOCS)*, pages 80–91, Chicago, IL, 1982. IEEE.

The Two Faces of Lattices in Cryptology

Phong Q. Nguyen and Jacques Stern

École Normale Supérieure, Département d'Informatique
45 rue d'Ulm, 75005 Paris, France
pnguyen@ens.fr
http://www.di.ens.fr/~pnguyen/
stern@di.ens.fr
http://www.di.ens.fr/~stern/

Abstract. Lattices are regular arrangements of points in n-dimensional space, whose study appeared in the 19th century in both number theory and crystallography. Since the appearance of the celebrated Lenstra-Lenstra-Lovász lattice basis reduction algorithm twenty years ago, lattices have had surprising applications in cryptology. Until recently, the applications of lattices to cryptology were only negative, as lattices were used to break various cryptographic schemes. Paradoxically, several positive cryptographic applications of lattices have emerged in the past five years: there now exist public-key cryptosystems based on the hardness of lattice problems, and lattices play a crucial rôle in a few security proofs. We survey the main examples of the two faces of lattices in cryptology.

1 Introduction

Lattices are discrete subgroups of $\mathbb{R}^n$. A lattice has infinitely many $\mathbb{Z}$-bases, but some are more useful than others. The goal of *lattice reduction* is to find interesting lattice bases, such as bases consisting of reasonably short and almost orthogonal vectors. From the mathematical point of view, the history of lattice reduction goes back to the reduction theory of quadratic forms developed by Lagrange [86], Gauss [55], Hermite [68], Korkine and Zolotarev [82,83], among others, and to Minkowski's geometry of numbers [103]. With the advent of algorithmic number theory, the subject had a revival in 1981 with Lenstra's celebrated work on integer programming (see [89,90]), which was, among others, based on a novel lattice reduction technique (which can be found in the preliminary version [89] of [90]). Lenstra's reduction technique was only polynomial-time for fixed dimension, which was however enough in [89]. That inspired Lovász to develop a polynomial-time variant of the algorithm, which computes a so-called *reduced* basis of a lattice. The algorithm reached a final form in the seminal paper [88] where Lenstra, Lenstra and Lovász applied it to factor rational polynomials in polynomial time (back then, a famous problem), from which the name LLL comes. Further refinements of the LLL algorithm were later proposed, notably by Schnorr [121,122].

Those algorithms have proved invaluable in many areas of mathematics and computer science (see [91,78,132,64,36,84]). In particular, their relevance to

J.H. Silverman (Ed.): CaLC 2001, LNCS 2146, pp. 146–180, 2001.

cryptology was immediately understood, and they were used to break schemes based on the knapsack problem (see [119,29]), which were early alternatives to the RSA cryptosystem [120]. The success of reduction algorithms at breaking various cryptographic schemes over the past twenty years (see [75]) have arguably established lattice reduction techniques as the most popular tool in public-key cryptanalysis. As a matter of fact, applications of lattices to cryptology have been mainly negative. Interestingly, it was noticed in many cryptanalytic experiments that LLL, as well as other lattice reduction algorithms, behave much more nicely than what was expected from the worst-case proved bounds. This led to a common belief among cryptographers, that lattice reduction is an easy problem, at least in practice.

That belief has recently been challenged by some exciting progress on the complexity of lattice problems, which originated in large part in two seminal papers written by Ajtai in 1996 and in 1997 respectively. Prior to 1996, little was known on the complexity of lattice problems. In his 1996 paper [3], Ajtai discovered a fascinating connection between the worst-case complexity and the average-case complexity of some well-known lattice problems. Such a connection is not known to hold for any other problem in NP believed to be outside P. In his 1997 paper [4], building on previous work by Adleman [2], Ajtai further proved the NP-hardness (under randomized reductions) of the most famous lattice problem, the shortest vector problem (SVP). The NP-hardness of SVP has been a long standing open problem. Ajtai's breakthroughs initiated a series of new results on the complexity of lattice problems, which are nicely surveyed by Cai [30,31].

Those complexity results opened the door to positive applications in cryptology. Indeed, several cryptographic schemes based on the hardness of lattice problems were proposed shortly after Ajtai's discoveries (see [5,61,69,32,99, 50]). Some have been broken, while others seem to resist state-of-the-art attacks, for now. Those schemes attracted interest for at least two reasons: on the one hand, there are very few public-key cryptosystems based on problems different from integer factorization or the discrete logarithm problem, and on the other hand, some of those schemes offered encryption/decryption rates asymptotically higher than classical schemes. Besides, one of those schemes, by Ajtai and Dwork [5], enjoyed a surprising security proof based on worst-case (instead of average-case) hardness assumptions.

Independently of those developments, there has been renewed cryptographic interest in lattice reduction, following a beautiful work by Coppersmith [38] in 1996. Coppersmith showed, by means of lattice reduction, how to solve rigorously certain problems, apparently non-linear, related to the question of finding small roots of low-degree polynomial equations. In particular, this has led to surprising attacks on the RSA [120] cryptosystem in special settings such as low public or private exponent, but curiously, also to new security proofs [128,18]. Coppersmith's results differ from "traditional" applications of lattice reduction in cryptanalysis, where the underlying problem is already linear, and the attack often heuristic by requiring (at least) that current lattice reduction algorithms behave ideally, as opposed to what is theoretically guaranteed. The use of lattice reduction techniques to solve polynomial equations goes back to the eighties [66, 133]. The first result of that kind, the broadcast attack on low-exponent RSA due to Håstad [66], can be viewed as a weaker version of Coppersmith's theorem on univariate modular polynomial equations.

A shorter version of this survey previously appeared in [118]. The rest of the paper is organized as follows. In Section 2, we give basic definitions and results on lattices and their algorithmic problems. In Section 3, we survey an old application of lattice reduction in cryptology: finding small solutions of multivariate linear equations, which includes the well-known subset sum or knapsack problem as a special case. In Section 4, we review a related problem: the hidden number problem. In Section 5, we discuss lattice-based cryptography, somehow a revival for knapsack-based cryptography. In Section 6, we discuss developments on the problem of finding small roots of polynomial equations, inspired by Coppersmith's discoveries in 1996. In Section 7, we survey the surprising links between lattice reduction, the RSA cryptosystem, and integer factorization.

2 Lattice Problems

2.1 Definitions

Recall that a *lattice* is a discrete (additive) subgroup of $\mathbb{R}^n$. In particular, any subgroup of $\mathbb{Z}^n$ is a lattice, and such lattices are called *integer lattices*. An equivalent definition is that a lattice consists of all integral linear combinations of a set of linearly independent vectors, that is,

$$L = \left\{ \sum_{i=1}^{d} n_i \mathbf{b}_i \mid n_i \in \mathbb{Z} \right\},$$

where the $\mathbf{b}_i$'s are linearly independent over $\mathbb{R}$. Such a set of vectors $\mathbf{b}_i$'s is called a lattice *basis*. All the bases have the same number $\dim(L)$ of elements, called the *dimension* (or *rank*) of the lattice since it matches the dimension of the vector subspace $\text{span}(L)$ spanned by L.

There are infinitely many lattice bases when $\dim(L) \geq 2$. Any two bases are related to each other by some unimodular matrix (integral matrix of determinant ± 1), and therefore all the bases share the same Gramian determinant $\det_{1 \leq i,j \leq d} \langle \mathbf{b}_i, \mathbf{b}_j \rangle$. The *volume* $\text{vol}(L)$ (or *determinant*) of the lattice is by definition the square root of that Gramian determinant, thus corresponding to the d-dimensional volume of the parallelepiped spanned by the $\mathbf{b}_i$'s. In the important case of full-dimensional lattices where $\dim(L) = n$, the volume is equal to the absolute value of the determinant of any lattice basis (hence the name determinant). If the lattice is further an integer lattice, then the volume is also equal to the index $[\mathbb{Z}^n : L]$ of L in $\mathbb{Z}^n$.

Since a lattice is discrete, it has a shortest non-zero vector: the Euclidean norm of such a vector is called the lattice *first minimum*, denoted by $\lambda_1(L)$ or

$\|L\|$. Of course, one can use other norms as well: we will use $\|L\|_\infty$ to denote the first minimum for the infinity norm. More generally, for all $1 \leq i \leq \dim(L)$, Minkowski's *i-th minimum* $\lambda_i(L)$ is defined as the minimum of $\max_{1 \leq j \leq i} \|\mathbf{v}_j\|$ over all i linearly independent lattice vectors $\mathbf{v}_1, \ldots, \mathbf{v}_i \in L$. There always exist linearly independent lattice vectors $\mathbf{v}_1, \ldots, \mathbf{v}_d$ reaching the minima, that is $\|\mathbf{v}_i\| = \lambda_i(L)$. However, surprisingly, as soon as $\dim(L) \geq 4$, such vectors do not necessarily form a lattice basis, and when $\dim(L) \geq 5$, there may not even exist a lattice basis reaching the minima. This is one of the reasons why there exist several notions of basis reduction in high dimension, without any "optimal" one. It will be convenient to define the *lattice gap* as the ratio $\lambda_2(L)/\lambda_1(L)$ between the first two minima.

Minkowski's Convex Body Theorem guarantees the existence of short vectors in lattices: a careful application shows that any d-dimensional lattice L satisfies $\|L\|_\infty \leq \mathrm{vol}(L)^{1/d}$, which is obviously the best possible bound. It follows that $\|L\| \leq \sqrt{d}\mathrm{vol}(L)^{1/d}$, which is not optimal, but shows that the value $\lambda_1(L)/\mathrm{vol}(L)^{1/d}$ is bounded when L runs over all d-dimensional lattices. The supremum of $\lambda_1(L)^2/\mathrm{vol}(L)^{2/d}$ is denoted by γ_d, and called Hermite's constant[1] of dimension d, because Hermite was the first to establish its existence in the language of quadratic forms. The exact value of Hermite's constant is only known for $d \leq 8$. The best asymptotic bounds known for Hermite's constant are the following ones (see [102, Chapter II] for the lower bound, and [37, Chapter 9] for the upper bound):

$$\frac{d}{2\pi e} + \frac{\log(\pi d)}{2\pi e} + o(1) \leq \gamma_d \leq \frac{1.744d}{2\pi e}(1 + o(1)).$$

Minkowski proved more generally:

Theorem 1 (Minkowski). *For all d-dimensional lattices L and all $r \leq d$:*

$$\prod_{i=1}^{r} \lambda_i(L) \leq \sqrt{\gamma_d^r}\mathrm{vol}(L)^{r/d}.$$

A general principle, dating back to Gauss, estimates the number of lattice points (in a full-rank lattice) in nice sets of $\mathbb{R}^n$ by the volume of the set divided by the volume of the lattice, with a small error term. This approach can be proved to be rigorous in certain settings, such as when the lattice dimension is fixed and the set is the ball centered at the origin with radius growing to infinity. Thus, one often heuristically approximates the successive minima of a d-dimensional lattice L by $\sqrt{\frac{d}{2\pi e}}\mathrm{vol}(L)^{1/d}$. This is of course only an intuitive estimate, which may be far away from the truth.

For any lattice L of $\mathbb{R}^n$, one defines the *dual lattice* (also called *polar lattice*) of L as:

$$L^* = \{\mathbf{x} \in \mathrm{span}(L) : \forall \mathbf{y} \in L, \langle \mathbf{x}, \mathbf{y} \rangle \in \mathbb{Z}\}.$$

[1] For historical reasons, Hermite's constant refers to $\max \lambda_1(L)^2/\mathrm{vol}(L)^{2/d}$ and not to $\max \lambda_1(L)/\mathrm{vol}(L)^{1/d}$.

If $(\mathbf{b}_1, \ldots, \mathbf{b}_d)$ is a basis of L, then the dual family $(\mathbf{b}_1^*, \ldots, \mathbf{b}_d^*)$ is a basis of L^* (the dual family is the unique linearly independent family of span(L) such that $\langle \mathbf{b}_i^*, \mathbf{b}_j \rangle$ is equal to 1 if $i = j$, and to 0 otherwise). Thus, $(L^*)^* = L$ and $\mathrm{vol}(L)\mathrm{vol}(L^*) = 1$. The so-called transference theorems relate the successive minima of a lattice and its dual lattice. The first transference theorem follows from the definition of Hermite's constant:

$$\lambda_1(L)\lambda_1(L^*) \leq \gamma_d.$$

A more difficult transference theorem (see [9]) ensures that for all $1 \leq r \leq d$:

$$\lambda_r(L)\lambda_{d-r+1}(L^*) \leq d.$$

Both these transference bounds are optimal up to a constant. More information on lattice theory can be found in numerous textbooks, such as [65, 131, 92].

2.2 Algorithmic Problems

In the rest of this section, we assume implicitly that lattices are rational lattices (lattices in $\mathbb{Q}^n$), and d will denote the lattice dimension.

The most famous lattice problem is the *shortest vector problem* (SVP): given a basis of a lattice L, find $\mathbf{u} \in L$ such that $\|\mathbf{u}\| = \|L\|$ (recall that $\|L\| = \lambda_1(L)$). SVP_∞ will denote the analogue for the infinity norm. One defines approximate short vector problems by asking a non-zero $\mathbf{u} \in L$ with norm bounded by some approximation factor: $\|\mathbf{u}\| \leq f(d)\|L\|$.

The *closest vector problem* (CVP), also called the *nearest lattice point problem*, is a non-homogeneous version of the shortest vector problem: given a basis of a lattice L and a vector $\mathbf{v} \in \mathbb{R}^n$ (it does not matter whether $\mathbf{v} \in$ span(L)), find a lattice vector minimizing the distance to $\mathbf{v}$. Again, one defines approximate closest vector problems by asking $\mathbf{u} \in L$ such that for all $\mathbf{w} \in L$, $\|\mathbf{u} - \mathbf{v}\| \leq f(d)\|\mathbf{w} - \mathbf{v}\|$.

Another problem is the *smallest basis problem* (SBP), which has many variants depending on the exact meaning of "smallest". The variant currently in vogue (see [3, 14]) is the following: find a lattice basis minimizing the maximum of the lengths of its elements. A more geometric variant asks instead to minimize the product of the lengths (see [64]), since the product is always larger than the lattice volume, with equality if and only if the basis is orthogonal.

2.3 Complexity Results

We refer to Cai [30, 31] for an up-to-date survey of complexity results. Ajtai [4] recently proved that SVP is NP-hard under randomized reductions. Micciancio [98, 97] simplified and improved the result by showing that approximating SVP to within a factor $< \sqrt{2}$ is also NP-hard under randomized reductions. The NP-hardness of SVP under deterministic (Karp) reductions remains an open problem.

CVP seems to be a more difficult problem. Goldreich *et al.* [62] recently noticed that CVP cannot be easier than SVP: given an oracle that approximates CVP to within a factor $f(d)$, one can approximate SVP in polynomial time to within the same factor $f(d)$. Reciprocally, Kannan proved in [78, Section 7] that any algorithm approximating SVP to within a non-decreasing function $f(d)$ can be used to approximate CVP to within $d^{3/2}f(d)^2$. CVP was shown to be NP-hard as early as in 1981 [49] (for a much simpler "one-line" proof using the knapsack problem, see [100]). Approximating CVP to within a quasi-polynomial factor $2^{\log^{1-\varepsilon} d}$ is NP-hard [7,45].

However, NP-hardness results for SVP and CVP have limits. Goldreich and Goldwasser [58] showed that approximating SVP or CVP to within $\sqrt{d/\log d}$ is not NP-hard, unless the polynomial-time hierarchy collapses.

Interestingly, SVP and CVP problems seem to be more difficult with the infinity norm. It was shown that SVP_∞ and CVP_∞ are NP-hard in 1981 [49]. In fact, approximating $\mathrm{SVP}_\infty/\mathrm{CVP}_\infty$ to within an almost-polynomial factor $d^{1/\log\log d}$ is NP-hard [44]. On the other hand, Goldreich and Goldwasser [58] showed that approximating $\mathrm{SVP}_\infty/\mathrm{CVP}_\infty$ to within $d/\log d$ is not NP-hard, unless the polynomial-time hierarchy collapses.

We will not discuss Ajtai's worst-case/average-case equivalence [3,33], which refers to special versions of SVP and SBP (see [30,31,14]) such as SVP when the lattice gap λ_2/λ_1 is at least polynomial in the dimension.

2.4 Algorithmic Results

The main algorithmic results are surveyed in [91,78,132,64,36,84,30,109]. No polynomial-time algorithm is known for approximating either SVP, CVP or SBP to within a polynomial factor in the dimension d. In fact, the existence of such algorithms is an important open problem. The best polynomial-time algorithms achieve only slightly subexponential factors, and are based on the LLL algorithm [88], which can approximate SVP and SBP. However, it should be emphasized that these algorithms typically perform much better than is theoretically guaranteed, on instances of practical interest. Given as input any basis of a lattice L, LLL provably outputs in polynomial time a basis $(\mathbf{b}_1,\ldots,\mathbf{b}_d)$ satisfying:

$$\|\mathbf{b}_1\| \le 2^{(d-1)/4}\mathrm{vol}(L)^{1/d},\ \|\mathbf{b}_i\| \le 2^{(d-1)/2}\lambda_i(L) \text{ and } \prod_{i=1}^{d}\|\mathbf{b}_i\| \le 2^{\binom{d}{2}/2}\mathrm{vol}(L).$$

Thus, LLL can approximate SVP to within $2^{(d-1)/2}$. Schnorr[2] [121] improved the bound to $2^{O(d(\log\log d)^2/\log d)}$, and Ajtai *et al.* [6] recently further improved it to $2^{O(d\log\log d/\log d)}$ in randomized polynomial time thanks to a new randomized algorithm to find the shortest vector. In fact, Schnorr defined an LLL-based

[2] Schnorr's result is usually cited in the literature as an approximation algorithm to within $(1+\varepsilon)^n$ for any constant $\varepsilon > 0$. However, Goldreich and Håstad noticed about two years ago that one can choose some $\varepsilon = o(1)$ and still have polynomial running time, for instance using the blocksize $k = \log d/\log\log d$ in [121].

family of algorithms [121] (named BKZ for blockwise Korkine-Zolotarev) whose performances depend on a parameter called the blocksize. These algorithms use some kind of exhaustive search super-exponential in the blocksize. So far, the best reduction algorithms in practice are variants [124,125] of those BKZ-algorithms, which apply a heuristic to reduce exhaustive search. But little is known on the average-case (and even worst-case) complexity of reduction algorithms.

Babai's nearest plane algorithm [8] uses LLL to approximate CVP to within $2^{d/2}$, in polynomial time (see also [80]). Using Schnorr's algorithm [121], this can be improved to $2^{O(d(\log\log d)^2/\log d)}$ in polynomial time, and even further to $2^{O(d\log\log d/\log d)}$ in randomized polynomial time using [6], due to Kannan's link between CVP and SVP (see previous section). In practice however, the best strategy seems to be the *embedding method* (see [61,108]), which uses the previous algorithms for SVP and a simple heuristic reduction from CVP to SVP. Namely, given a lattice basis $(\mathbf{b}_1,\ldots,\mathbf{b}_d)$ and a vector $\mathbf{v}\in\mathbb{R}^n$, the embedding method builds the $(d+1)$-dimensional lattice (in $\mathbb{R}^{n+1}$) spanned by the row vectors $(\mathbf{b}_i,0)$ and $(\mathbf{v},1)$. Depending on the lattice, one should choose a coefficient different than 1 in $(\mathbf{v},1)$. It is hoped that a shortest vector of that lattice is of the form $(\mathbf{v}-\mathbf{u},1)$ where $\mathbf{u}$ is a closest vector (in the original lattice) to $\mathbf{v}$, whenever the distance to the lattice is smaller than the lattice first minimum. This heuristic may fail (see for instance [97] for some simple counterexamples), but it can also sometimes be proved, notably in the case of lattices arising from low-density knapsacks.

For exact SVP, the best algorithm known (in theory) is the recent randomized $2^{O(d)}$-time algorithm by Ajtai *et al.* [6], which improved Kannan's super-exponential algorithm [77,79] (see also [67]). For exact CVP, the best algorithm remains Kannan's super-exponential algorithm [77,79], with running time $2^{O(d\log d)}$ (see also [67] for an improved constant).

3 Finding Small Roots of Multivariate Linear Equations

One of the early and most natural applications of lattice reduction in cryptology was to find small roots of multivariate linear equations, where the equations are either integer equations or modular equations.

3.1 Knapsacks

Cryptology and lattices share a long history with the *knapsack* (also called *subset sum*) problem, a well-known NP-hard problem considered by Karp, and a particular case of multivariate linear equation: given a set $\{a_1,a_2,\ldots,a_n\}$ of positive integers and a sum $s=\sum_{i=1}^n x_i a_i$, where $x_i\in\{0,1\}$, recover the x_i's.

In 1978, Merkle and Hellman [96] invented one of the first public-key cryptosystems, by converting some easy knapsacks into what they believed were hard knapsacks. It was basically the unique alternative to RSA until 1982, when Shamir [126] proposed a (heuristic) attack against the simplest version of the Merkle-Hellman scheme. Shamir used Lenstra's integer programming algorithm [89,90] but, the same year, Adleman [1] showed how to use LLL instead, making experiments much easier. Brickell [27,28] later extended the attacks to the more general "iterated" Merkle-Hellman scheme, and showed that Merkle-Hellman was insecure for all realistic parameters. The cryptanalysis of Merkle-Hellman schemes was the first application of lattice reduction in cryptology.

Despite the failure of Merkle-Hellman cryptosystems, researchers continued to search for knapsack cryptosystems because such systems are very easy to implement and can attain very high encryption/decryption rates. But basically, all knapsack cryptosystems have been broken (for a survey, see [119]), either by specific (often lattice-based) attacks or by the low-density attacks. The last significant candidate to survive was the Chor-Rivest cryptosystem [35], broken by Vaudenay [135] in 1997 with algebraic (not lattice) methods.

3.2 Low-Density Attacks on Knapsacks

We only describe the basic link between lattices and knapsacks. Note that Ajtai's original proof [4] for the NP-hardness (under randomized reductions) of SVP used a connection between the subset sum problem and SVP.

Solving the knapsack problem amounts to find a 0, 1-solution of an inhomogeneous linear equation, which can be viewed as a closest vector problem in a natural way, by considering the corresponding homogeneous linear equation, together with an arbitrary solution of the inhomogeneous equation. Indeed, let $s = \sum_{i=1}^{n} x_i a_i$ be a knapsack instance. One can compute in polynomial time integers $y_1, \dots, y_n$ such that $s = \sum_{i=1}^{n} y_i a_i$, using for instance an extended gcd algorithm. Then the vector $(y_1 - x_1, \dots, y_n - x_n)$ belongs to the $(n-1)$-dimensional lattice L formed by all the solutions of the homogeneous equation, that is the vectors $(z_1, \dots, z_n) \in \mathbb{Z}^n$ such that:

$$z_1 a_1 + \dots + z_n a_n = 0.$$

And this lattice vector is fairly close to the vector $(y_1, \dots, y_n)$, since the distance is roughly $\sqrt{n/2}$. But because $x_i \in \{0, 1\}$, the lattice vector is even closer to the vector $\mathbf{y} = (y_1 - 1/2, \dots, y_n - 1/2)$ for which the distance is exactly $\sqrt{n/4}$. In fact, it is easy to see that $\mathbf{x} = (y_1 - x_1, \dots, y_n - x_n)$ is a closest vector to $\mathbf{y}$ in the lattice L, and that any lattice vector whose distance to $\mathbf{y}$ is exactly $\sqrt{n/4}$ is necessarily of the form $(y_1 - x'_1, \dots, y_n - x'_n)$ where $s = \sum_{i=1}^{n} x'_i a_i$ and $x'_i \in \{0, 1\}$. This gives a deterministic polynomial-time reduction from the knapsack problem to CVP (this reduction appeared in [100] with a slightly different lattice).

One can derive from this reduction a provable method to solve the knapsack problem in polynomial time with high probability when the knapsack *density* defined as $d = n / \max_{1 \le i \le n} \log_2 a_i$ is low (see [85,51,54]). Indeed, if $\|\mathbf{x} - \mathbf{y}\| = \sqrt{n/4}$ is strictly less than $2^{-(n-1)/2-1} \|L\|$, then by applying Babai's nearest plane CVP approximation algorithm to L and $\mathbf{y}$, one obtains $\mathbf{z} \in L$ such that $\|\mathbf{z} - \mathbf{y}\| < 2^{n/2} \|\mathbf{x} - \mathbf{y}\| < \|L\|/2$, and thus $\|\mathbf{z} - \mathbf{x}\| < \|L\|$ where $\mathbf{z} - \mathbf{x} \in L$, which implies that $\mathbf{z} = \mathbf{x}$, disclosing the x_i's. It remains to estimate the first minimum $||L||$. With high probability, the a_i's are coprime, and then:

$$\text{vol}(L) = \left(\sum_{i=1}^{n} a_i^2\right)^{1/2} \approx 2^{n/d}\sqrt{n}.$$

Thus, one expects $||L|| \approx 2^{1/d}\sqrt{\frac{n}{2\pi e}}$. It follows that the method should work whenever

$$\sqrt{\frac{n}{4}} < 2^{-(n-1)/2-1} 2^{1/d}\sqrt{\frac{n}{2\pi e}},$$

that is, roughly $d \leq 2/n$. This volume argument can be made rigorous because the probability that a fixed non-zero vector belongs to L is less than $1/A$ when the a_i's are chosen uniformly at random from $[0, A]$. One deduces that most knapsacks of density roughly less than $2/n$ are solvable in polynomial time (see [85, 51,54]).

One does not know how to provably solve the knapsack problem in polynomial time when the density lies between $2/n$ and 1, which is typically the case for cryptographic knapsacks (where the density should be less than 1, otherwise heuristically, there would be several solutions, causing decryption troubles). However, one can hope that the embedding method that heuristically reduces CVP to SVP works, as while as the distance to the lattice (which is $\sqrt{n/4}$) is smaller than the first minimum $||L||$. By the previous reasoning, this should happen when

$$\sqrt{\frac{n}{4}} \leq 2^{1/d}\sqrt{\frac{n}{2\pi e}},$$

that is,

$$d \leq \frac{1}{\log_2 \sqrt{\pi e/2}} \approx 0.955\ldots$$

This heuristic bound turns out to be not too far away from the truth. Indeed, one can show that the target vector $(x_1 - 1/2, \ldots, x_n - 1/2, 1)$ is with high probability (over the choice of the a_i's) the shortest vector in the embedding lattice, when the density $d \leq 0.9408\ldots$ (see [41] who used a slightly different lattice, but the proof carries through). This is done by enumerating all possible short vectors, and using bounds on the number of integral points in high-dimensional spheres [93]. The result improved the earlier bound of $0.6463\ldots$ from Lagarias and Odlyzko [85], which was essentially obtained by approximating the vector $(y_1, \ldots, y_n)$ in the lattice L, instead of $(y_1 - 1/2, \ldots, y_n - 1/2)$. This rigorous bound of $0.6463\ldots$ matches the heuristic bound obtained by a volume argument on the corresponding embedding lattice:

$$\sqrt{\frac{n}{2}} \leq 2^{1/d}\sqrt{\frac{n}{2\pi e}}.$$

To summarize, the subset sum problem can always be efficiently reduced to CVP, and this reduction leads to an efficient probabilistic reduction to SVP in low density, and to a polynomial-time solution in extremely low density. In the light of recent results on the complexity of SVP, those reductions from knapsack to SVP may seem useless. Indeed, the NP-hardness of SVP under randomized reductions suggests that there is no polynomial-time algorithm that solves SVP. However, it turns out that in practice, one can hope that standard lattice reduction algorithms behave like SVP-oracles, up to reasonably high dimensions. Experiments carried out in [85, 124, 125] show the effectiveness of such an approach for solving low-density subset sums, up to n about the range of 100–200. It does not prove nor disprove that one can solve, in theory or in practice, low-density knapsacks with n over several hundreds. But it was sufficient to show that knapsack cryptography was impractical: indeed, the keysize of knapsack schemes grows in general at least quadratically with n, so that high values of n (as required by lattice attacks) are not practical.

Thus, lattice methods to solve the subset sum problem are mainly heuristic. And lattice attacks against knapsack cryptosystems are somehow even more heuristic, because the reductions from knapsack to SVP assume a uniform distribution of the weights a_i's, which is in general not necessarily satisfied by knapsacks arising from cryptosystems.

3.3 The Orthogonal Lattice

Recently, Nguyen and Stern proposed in [113] a natural generalization of the lattices arising from the knapsack problem. More precisely, they defined for any integer lattice L in $\mathbb{Z}^n$, the *orthogonal lattice* $L^\perp$ as the set of integral vectors orthogonal to L, that is, the set of $\mathbf{x} \in \mathbb{Z}^n$ such that the dot product $\langle \mathbf{x}, \mathbf{y} \rangle = 0$ for all $\mathbf{y} \in L$. Note that the lattice $L^\perp$ has dimension $n - \dim(L)$, and can be computed in polynomial time from L (see [36]). Interestingly, the links between duality and orthogonality (see Martinet's book [92, pages 34–35]) enable to prove that the volume of $L^\perp$ is equal to the volume of the lattice $\mathrm{span}(L) \cap \mathbb{Z}^n$ which we denote by $\bar{L}$. Thus, if a lattice in $\mathbb{Z}^n$ is low-dimensional, its orthogonal lattice is high-dimensional with a volume at most equal: the successive minima of the orthogonal lattice are likely to be much shorter than the ones of the original lattice. That property of orthogonal lattices has led to effective (though heuristic) lattice-based attacks on various cryptographic schemes [113, 115, 116, 114, 117]. We refer to [109] for more information. In particular, it was used in [117] to solve the *hidden subset sum problem* (used in [26]) in low density. The hidden subset sum problem was apparently a non-linear version of the subset sum problem: given M and n in $\mathbb{N}$, and $b_1, \ldots, b_m \in \mathbb{Z}_M$, find $\alpha_1, \ldots, \alpha_n \in \mathbb{Z}_M$ such that each b_i is some subset sum modulo M of $\alpha_1, \ldots, \alpha_n$.

We sketch the solution of [117] to give a flavour of cryptanalyses based on orthogonal lattices. We first restate the hidden subset sum problem in terms of vectors. We are given an integer M, and a vector $\mathbf{b} = (b_1, \ldots, b_m) \in \mathbb{Z}^m$ with entries in $[0..M-1]$ such that there exist integers $\alpha_1, \ldots, \alpha_n \in [0..M-1]$, and vectors $\mathbf{x}_1, \ldots, \mathbf{x}_n \in \mathbb{Z}^m$ with entries in $\{0, 1\}$ satisfying:

$$\mathbf{b} = \alpha_1 \mathbf{x}_1 + \alpha_2 \mathbf{x}_2 + \cdots + \alpha_n \mathbf{x}_n \pmod{M}.$$

We want to determine the α_i's. There exists a vector $\mathbf{k} \in \mathbb{Z}^m$ such that:

$$\mathbf{b} = \alpha_1 \mathbf{x}_1 + \alpha_2 \mathbf{x}_2 + \cdots + \alpha_n \mathbf{x}_n + M\mathbf{k}.$$

Notice that if $\mathbf{u}$ in $\mathbb{Z}^n$ is orthogonal to $\mathbf{b}$, then $\mathbf{p_u} = (\langle \mathbf{u}, \mathbf{x}_1 \rangle, \ldots, \langle \mathbf{u}, \mathbf{x}_n \rangle, \langle \mathbf{u}, \mathbf{k} \rangle)$ is orthogonal to the vector $\mathbf{v}_\alpha = (\alpha_1, \ldots, \alpha_n, M)$. But $\mathbf{v}_\alpha$ is independent of m, and so is the n-dimensional lattice $\mathbf{v}_\alpha^\perp$. On the other hand, as m grows for a fixed M, most of the vectors of any reduced basis of the $(m-1)$-dimensional lattice $\mathbf{b}^\perp$ should get shorter and shorter, because they should have norm close to $\text{vol}(\mathbf{b}^\perp)^{1/(m-1)} \leq \text{vol}(\mathbf{b})^{1/(m-1)} = \|\mathbf{b}\|^{1/(m-1)} \approx (M\sqrt{m})^{1/(m-1)}$. For such vectors $\mathbf{u}$, the corresponding vectors $\mathbf{p}_u$ also get shorter and shorter. But if $\mathbf{p}_u$ gets smaller than $\lambda_1(\mathbf{v}_\alpha^\perp)$ (which is independent of m), then it is actually zero, that is, $\mathbf{u}$ is orthogonal to all the $\mathbf{x}_j$'s and $\mathbf{k}$. Note that one expects $\lambda_1(\mathbf{v}_\alpha^\perp)$ to be of the order of $\|\mathbf{v}_\alpha\|^{1/n} \approx (M\sqrt{n})^{1/n}$.

This suggests that if $(\mathbf{u}_1, \ldots, \mathbf{u}_{m-1})$ is a sufficiently reduced basis of $\mathbf{b}^\perp$, then the first $m-(n+1)$ vectors $\mathbf{u}_1, \ldots, \mathbf{u}_{m-(n+1)}$ should heuristically be orthogonal to all the $\mathbf{x}_j$'s and $\mathbf{k}$. One cannot expect that more than $m-(n+1)$ vectors are orthogonal because the lattice L_x spanned by the $\mathbf{x}_j$'s and $\mathbf{k}$ is likely to have dimension $(n+1)$. From the previous discussion, one can hope that the heuristic condition is satisfied when the density $n/\log(M)$ is very small (so that $\lambda_1(\mathbf{v}_\alpha^\perp)$ is not too small), and m is sufficiently large. And if the heuristic condition is satisfied, the lattice $\bar{L}_x$ is disclosed, because it is then equal to the orthogonal lattice $(\mathbf{u}_1, \ldots, \mathbf{u}_{m-(n+1)})^\perp$. Once $\bar{L}_x$ is known, it is not difficult to recover (heuristically) the vectors $\mathbf{x}_j$'s by lattice reduction, because they are very short vectors. One eventually determines the coefficients α_j's from a linear modular system. The method is quite heuristic, but it works in practice for small parameters in low density (see [117] for more details).

3.4 Multivariate Modular Linear Equations

The technique described in Section 3.2 to solve the knapsack problem can easily be extended to find small solutions of a system of multivariate linear equations over the integers: one views the problem as a closest vector problem in the lattice corresponding to the homogenized equations, which is an orthogonal lattice. Naturally, a similar method can be applied to a system of multivariate linear modular equations, except that in this case, the corresponding lattice is not an orthogonal lattice.

Let $A = (a_{i,j})$ be an $\ell \times k$ integral matrix, $\mathbf{c} \in \mathbb{Z}^\ell$ be a column vector and q be a prime number. The problem is to find a short column vector $\mathbf{x} \in \mathbb{Z}^k$ such that:

$$A\mathbf{x} \equiv \mathbf{c} \pmod{q}.$$

The interesting case is when the number of unknowns k is larger than the number of equations ℓ. Following Section 3, one computes an arbitrary solution $\mathbf{y} \in \mathbb{Z}^k$ such that $A\mathbf{y} \equiv \mathbf{c} \pmod{q}$, for instance by finding a solution of a solvable system of linear equations over the integers (if the system is not solvable, then the original problem has no solution). And one computes a basis of the full-dimensional lattice L of all column vectors $\mathbf{z} \in \mathbb{Z}^k$ such that

$$A\mathbf{z} \equiv 0 \pmod{q}.$$

Then any short solution $\mathbf{x}$ to $A\mathbf{x} \equiv \mathbf{c} \pmod{q}$ corresponds to a lattice vector $\mathbf{y} - \mathbf{x} \in L$ close to $\mathbf{y}$. Thus, there is at most one $\mathbf{x} \in \mathbb{Z}^k$ such that $A\mathbf{x} \equiv \mathbf{c}$ (mod q) and $\|\mathbf{x}\| < \|L\|/2$. And if ever there is an unusually short vector $\mathbf{x} \in \mathbb{Z}^k$ such that $A\mathbf{x} \equiv \mathbf{c} \pmod{q}$ and

$$\|\mathbf{x}\| < \|L\| 2^{-k/2-1},$$

then Babai's CVP approximation algorithm will disclose it, as in Section 3. It remains to lower bound the first minimum of the lattice.

One can see that the volume of L is an integer dividing q^ℓ, because it is the index of L in $\mathbb{Z}^k$. In fact, for most matrices A, one expects the volume to be exactly q^ℓ, so that:

$$\|L\| \approx \sqrt{\frac{k}{2\pi e}} q^{\ell/k}.$$

This estimate is not far from the truth, since for any fixed non-zero vector $\mathbf{z} \in \mathbb{Z}^k$ such that $\|\mathbf{x}\|_\infty < q$, the probability that $\mathbf{z} \in L$ (when A is uniformly distributed) is exactly $q^{-\ell}$. It follows that for most matrices, if ever there exists $\mathbf{x} \in \mathbb{Z}^k$ such that $A\mathbf{x} \equiv \mathbf{c} \pmod{q}$ and $\|\mathbf{x}\|$ roughly less than $q^{\ell/k} 2^{-k/2-1}$, then one can find such an $\mathbf{x}$ in polynomial time. For a precise statement, we refer to [52] who actually used a dual approach requiring transference theorems (which we do not need here). An interesting application is that if we know a few bits of each entry of an arbitrary solution $\mathbf{x}$ of a system of linear modular equations, then we can recover all of $\mathbf{x}$, because if the number of bits is sufficiently large, the problem is reduced to finding an unusually short solution of a system of linear modular equations. This was used to show the insecurity of truncated linear congruential pseudo-random number generators in [52].

The result can in fact be extended to a wider class of parameters, when the modulus q is not necessarily prime (see [52]), and when the equations may have different modulus (see [10]). We note that the exponent $-k/2$ can be suppressed when a CVP-oracle is available, which is the case when k is fixed. Furthermore, the previous reasoning not only shows how to find unusually short solutions, it also shows how to find reasonably short solutions when the matrix A is random. Indeed, a tighter analysis then shows that all the minima of the lattice L are in fact not too far away from $\sqrt{k/(2\pi e)} q^{\ell/k}$, so that all points are reasonably close to the lattice. In this case, one can find in polynomial time a vector $\mathbf{x} \in \mathbb{Z}^k$ such that $A\mathbf{x} \equiv \mathbf{c} \pmod{q}$ and $\|\mathbf{x}\|$ is very roughly less than $\sqrt{k/(2\pi e)} q^{\ell/k} 2^{k/2}$. This was used to attack certain RSA padding signature schemes in which the messages have a fixed pattern (see [104, 57]), and it was also used to complete the proof of security of the RSA–OAEP encryption scheme (see [53]).

However, the previous results are weak in a certain sense. First, the results depend strongly on the distribution of the coefficients of the linear equations.

More precisely, the first minimum of the lattice can be arbitrary small, leading to possibly much weaker bounds: hence, one must perform a new analysis of the lattice for any system of equations which is not uniformly distributed. This was the case in [52] where linear congruential generators gave rise to special systems of equations. Furthermore, the exponential or slightly subexponential factors in the polynomial-time approximation of CVP imply that the bounds obtained are rather weak as the number k of unknowns increases. The situation is somewhat similar to that of knapsacks for which only knapsacks of very low density can provably be solved. This is one of the reasons why the attack of [104] was only heuristic. On the other hand, k was as small as 2 in [53], making provable results useful. We will see in the next section a particular case of a system of linear modular equations for which the generic method can be replaced by another lattice-based method.

4 The Hidden Number Problem

4.1 Hardness of Diffie-Hellman Bits

In [24], Boneh and Venkatesan used the LLL algorithm to solve the *hidden number problem*, which enables to prove the hardness of the most significant bits of secret keys in Diffie-Hellman and related schemes in prime fields. This was the first positive application of LLL in cryptology. Recall the Diffie-Hellman key exchange protocol [43]: Alice and Bob fix a finite cyclic G and a generator g. They respectively pick random $a, b \in [1, |G|]$ and exchange g^a and g^b. The secret key is g^{ab}. Proving the security of the protocol under "reasonable" assumptions has been a challenging problem in cryptography (see [15]). Computing the most significant bits of g^{ab} is as hard as computing g^{ab} itself, in the case of prime fields:

Theorem 2 (Boneh-Venkatesan). *Let q be an n-bit prime and g be a generator of $\mathbb{Z}_q^*$. Let $\varepsilon > 0$ be fixed, and set $\ell = \ell(n) = \lceil \varepsilon\sqrt{n} \rceil$. Suppose there exists an expected polynomial time (in n) algorithm $\mathcal{A}$, that on input q, g, g^a and g^b, outputs the ℓ most significant bits of g^{ab}. Then there is also an expected polynomial time algorithm that on input q, g, g^a, g^b and the factorization of $q-1$, computes all of g^{ab}.*

The above result is slightly different from [24], due to an error in the proof of [24] spotted by [63]. The same result holds for the least significant bits. For a more general statement when g is not necessarily a generator, and the factorization of $q-1$ is unknown, see [63]. For analogous results in other groups, we refer to [136] for finite fields and to [23] for the elliptic curve case.

The proof goes as follows. We are given some g^a and g^b, and want to compute g^{ab}. We repeatedly pick a random r until g^{a+r} is a generator of $\mathbb{Z}_q^*$ (testing is easy thanks to the factorization of $q-1$). For each r, the probability of success is $\phi(q-1)/(q-1) \geq C/\log\log q$. Next, we apply $\mathcal{A}$ to the points g^{a+r} and g^{b+t} for many random values of t, so that we learn the most significant bits of $g^{(a+r)b}g^{(a+r)t}$, where $g^{(a+r)t}$ is a random element of $\mathbb{Z}_q^*$ since g^{a+r} is a generator. Note that one can easily recover g^{ab} from $\alpha = g^{(a+r)b}$. The problem becomes the *hidden number problem* (HNP): given $t_1, \ldots, t_d$ chosen uniformly and independently at random in $\mathbb{Z}_q^*$, and $\mathrm{MSB}_\ell(\alpha t_i \bmod q)$ for all i, recover $\alpha \in \mathbb{Z}_q$. Here, $\mathrm{MSB}_\ell(x)$ for $x \in \mathbb{Z}_q$ denotes any integer z satisfying $|x - z| < q/2^{\ell+1}$.

To achieve the proof, Boneh and Venkatesan presented a simple solution to HNP when ℓ is not too small, by reducing HNP to a lattice closest vector problem. We sketch this solution in the next section.

4.2 Solving the Hidden Number Problem by Lattice Reduction

Consider an HNP-instance: let $t_1, \ldots, t_d$ be chosen uniformly and independently at random in $\mathbb{Z}_q^*$, and $a_i = \mathrm{MSB}_\ell(\alpha t_i \bmod q)$ where $\alpha \in \mathbb{Z}_q$ is hidden. Clearly, the vector $\mathbf{t} = (t_1\alpha \bmod q, \ldots t_d\alpha \bmod q, \alpha/2^{\ell+1})$ belongs to the $(d+1)$-dimensional lattice $L = L(q, \ell, t_1, \ldots, t_d)$ spanned by the rows of the following matrix:

$$\begin{pmatrix} q & 0 & \cdots & 0 & 0 \\ 0 & q & \ddots & \vdots & \vdots \\ \vdots & \ddots & \ddots & 0 & \vdots \\ 0 & \ldots & 0 & q & 0 \\ t_1 & \ldots & \ldots & t_d & 1/2^{\ell+1} \end{pmatrix}$$

The vector $\mathbf{a} = (a_1, \ldots, a_d, 0)$ is very close to L, because it is very close to $\mathbf{t}$. Indeed, $\|\mathbf{t} - \mathbf{a}\| \leq q\sqrt{d+1}/2^{\ell+1}$. It is not difficult to show that any lattice point sufficiently close to $\mathbf{a}$ discloses the hidden number α, because sufficiently short lattice vectors must have their first d coordinates equal to zero (see [24, Theorem 5] or [110, 112]):

Lemma 3 (Uniqueness). *Set $d = 2\lceil\sqrt{\log q}\rceil$ and $\mu = \frac{1}{2}\sqrt{\log q} + 3$. Let α be in $\mathbb{Z}_q^*$. Choose integers $t_1, \ldots, t_d$ uniformly and independently at random in $\mathbb{Z}_q^*$. Let $\mathbf{a} = (a_1, \ldots, a_d, 0)$ be such that $|(\alpha t_i \bmod q) - a_i| < q/2^\mu$. Then with probability at least $\frac{1}{2}$, all $\mathbf{u} \in L$ with $\|\mathbf{u} - \mathbf{a}\| < \frac{q}{2^\mu}$ are of the form:*

$$\mathbf{u} = (t_1\beta \bmod q, \ldots t_d\beta \bmod q, \beta/2^{\ell+1}) \text{ where } \alpha \equiv \beta \pmod{q}.$$

Since $\mathbf{a}$ is close enough to L, Babai's nearest plane CVP approximation algorithm [8] yields a lattice point sufficiently close to $\mathbf{a}$, which leads to:

Theorem 4 (Boneh-Venkatesan). *Let α be in $\mathbb{Z}_q^*$. Let $\mathcal{O}$ be a function defined by $\mathcal{O}(t) = \mathrm{MSB}_\ell(\alpha t \bmod q)$ with $\ell = \lceil\sqrt{\log q}\rceil + \lceil\log\log q\rceil$. There exists a deterministic polynomial time algorithm $\mathcal{A}$ which, on input $t_1, \ldots, t_d$, $\mathcal{O}(t_1), \ldots, \mathcal{O}(t_d)$, outputs α with probability at least $1/2$ over $t_1, \ldots, t_d$ chosen uniformly and independently at random from $\mathbb{Z}_q^*$, where $d = 2\lceil\sqrt{\log q}\rceil$.*

Thus, the hidden number problem can be solved using $\ell = \sqrt{\log q} + \log\log q$ bits. Using the best polynomial-time CVP approximation algorithm known, this can be asymptotically improved to $O(\sqrt{\log q}\log\log\log q/\log\log q)$. Theorem 2 is a simple consequence.

We note that Theorem 4 could have alternatively be obtained from the generic method described in Section 3.4. Indeed, the hidden number problem can be viewed as a system of d modular linear equations in the $d+1$ unknowns α and $(\alpha t_i \bmod q) - \mathrm{MSB}_\ell(\alpha t_i \bmod q)$ where $1 \leq i \leq d$. Among these $d+1$ unknowns, only α may be large. One may transform the system to eliminate the possibly large unknown α. One then obtains a new system of $d-1$ modular linear equations in the d unknowns $(\alpha t_i \bmod q) - \mathrm{MSB}_\ell(\alpha t_i \bmod q)$ all smaller than $q/2^{\ell+1}$ in absolute value. Although this system does not correspond to a uniformly distributed matrix, one can easily obtain the same lower bound on the first minimum of the lattice as in the random case (see Section 3.4). It follows that one can find the (unique) small solution of the system in polynomial time (and thus, α) provided that roughly:

$$\frac{q}{2^{\ell+1}} \leq q^{(d-1)/d} 2^{-d/2-1},$$

that is $\ell \geq d/2 + 1 + \log(q)/d$, where the right-hand term is minimized for $d \approx \sqrt{2\log q}$, leading to ℓ larger than roughly $\sqrt{2\log q}$. Thus, one can obtain essentially the same bounds.

4.3 Variants of the Hidden Number Problem

It was recently realized that the condition that the t_i's are uniformly distributed is often too restrictive for applications. The previous solution to the hidden number problem can in fact be extended to cases where the distribution of the t_i's is not necessarily perfectly uniform (see [63,111]). A precise definition of this relaxed uniformity property can be made with the classical notion of discrepancy (see [111] for more details). To apply the solution to this generalized hidden number problem, it suffices to show that the distribution of the t_i's is sufficiently uniform, which is usually obtained by exponential sum techniques (see [63,111, 112,48,130,129] for some examples).

One may also extend the solution to the hidden number problem to the case when an oracle for CVP (in the Euclidean norm or the infinity norm) is available, which significantly decreases the number of necessary bits (see [110, 111]). This is useful to estimate what can be achieved in practice, especially when the lattice dimension is small. It turns out that the required number of bits becomes $O(\log\log q)$ and 2 respectively, with a CVP-oracle and a CVP_∞-oracle.

One may also study the hidden number problem with arbitrary bits instead of most significant bits. It is easy to see that the HNP with ℓ least significant bits can be reduced to the original HNP with ℓ most significant bits, but the situation worsens with arbitrary bits. By multiplying the t_i's with an appropriate number independent of the t_i's (see [111]), one obtains a deterministic polynomial-time reduction from the HNP with ℓ consecutive bits at a known position to the original HNP with $\ell/2$ most significant bits (the prime field $\mathbb{Z}_q$ and the number of random multipliers remain the same). This appropriate number can be found either by continued fractions or lattice reduction in dimension 2. More generally, by using high-dimensional lattice reduction, it is not difficult to show that there is a deterministic polynomial-time reduction from the HNP with ℓ arbitrary bits at known positions such that the number of blocks of consecutive unknown bits is m, to the original HNP with $\ell/m + 1 - \log m$ most significant bits. Thus, the HNP with arbitrary bits seems to be harder, especially when there are many blocks of consecutive unknown bits.

Finally, variants of the hidden number problem in settings other than prime fields have been studied in [130, 129, 23].

4.4 Lattice Attacks on DSA

Interestingly, the previous solution of the hidden number problem also has a dark side: it leads to a simple attack against the Digital Signature Algorithm [106, 95] (DSA) in special settings (see [73, 110]). Recall that the DSA uses a public element $g \in \mathbb{Z}_p$ of order q, a 160-bit prime dividing $p-1$ where p is a large prime (at least 512 bits). The signer has a secret key $\alpha \in \mathbb{Z}_q^*$ and a public key $\beta = g^\alpha \bmod p$. The DSA signature of a message m is $(r,s) \in \mathbb{Z}_q^2$ where $r = (g^k \bmod p) \bmod q$, $s = k^{-1}(h(m) + \alpha r) \bmod q$, h is SHA-1 hash function and k is a random element in $\mathbb{Z}_q^*$ chosen at each signature.

It is well-known that the secret key α can easily be recovered if the random nonce k is disclosed, or if k is produced by a cryptographically weak pseudo-random generator such as a linear congruential generator with known parameters [10] and a few signatures are available. Recently, Howgrave-Graham and Smart [73] noticed that Babai's nearest plane CVP algorithm could heuristically recover α, provided that sufficiently many signatures and sufficiently many bits of the corresponding nonces k are known. This is not surprising, because the underlying problem is in fact a generalized hidden number problem.

Indeed, assume that for d signatures (r_i, s_i) of messages m_i, the ℓ least significant bits of the random nonce k_i are known to the attacker: one knows $a_i < 2^\ell$ such that $k_i - a_i$ is of the form $2^\ell b_i$. Then $\alpha r_i \equiv s_i(a_i + b_i 2^\ell) - h(m_i) \pmod q$, which can be rewritten as: $\alpha r_i 2^{-\ell} s_i^{-1} \equiv (a_i - s_i^{-1} h(m_i)) \cdot 2^{-\ell} + b_i \pmod q$. Letting $t_i = r_i 2^{-\ell} s_i^{-1} \bmod q$, one sees that $\mathrm{MSB}_\ell(\alpha t_i \bmod q)$ is known. Recovering the secret key α is therefore a generalized hidden number problem in which the t_i's are not assumed to be independent and uniformly distributed over $\mathbb{Z}_q$, but are of the form $r_i 2^{-\ell} s_i^{-1}$ where the underlying k_i's are independent and uniformly distributed over $\mathbb{Z}_q^*$. Nguyen and Shparlinski [111] proved that under a reasonable assumption on the hash function, the t_i's are sufficiently uniform to make the corresponding hidden number problem provably tractable with the same number of bits as in Theorem 4, that is, essentially $\sqrt{\log q}$. Since lattice reduction algorithms can behave much better than theoretically expected, one may even hope to solve CVP exactly, yielding better bounds to Theorem 4. For the case of a 160-bit prime q as in DSA, one obtains that the DSA–HNP can be solved using respectively $\ell = 2$ bits and $d = 160$, or $\ell = 6$ bits and $d = 100$ respectively, when an oracle for CVP_∞ or CVP is available (see [110, 112]). In fact, the bounds are better in practice. It turns out that using standard lattice reduction algorithms implemented in Shoup's NTL library [127], one can often solve HNP for a 160-bit prime q using $\ell = 3$ bits and $d = 100$ (see [110, 112]).

Naturally, this attack can also be applied to similar signature algorithms (see [111]), such as the elliptic curve variant of DSA (see [112]), or the Nyberg-Rueppel signature scheme and related schemes (see [48]). The only difference is that one needs to establish the uniformity of different types of multipliers. This usually requires different kinds of exponential sums.

5 Lattice-Based Cryptography

We review state-of-the-art results on the main lattice-based cryptosystems. To keep the presentation simple, descriptions of the schemes are intuitive, referring to the original papers for more details. Only one of these schemes (the GGH cryptosystem [61]) explicitly works with lattices.

5.1 The Ajtai-Dwork Cryptosystem

Description. The Ajtai-Dwork cryptosystem [5] (AD) works in $\mathbb{R}^n$, with some finite precision depending on n. Its security is based on a variant of SVP.

The private key is a uniformly chosen vector u in the n-dimensional unit ball. One then defines a distribution $\mathcal{H}_u$ of points $\mathbf{a}$ in a large n-dimensional cube such that the dot product $\langle \mathbf{a}, \mathbf{u} \rangle$ is very close to $\mathbb{Z}$.

The public key is obtained by picking $\mathbf{w}_1, \ldots, \mathbf{w}_n, \mathbf{v}_1, \ldots, \mathbf{v}_m$ (where $m = n^3$) independently at random from the distribution $\mathcal{H}_u$, subject to the constraint that the parallelepiped w spanned by the $\mathbf{w}_i$'s is not flat. Thus, the public key consists of a polynomial number of points close to a collection of parallel affine hyperplanes, which is kept secret.

The scheme is mainly of theoretical purpose, as encryption is bit-by-bit. To encrypt a '0', one randomly selects $b_1, \ldots, b_m$ in $\{0, 1\}$, and reduces $\sum_{i=1}^{m} b_i \mathbf{v}_i$ modulo the parallelepiped w. The vector obtained is the ciphertext. The ciphertext of '1' is just a randomly chosen vector in the parallelepiped w. To decrypt a ciphertext $\mathbf{x}$ with the private key $\mathbf{u}$, one computes $\tau = \langle \mathbf{x}, \mathbf{u} \rangle$. If τ is sufficiently close to $\mathbb{Z}$, then $\mathbf{x}$ is decrypted as '0', and otherwise as '1'. Thus, an encryption of '0' will always be decrypted as '0', and an encryption of '1' has a small probability to be decrypted as '0'. These decryption errors can be removed (see [60]).

Security. The Ajtai-Dwork [5] cryptosystem received wide attention due to a surprising security proof based on worst-case assumptions. Indeed, it was shown that any probabilistic algorithm distinguishing encryptions of a '0' from encryptions of a '1' with some polynomial advantage can be used to solve SVP in any n-dimensional lattice with gap λ_2/λ_1 larger than n^8. There is a converse, due to Nguyen and Stern [115]: one can decrypt in polynomial time with high probability, provided an oracle that approximates SVP to within $n^{0.5-\varepsilon}$, or one that approximates CVP to within $n^{1.33}$. It follows that the problem of decrypting ciphertexts is unlikely to be NP-hard, due to the result of Goldreich-Goldwasser [58].

Nguyen and Stern [115] further presented a heuristic attack to recover the secret key. Experiments suggest that the attack is likely to succeed up to at least $n = 32$. For such parameters, the system is already impractical, as the public key requires 20 megabytes and the ciphertext for each bit has bit-length 6144. This shows that unless major improvements[3] are found, the Ajtai-Dwork cryptosystem is only of theoretical importance.

Cryptanalysis Overview. At this point, the reader might wonder how lattices come into play, since the description of AD does not involve lattices. Any ciphertext of '0' is a sum of $\mathbf{v}_i$'s minus some integer linear combination of the $\mathbf{w}_i$'s. Since the parallelepiped spanned by the $\mathbf{w}_i$'s is not too flat, the coefficients of the linear combination are relatively small. On the other hand, any linear combination of the $\mathbf{v}_i$'s and the $\mathbf{w}_i$'s with small coefficients is close to the hidden hyperplanes. This enables to build a particular lattice of dimension $n+m$ such that any ciphertext of '0' is in some sense close to the lattice, and reciprocally, any point sufficiently close to the lattice gives rise to a ciphertext of '0'. Thus, one can decrypt ciphertexts provided an oracle that approximates CVP sufficiently well. The analogous version for SVP uses related ideas, but is technically more complicated. For more details, see [115].

The attack to recover the secret key can be described quite easily. One knows that each $\langle \mathbf{v}_i, \mathbf{u} \rangle$ is close to some unknown integer V_i. It can be shown that any sufficiently short linear combination of the $\mathbf{v}_i$'s give information on the V_i's. More precisely, if $\sum_i \lambda_i \mathbf{v}_i$ is sufficiently short and the λ_i's are sufficiently small, then $\sum_i \lambda_i V_i = 0$ (because it is a too small integer). Note that the V_i's are disclosed if enough such equations are found. And each V_i gives an approximate linear equation satisfied by the coefficients of the secret key $\mathbf{u}$. Thus, one can compute a sufficiently good approximation of $\mathbf{u}$ from the V_i's. To find the V_i's, we produce many short combinations $\sum_i \lambda_i \mathbf{v}_i$ with small λ_i's, using lattice reduction. Heuristic arguments can justify that there exist enough such combinations. Experiments showed that the assumption was reasonable in practice.

5.2 The Goldreich-Goldwasser-Halevi Cryptosystem

The Goldreich-Goldwasser-Halevi cryptosystem [61] (GGH) can be viewed as a lattice-analog to the McEliece [94] cryptosystem based on algebraic coding theory. In both schemes, a ciphertext is the addition of a random noise vector to a vector corresponding to the plaintext. The public key and the private key are two representations of the same object (a lattice for GGH, a linear code for McEliece). The private key has a particular structure allowing to cancel noise

[3] A variant of AD with less message expansion was proposed in [32], however without any security proof. It mixes AD with a knapsack.

vectors up to a certain bound. However, the domains in which all these operations take place are quite different.

Description. The GGH scheme works in $\mathbb{Z}^n$. The private key is a non-singular $n \times n$ integral matrix R, with very short row vectors[4] (entries polynomial in n). The lattice L is the full-dimensional lattice in $\mathbb{Z}^n$ spanned by the rows of R. The basis R is then transformed to a non-reduced basis B, which will be public. In the original scheme, B is the multiplication of R by sufficiently many small unimodular matrices. Computing a basis as "good" as the private basis R, given only the non-reduced basis B, means approximating SBP.

The message space is a "large enough" cube in $\mathbb{Z}^n$. A message $\mathbf{m} \in \mathbb{Z}^n$ is encrypted into $\mathbf{c} = \mathbf{m}B + \mathbf{e}$ where $\mathbf{e}$ is an error vector uniformly chosen from $\{-\sigma, \sigma\}^n$, where σ is a security parameter. A ciphertext $\mathbf{c}$ is decrypted as $\lfloor \mathbf{c}R^{-1} \rceil RB^{-1}$ (note: this is Babai's round method [8] to solve CVP). But an eavesdropper is left with the CVP-instance defined by $\mathbf{c}$ and B. The private basis R is generated in such a way that the decryption process succeeds with high probability. The larger σ is, the harder the CVP-instances are expected to be. But σ must be small for the decryption process to succeed.

Improvements. In the original scheme, the public matrix B is the multiplication of the secret matrix by sufficiently many unimodular matrices. This means that without appropriate precaution, the public matrix may be as large as $O(n^3 \log n)$ bits. Micciancio [99, 101] therefore suggested to define instead B as the Hermite normal form (HNF) of R. Recall that the HNF of an integer square matrix R in row notation is the unique lower triangular matrix with coefficients in $\mathbb{N}$ such that: the rows span the same lattice as R, and any entry below the diagonal is strictly less than the diagonal entry in its column. Here, one can see that the HNF of R is $O(n^2 \log n)$ bits, which is much better but still big. When using the HNF, one should encode messages into the error vector $\mathbf{e}$ instead of a lattice point, because the HNF is unbalanced. The ciphertext is defined as the reduction of $\mathbf{e}$ modulo the HNF, and hence uses less than $O(n \log n)$ bits. One can easily prove that the new scheme (which is now deterministic) cannot be less secure than the original GGH scheme (see [99, 101]).

Security. GGH has no proven worst-case/average-case property, but it is much more efficient than AD. Specifically, for security parameter n, key-size and encryption time can be $O(n^2 \log n)$ for GGH (note that McEliece is slightly better though), *vs.* at least $O(n^4)$ for AD. For RSA and El-Gamal systems, key size is $O(n)$ and computation time is $O(n^3)$. The authors of GGH argued that the increase in size of the keys was more than compensated by the decrease in computation time. To bring confidence in their scheme, they published on the Internet a series of five numerical challenges [59], in dimensions 200, 250, 300, 350 and

[4] A different construction for R based on tensor product was proposed in [50].

400. In each of these challenges, a public key and a ciphertext were given, and the challenge was to recover the plaintext.

The GGH scheme is now considered broken, at least in its original form, due to an attack recently developed by Nguyen [108]. As an application, using small computing power and Shoup's NTL library [127], Nguyen was able to solve all the GGH challenges, except the last one in dimension 400. But already in dimension 400, GGH is not very practical: in the 400-challenge, the public key takes 1.8 Mbytes without HNF or 124 Kbytes using the HNF.[5]

Nguyen's attack used two "qualitatively different" weaknesses of GGH. The first one is inherent to the GGH construction: the error vectors used in the encryption process are always much shorter[6] than lattice vectors. This makes CVP-instances arising from GGH easier than general CVP-instances. The second weakness is the particular form of the error vectors in the encryption process. Recall that $\mathbf{c} = \mathbf{m}B + \mathbf{e}$ where $\mathbf{e} \in \{\pm\sigma\}^n$. The form of $\mathbf{e}$ was apparently chosen to maximize the Euclidean norm under requirements on the infinity norm. However, if we let $\mathbf{s} = (\sigma, \ldots, \sigma)$ then $\mathbf{c} + \mathbf{s} \equiv \mathbf{m}B \pmod{2\sigma}$, which allows to guess $\mathbf{m}$ mod 2σ. Then the original closest vector problem can be reduced to finding a lattice vector within (smaller) distance $\mathbf{e}/(2\sigma)$ from $(\mathbf{c} - (\mathbf{m} \bmod 2\sigma)B)/(2\sigma)$. The simplified closest vector problem happens to be within reach (in practice) of current lattice reduction algorithms, thanks to the embedding strategy that heuristically reduces CVP to SVP. We refer to [108] for more information.

It is easy to fix the second weakness by selecting the entries of the error vector $\mathbf{e}$ at random in $\{-\sigma, \ldots, +\sigma\}$ instead of $\{\pm\sigma\}$. However, one can argue that the resulting GGH system would still not be much practical, even using [99,101]. Indeed, Nguyen's experiments [108] showed that SVP could be solved in practice up to dimensions as high as 350, for (certain) lattices with gap as small as 10. To be competitive, the new GGH system would require the hardness (in lower dimensions due to the size of the public key, even using [99]) of SVP for certain lattices of only slightly smaller gap, which means a rather smaller improvement in terms of reduction. Note also that those experiments do not support the practical hardness of Ajtai's variant of SVP in which the gap is polynomial in the lattice dimension. Besides, it is not clear how to make decryption efficient without a huge secret key (Babai's rounding requires the storage of R^{-1} or a good approximation, which could be in [61] over 1 Mbytes in dimension 400).

5.3 The NTRU Cryptosystem

Description. The NTRU cryptosystem [69], proposed by Hoffstein, Pipher and Silverman, works in the ring $R = \mathbb{Z}[X]/(X^N - 1)$. An element $F \in R$ is seen as a polynomial or a row vector: $F = \sum_{i=0}^{N-1} F_i x^i = [F_0, F_1, \ldots, F_{N-1}]$. To select keys, one uses the set $\mathcal{L}(d_1, d_2)$ of polynomials $F \in R$ such that d_1 coefficients

[5] The challenges do not use the HNF, as they were proposed before [99]. Note that 124 Kbytes is about twice as large as McEliece for the recommended parameters.

[6] In all GGH-like constructions known, the error vector is always at least twice as short.

are equal to 1, d_2 coeffients are equal to -1, and the rest are zero. There are two small coprime moduli $p < q$: a possible choice is $q = 128$ and $p = 3$. There are also three integer parameters d_f, d_g and d_ϕ quite a bit smaller than the prime number N (which is around a few hundreds).

The private keys are $f \in \mathcal{L}(d_f, d_f - 1)$ and $g \in \mathcal{L}(d_g, d_g)$. With high probability, f is invertible mod q. The public key $h \in R$ is defined as $h = g/f \mod q$. A message $m \in \{-(p-1)/2 \cdots +(p-1)/2\}^N$ is encrypted into: $e = (p\phi * h + m) \mod q$, where ϕ is randomly chosen in $\mathcal{L}(d_\phi, d_\phi)$. The user can decrypt thanks to the congruence $e * f \equiv p\phi * g + m * f \pmod q$, where the reduction is centered (one takes the smallest residue in absolute value). Since ϕ, f, g and m all have small coefficients and many zeroes (except possibly m), that congruence is likely to be a polynomial equality over $\mathbb{Z}$. By further reducing $e * f$ modulo p, one thus recovers $m * f \mod q$, hence m.

Security. The best attack known against NTRU is based on lattice reduction, but this does not mean that lattice reduction is necessary to break NTRU. The simplest lattice-based attack can be described as follows. Coppersmith and Shamir [40] noticed that the target vector $f||g \in \mathbb{Z}^{2N}$ (the symbol || denotes vector concatenation) belongs to the following natural lattice:

$$L_{CS} = \{F||G \in \mathbb{Z}^{2N} \mid F \equiv h * G \mod q \text{ where } F, G \in R\}.$$

It is not difficult to see that L_{CS} is a full-dimensional lattice in $\mathbb{Z}^{2N}$, with volume q^N. The volume suggests that the target vector is a shortest vector of L_{CS} (but with small gap), so that a SVP-oracle should heuristically output the private keys f and g. However, based on numerous experiments with Shoup's NTL library [127], the authors of NTRU claimed in [69] that all such attacks are exponential in N, so that even reasonable choices of N ensure sufficient security. The parameter N must be prime, otherwise the lattice attacks can be improved due to the factorization of $X^N - 1$ (see [56]). Note that the keysize of NTRU is only $O(N \log q)$, which makes NTRU the leading candidate among knapsack-based and lattice-based cryptosystems, and allows high lattice dimensions. It seems that better attacks or better lattice reduction algorithms are required in order to break NTRU. To date, none of the numerical challenges proposed in [69] has been solved. However, it is probably too early to tell whether or not NTRU is secure. Note that NTRU, like RSA, should only be used with appropriate preprocessing. Indeed, NTRU without padding cannot be semantically secure since $e(1) \equiv m(1) \pmod q$ as polynomials, and it is easily malleable using multiplications by X of polynomials (circular shifts). And there exist simple chosen ciphertext attacks [74] that can recover the secret key.

6 Finding Small Roots of Low-Degree Polynomial Equations

We survey an important application of lattice reduction found in 1996 by Coppersmith [38,39], and its developments. These results illustrate the power of linearization combined with lattice reduction.

6.1 Univariate Modular Equations

The general problem of solving univariate polynomial equations modulo some integer N of unknown factorization seems to be hard. Indeed, notice that for some polynomials, it is equivalent to the knowledge of the factorization of N. And the particular case of extracting e-th roots modulo N is the problem of decrypting ciphertexts in the RSA cryptosystem, for an eavesdropper. Curiously, Coppersmith [38] showed using LLL that the special problem of finding small roots is easy:

Theorem 5 (Coppersmith). *Let P be a monic polynomial of degree δ in one variable modulo an integer N of unknown factorization. Then one can find in time polynomial in $(\log N, \delta)$ all integers x_0 such that $P(x_0) \equiv 0 \pmod{N}$ and $|x_0| \leq N^{1/\delta}$.*

Related (but weaker) results appeared in the eighties [66,133].[7] Incidentally, the result implies that the number of roots less than $N^{1/\delta}$ is polynomial, which was also proved in [81] (using elementary techniques).

We sketch a proof of Theorem 5, in the spirit of Howgrave-Graham [70], who simplified Coppersmith's original proof (see also [76]) by working in the dual lattice of the lattice originally considered by Coppersmith. More details can be found in [39]. Coppersmith's method reduces the problem of finding small modular roots to the (easy) problem of solving polynomial equations over $\mathbb{Z}$. More precisely, it applies lattice reduction to find an integral polynomial equation satisfied by all small modular roots of P. The intuition is to linearize all the equations of the form $x^i P(x)^j \equiv 0 \pmod{N^j}$ for appropriate integral values of i and j. Such equations are satisfied by any solution of $P(x) \equiv 0 \pmod{N}$. Small solutions x_0 give rise to unusually short solutions to the resulting linear system. To transform modular equations into integer equations, the following elementary lemma[8] is used, with the notation $\|r(x)\| = \sqrt{\sum a_i^2}$ for any polynomial $r(x) = \sum a_i x^i \in \mathbb{Q}[x]$:

Lemma 6. *Let $r(x) \in \mathbb{Q}[x]$ be a polynomial of degree $< n$ and let X be a positive integer. Suppose $\|r(xX)\| < 1/\sqrt{n}$. If $r(x_0) \in \mathbb{Z}$ with $|x_0| < X$, then $r(x_0) = 0$ holds over the integers.*

[7] Håstad [66] presented his result in terms of system of low-degree modular equations, but he actually studies the same problem, and his approach achieves the weaker bound $N^{2/(\delta(\delta+1))}$.

[8] A similar lemma is used in [66]. Note also the resemblance with [88, Prop. 2.7].

This is just because any sufficiently small integer must be zero. Now the trick is to, given a parameter h, consider the $n = (h+1)\delta$ polynomials $q_{u,v}(x) = x^u(P(x)/N)^v$, where $0 \leq u \leq \delta - 1$ and $0 \leq v \leq h$. Notice that the degree of $q_{u,v}(x)$ is strictly less than n, and that the evaluation of $q_{u,v}(x)$ at any root x_0 of $P(x)$ modulo N is necessarily an integer. The same is true for any integral linear combination $r(x)$ of the $q_{u,v}(x)$'s. If such a combination $r(x)$ further satisfies $\|r(xX)\| < 1/\sqrt{n}$, then by Lemma 6, solving the equation $r(x) = 0$ over $\mathbb{Z}$ yields all roots of $P(x)$ modulo N less than X in absolute value. This suggests to look for a short vector in the lattice corresponding to the $q_{u,v}(xX)$'s. More precisely, define the $n \times n$ matrix M whose i-th row consists of the coefficients of $q_{u,v}(xX)$, starting by the low-degree terms, where $v = \lfloor (i-1)/\delta \rfloor$ and $u = (i-1) - \delta v$. Notice that M is lower triangular, and a simple calculation leads to $\det(M) = X^{n(n-1)/2}N^{-nh/2}$. We apply an LLL-reduction to the full-dimensional lattice spanned by the rows of M. The first vector of the reduced basis corresponds to a polynomial of the form $r(xX)$, and has Euclidean norm $\|r(xX)\|$. The theoretical bounds of the LLL algorithm ensure that:

$$\|r(xX)\| \leq 2^{(n-1)/4}\det(M)^{1/n} = 2^{(n-1)/4}X^{(n-1)/2}N^{-h/2}.$$

Recall that we need $\|r(xX)\| \leq 1/\sqrt{n}$ to apply the lemma. Hence, for a given h, the method is guaranteed to find modular roots up to X if:

$$X \leq \frac{1}{\sqrt{2}}N^{h/(n-1)}n^{-1/(n-1)}.$$

The limit of the upper bound, when h grows to ∞, is $\frac{1}{\sqrt{2}}N^{1/\delta}$. Theorem 5 follows from an appropriate choice of h. This result is practical (see [42,71] for experimental results) and has many applications. It can be used to attack RSA encryption when a very low public exponent is used (see [16] for a survey). Boneh *et al.* [21] applied it to factor efficiently numbers of the form $N = p^r q$ for large r. Boneh [17] used a variant to find smooth numbers in short interval. See also [13] for an application to Chinese remaindering in the presence of noise, and [72] to find approximate integer common divisors. Curiously, Coppersmith's theorem was also recently used in security proofs of factoring-based schemes (see [128, 18]).

Remarks. Theorem 5 is trivial if $P(x) = x^\delta + c$. Note also that one cannot hope to improve the (natural) bound $N^{1/\delta}$ for all polynomials and all moduli N. Indeed, for the polynomial $P(x) = x^\delta$ and $N = p^\delta$ where p is prime, the roots of P mod N are the multiples of p. Thus, one cannot hope to find all the small roots (slightly) beyond $N^{1/\delta} = p$, because there are too many of them. This suggests that even an SVP-oracle (instead of LLL) should not help Theorem 5 in general, as evidenced by the value of the lattice volume (the fudge factor $2^{(n-1)/4}$ yielded by LLL is negligible compared to $\det(M)^{1/n}$). It was recently noticed in [13] that if one only looks for the smallest root mod N, an SVP-oracle can improve the bound $N^{1/\delta}$ for very particular moduli (namely, squarefree N of known factorization, without too small factors). Note that in such cases, finding modular roots can still be difficult, because the number of modular roots can be exponential in the number of prime factors of N. Coppersmith discusses potential improvements in [39].

6.2 Multivariate Modular Equations

Interestingly, Theorem 5 can heuristically extend to multivariate polynomial modular equations. Assume for instance that one would like to find all small roots of $P(x, y) \equiv 0 \pmod N$, where $P(x, y)$ has total degree δ and has at least one monic monomial $x^\alpha y^{\delta-\alpha}$ of maximal total degree. If one could obtain two algebraically independent integral polynomial equations satisfied by all sufficiently small modular roots (x, y), then one could compute (by resultant) a univariate integral polynomial equation satisfied by x, and hence find efficiently all small (x, y). To find such equations, one can use an analogue of lemma 6 to bivariate polynomials, with the (natural) notation $\|r(x,y)\| = \sqrt{\sum_{i,j} a_{i,j}^2}$ for $r(x, y) = \sum_{i,j} a_{i,j} x^i y^j$:

Lemma 7. *Let $r(x,y) \in \mathbb{Q}[x,y]$ be a sum of at most w monomials. Assume $\|r(xX, yY)\| < 1/\sqrt{w}$ for some $X, Y \geq 0$. If $r(x_0, y_0) \in \mathbb{Z}$ with $|x_0| < X$ and $|y_0| < Y$, then $r(x_0, y_0) = 0$ holds over the integers.*

By analogy, one chooses a parameter h and select $r(x, y)$ as a linear combination of the polynomials $q_{u_1,u_2,v}(x, y) = x^{u_1} y^{u_2} (P(x, y)/N)^v$, where $u_1 + u_2 + \delta v \leq h\delta$ and $u_1, u_2, v \geq 0$ with $u_1 < \alpha$ or $u_2 < \delta - \alpha$. Such polynomials have total degree less than $h\delta$, and therefore are linear combinations of the $n = (h\delta + 1)(h\delta + 2)/2$ monic monomials of total degree $\leq \delta h$. Due to the condition $u_1 < \alpha$ or $u_2 < \delta - \alpha$, such polynomials are in bijective correspondence with the n monic monomials (associate to $q_{u_1,u_2,v}(x, y)$ the monomial $x^{u_1+v\alpha} y^{u_2+v(\delta-\alpha)}$). One can represent the polynomials as n-dimensional vectors in such a way that the $n \times n$ matrix consisting of the $q_{u_1,u_2,v}(xX, yY)$'s (for some ordering) is lower triangular with coefficients $N^{-v} X^{u_1+v\delta} Y^{u_2+v(\delta-\alpha)}$ on the diagonal.

Now consider the first two vectors $r_1(xX, yY)$ and $r_2(xX, yY)$ of an LLL-reduced basis of the lattice spanned by the rows of that matrix. Since the rational $q_{u_1,u_2,v}(x_0, y_0)$ is actually an integer for any root (x_0, y_0) of $P(x, y)$ modulo N, we need $\|r_1(xX, yY)\|$ and $\|r_2(xX, yY)\|$ to be less than $1/\sqrt{n}$ to apply Lemma 7. A (tedious) computation of the triangular matrix determinant enables to prove that $r_1(x, y)$ and $r_2(x, y)$ satisfy that bound when $XY < N^{1/\delta - \varepsilon}$ and h is sufficiently large (see [76]). Thus, one obtains two integer polynomial bivariate equations satisfied by all small modular roots of $P(x, y)$.

The problem is that, although such polynomial equations are linearly independent as vectors, they might be algebraically dependent, making the method heuristic. This heuristic assumption is unusual: many lattice-based attacks are heuristic in the sense that they require traditional lattice reduction algorithms to behave like SVP-oracles. An important open problem is to find sufficient conditions to make Coppersmith's method provable for bivariate (or multivariate) equations. Note that the method cannot work all the time. For instance, the polynomial $x - y$ has clearly too many roots over $\mathbb{Z}^2$ and hence too many roots mod any N (see [38] for more general counterexamples).

Such a result may enable to prove several attacks which are for now, only heuristic. Indeed, there are applications to the security of the RSA encryption scheme when a very low public exponent or a low private exponent is used (see [16] for a survey), and related schemes such as the KMOV cryptosystem (see [12]). In particular, the experimental evidence of [19, 12, 46] shows that the method is very effective in practice for certain polynomials.

Remarks. In the case of univariate polynomials, there was basically no choice over the polynomials $q_{u,v}(x) = x^u(P(x)/N)^v$ used to generate the appropriate univariate integer polynomial equation satisfied by all small modular roots. There is much more freedom with bivariate modular equations. Indeed, in the description above, we selected the indices of the polynomials $q_{u_1,u_2,v}(x, y)$ in such a way that they corresponded to all the monomials of total degree $\leq h\delta$, which form a triangle in $\mathbb{Z}^2$ when a monomial $x^i y^j$ is represented by the point (i, j). This corresponds to the general case where a polynomial may have several monomials of maximal total degree. However, depending on the shape of the polynomial $P(x, y)$ and the bounds X and Y, other regions of (u_1, u_2, v) might lead to better bounds.

Assume for instance $P(x, y)$ is of the form $x^{\delta_x} y^{\delta_y}$ plus a linear combination of $x^i y^j$'s where $i \leq \delta_x$, $j \leq \delta_y$ and $i + j < \delta_x + \delta_y$. Intuitively, it is better to select the (u_1, u_2, v)'s to cover the rectangle of sides $h\delta_x$ and $h\delta_y$ instead of the previous triangle, by picking all $q_{u_1,u_2,v}(x, y)$ such that $u_1 + v\delta_x \leq h\delta_x$ and $u_2 + v\delta_y \leq h\delta_y$, with $u_1 < \delta_x$ or $u_2 < \delta_y$. One can show that the polynomials $r_1(x, y)$ and $r_2(x, y)$ obtained from the first two vectors of an LLL-reduced basis of the appropriate lattice satisfy Lemma 7, provided that h is sufficiently large, and the bounds satisfy $X^{\delta_x} Y^{\delta_y} \leq N^{2/3-\varepsilon}$. Boneh and Durfee [19] applied similar and other tricks to a polynomial of the form $P(x, y) = xy + ax + b$. This allowed better bounds than the generic bound, leading to improved attacks on RSA with low secret exponent (see also [46] for an extension to the trivariate case, useful when the RSA primes are unbalanced).

6.3 Multivariate Integer Equations

The general problem of solving multivariate polynomial equations over $\mathbb{Z}$ is also hard, as integer factorization is a special case. Coppersmith [38] showed that a similar[9] lattice-based approach can be used to find small roots of bivariate polynomial equations over $\mathbb{Z}$:

Theorem 8 (Coppersmith). *Let $P(x, y)$ be a polynomial in two variables over $\mathbb{Z}$, of maximum degree δ in each variable separately, and assume the coefficients*

[9] However current proofs are somehow more technical than for Theorem 5. A simplification analogue to what has been obtained for Theorem 5 would be useful.

of f are relatively prime as a set. Let X, Y be bounds on the desired solutions x_0, y_0. Define $\hat{P}(x,y) = P(Xx, Yy)$ and let D be the absolute value of the largest coefficient of $\hat{P}$. If $XY < D^{2/(3\delta)}$, then in time polynomial in $(\log D, \delta)$, we can find all integer pairs (x_0, y_0) such that $P(x_0, y_0) = 0$, $|x_0| < X$ and $|y_0| < Y$.

Again, the method extends heuristically to more than two variables, and there can be improved bounds depending on the shape[10] of the polynomial (see [38]). Theorem 8 was introduced to factor in polynomial time an RSA–modulus[11] $N = pq$ provided that half of the (either least or most significant) bits of either p or q are known (see [38,17,20]). This was sufficient to break an ID-based RSA encryption scheme proposed by Vanstone and Zuccherato [134]. Boneh *et al.* [20] provide another application, for recovering the RSA secret key when a large fraction of the bits of the secret exponent is known. Curiously, none of the applications cited above happen to be "true" applications of Theorem 8. It was later realized in [71,21] that those results could alternatively be obtained from a (simple) variant of the univariate modular case (Theorem 5).

7 Lattices and RSA

Section 3 and 6 suggest to clarify the links existing between lattice reduction and RSA [120], the most famous public-key cryptosystem. We refer to [95] for an exposition of RSA, and to [16] for a survey of attacks on RSA encryption. Recall that in RSA, one selects two prime numbers p and q of approximately the same size. The number $N = pq$ is public. One selects an integer d coprime with $\phi(N) = (p-1)(q-1)$. The integer d is the private key, and is called the RSA *secret exponent*. The *public exponent* is the inverse e of d modulo $\phi(N)$.

7.1 Lattice Attacks on RSA Encryption

Small Public Exponent. When the public exponent e is very small, such as 3, one can apply Coppersmith's method (seen in the previous section) for univariate polynomials in various settings (see [16,38,42] for exact statements):

- An attacker can recover the plaintext of a given ciphertext, provided a large part of the plaintext is known.
- If a message is randomized before encryption, by simply padding random bits at a known place, an attacker can recover the message provided the amount of randomness is small.
- Håstad [66] attacks can be improved. An attacker can recover a message broadcasted (by RSA encryption and known affine transformation) to sufficiently many participants, each holding a different modulus N. This precisely happens if one sends a similar message with different known headers or time-stamps which are part of the encryption block.

[10] The coefficient 2/3 is natural from the remarks at the end of the previous section for the bivariate modular case. If we had assumed P to have total degree δ, the bound would be $XY < D^{1\delta}$.

[11] p and q are assumed to have similar size.

None of the attacks recover the secret exponent d: they can only recover the plaintext. The attacks do not work if appropriate padding is used (see current standards and [95]), or if the public exponent is not too small. For instance, the popular choice $e = 65537$ is not threatened by these attacks.

Small Private Exponent. When $d \leq N^{0.25}$, an old result of Wiener [137] shows that one can easily recover the secret exponent d (and thus the factorization of N) from the continued fractions algorithm. Boneh and Durfee [19] recently improved the bound to $d \leq N^{0.292}$, by applying Coppersmith's technique to bivariate modular polynomials and improving the generic bound. Note that the attack is heuristic (see Section 6), but experiments showed that it works well in practice (no counterexample has ever been found). This bound holds when the RSA primes are balanced: Durfee and Nguyen [46] improved the bound when the primes are unbalanced, using an extension to trivariate modular polynomials. All those attacks on RSA with small private exponent also hold against the RSA signature scheme, since they only use the public key. A related result (using Coppersmith's technique for either bivariate integer or univariate modular polynomials) is an attack [20] to recover d when a large portion of the bits of d is known (see [16]).

7.2 Lattice Attacks on RSA Signature

The RSA cryptosystem is often used as a digital signature scheme. To prevent various attacks, one must apply a preprocessing scheme to the message, prior to signature. The recommended solution is to use hash functions and appropriate padding (see current standards and [95]). However, several alternative simple solutions not involving hashing have been proposed, and sometimes accepted as standards. Today, all such solutions have been broken (see [57]), some of them by lattice reduction techniques (see [104, 57]). Those lattice attacks are heuristic but work well in practice. They apply lattice reduction algorithms to find small solutions to modular linear systems, which leads to signature forgeries for certain proposed RSA signature schemes. Finding such small solutions is viewed as a closest vector problem for some norm, as seen in Section 3.4.

7.3 Security of RSA–OAEP

Although no efficient method is known to invert the RSA encryption function in general, it is widely accepted that the RSA encryption scheme should not be directly used as such, because it does not satisfy strong security notions (see for instance [22, 95] for a simple explanation): a preprocessing function should be applied to the message prior to encryption. The most famous preprocessing scheme for RSA is OAEP proposed by Bellare and Rogaway [11], which is standardized in PKCS. The RSA–OAEP scheme was only recently proved to be strongly secure (semantic security against adaptive chosen-ciphertext attacks), under the assumption that the RSA function is hard to invert and the random oracle model. This was first proved by Shoup [128] for the particular case of public exponent 3 using Coppersmith's theorem on univariate polynomial equations, and later extended to any exponent by Fujisaki *et al.* [53]. Interestingly, the last part of the proof of [53] relied on lattices (in dimension 2) to find a small solution to a linear modular equation (see Section 3.4). Note however that the result could also have been obtained with continued fractions.

Boneh [18] recently proposed a simpler version of OAEP for the RSA and Rabin encryption functions. The proof for Rabin is based on Coppersmith's lattice-based theorem on univariate polynomial equations, while the proof for RSA uses lattices again to find small solutions of linear modular equations. It is somewhat surprising that lattices are used both to attack RSA in certain settings, and to prove the security of industrial uses of RSA.

7.4 Factoring and Lattice Reduction

In the general case, the best attack against RSA encryption or signature is integer factorization. Note that to prove (or disprove) the equivalence between integer factorization and breaking RSA encryption remains an important open problem in cryptology (latest results [25] suggest that breaking RSA encryption may actually be easier). We already pointed out that in some special cases, lattice reduction leads to efficient factorization: when the factors are partially known [38], or when the number to factor has the form $p^r q$ with large r [21].

Schnorr [123] was the first to establish a link between integer factorization and lattice reduction, which was later extended by Adleman [2]. Schnorr [123] proposed a heuristic method to factor general numbers, using lattice reduction to approximate the closest vector problem in the infinity or the L_1 norm. Adleman [2] showed how to use the Euclidean norm instead, which is more suited to current lattice reduction algorithms. Those methods use the same underlying ideas as sieving algorithms (see [36]): to factor a number n, they try to find many congruences of smooth numbers to produce random square congruences of the form $x^2 \equiv y^2 \pmod{n}$, after a linear algebra step. Heuristic assumptions are needed to ensure the existence of appropriate congruences. The problem of finding such congruences is seen as a closest vector problem. Still, it should be noted that those methods are theoretical, since they are not adapted to currently known lattice reduction algorithms. To be useful, they would require very good lattice reduction for lattices of dimension over at least several thousands.

We close this review by mentioning that current versions of the Number Field Sieve (NFS) (see [87,36]), the best algorithm known for factoring large integers, use lattice reduction. Indeed, LLL plays a crucial role in the last stage of NFS where one has to compute an algebraic square root of a huge algebraic number given as a product of hundreds of thousands of small ones. The best algorithm known to solve this problem is due to Montgomery (see [105,107]). It has been used in all recent large factorizations, notably the record factorization [34] of a 512-bit RSA-number of 155 decimal digits proposed in the RSA challenges. There, LLL is applied many times in low dimension (less than 10) to find nice algebraic integers in integral ideals. But the overall running time of NFS is dominated by other stages, such as sieving and linear algebra.

8 Conclusions

The LLL algorithm and other lattice basis reduction algorithms have proved invaluable in cryptology. They have become the most popular tool in public-key cryptanalysis. In particular, they play a crucial rôle in several attacks against the RSA cryptosystem. The past few years have seen new, sometimes provable, lattice-based methods for solving problems which were *a priori* not linear, and this definitely opens new fields of applications. Interestingly, several provable lattice-based results introduced in cryptanalysis have also recently been used in the area of security proofs. Paradoxically, at the same time, a series of complexity results on lattice reduction has emerged, giving rise to another family of cryptographic schemes based on the hardness of lattice problems. The resulting cryptosystems have enjoyed different fates, but it is probably too early to tell whether or not secure and practical cryptography can be built using hardness of lattice problems. Indeed, several questions on lattices remain open. In particular, we still do not know whether or not it is easy to approximate the shortest vector problem up to some polynomial factor, or to find the shortest vector when the lattice gap is larger than some polynomial in the dimension. Besides, only very few lattice basis reduction algorithms are known, and their behaviour (both complexity and output quality) is still not well understood. And so far, there has not been any massive computer experiment in lattice reduction comparable to what has been done for integer factorization or the elliptic curve discrete logarithm problem. Twenty years of lattice reduction yielded surprising applications in cryptology. We hope the next twenty years will prove as exciting.

Acknowledgements

We thank Dan Boneh, Don Coppersmith, Glenn Durfee, Arjen and Hendrik Lenstra, Lászlό Lovász, Daniele Micciancio, Igor Shparlinski and Joe Silverman for helpful discussions and comments.

References

1. L. M. Adleman. On breaking generalized knapsack publick key cryptosystems. In *Proc. of 15th STOC*, pages 402–412. ACM, 1983.
2. L. M. Adleman. Factoring and lattice reduction. Unpublished manuscript, 1995.
3. M. Ajtai. Generating hard instances of lattice problems. In *Proc. of 28th STOC*, pages 99–108. ACM, 1996. Available at [47] as TR96-007.
4. M. Ajtai. The shortest vector problem in L_2 is NP-hard for randomized reductions. In *Proc. of 30th STOC*. ACM, 1998. Available at [47] as TR97-047.
5. M. Ajtai and C. Dwork. A public-key cryptosystem with worst-case/average-case equivalence. In *Proc. of 29th STOC*, pages 284–293. ACM, 1997. Available at [47] as TR96-065.

6. M. Ajtai, R. Kumar, and D. Sivakumar. A sieve algorithm for the shortest lattice vector problem. In *Proc. 33rd STOC*, pages 601–610. ACM, 2001.
7. S. Arora, L. Babai, J. Stern, and Z. Sweedyk. The hardness of approximate optima in lattices, codes, and systems of linear equations. *Journal of Computer and System Sciences*, 54(2):317–331, 1997.
8. L. Babai. On Lovász lattice reduction and the nearest lattice point problem. *Combinatorica*, 6:1–13, 1986.
9. W. Banaszczyk. New bounds in some transference theorems in the geometry of numbers. *Mathematische Annalen*, 296:625–635, 1993.
10. M. Bellare, S. Goldwasser, and D. Micciancio. "Pseudo-random" number generation within cryptographic algorithms: The DSS case. In *Proc. of Crypto'97*, volume 1294 of *LNCS*. IACR, Springer-Verlag, 1997.
11. M. Bellare and P. Rogaway. Optimal asymmetric encryption. In *Proc. of Eurocrypt'94*, volume 950 of *LNCS*, pages 92–111. IACR, Springer-Verlag, 1995.
12. D. Bleichenbacher. On the security of the KMOV public key cryptosystem. In *Proc. of Crypto'97*, volume 1294 of *LNCS*, pages 235–248. IACR, Springer-Verlag, 1997.
13. D. Bleichenbacher and P. Q. Nguyen. Noisy polynomial interpolation and noisy Chinese remaindering. In *Proc. of Eurocrypt '00*, volume 1807 of *LNCS*. IACR, Springer-Verlag, 2000.
14. J. Blömer and J.-P. Seifert. On the complexity of computing short linearly independent vectors and short bases in a lattice. In *Proc. of 31st STOC*. ACM, 1999.
15. D. Boneh. The decision Diffie-Hellman problem. In *Algorithmic Number Theory – Proc. of ANTS-III*, volume 1423 of *LNCS*. Springer-Verlag, 1998.
16. D. Boneh. Twenty years of attacks on the RSA cryptosystem. *Notices of the AMS*, 46(2):203–213, 1999.
17. D. Boneh. Finding smooth integers in short intervals using CRT decoding. In *Proc. of 32nd STOC*. ACM, 2000.
18. D. Boneh. Simplified OAEP for the RSA and Rabin functions. In *Proc. of Crypto 2001*, LNCS. IACR, Springer-Verlag, 2001.
19. D. Boneh and G. Durfee. Cryptanalysis of RSA with private key d less than $N^{0.292}$. In *Proc. of Eurocrypt'99*, volume 1592 of *LNCS*, pages 1–11. IACR, Springer-Verlag, 1999.
20. D. Boneh, G. Durfee, and Y. Frankel. An attack on RSA given a small fraction of the private key bits. In *Proc. of Asiacrypt'98*, volume 1514 of *LNCS*, pages 25–34. Springer-Verlag, 1998.
21. D. Boneh, G. Durfee, and N. A. Howgrave-Graham. Factoring $n = p^r q$ for large r. In *Proc. of Crypto'99*, volume 1666 of *LNCS*. IACR, Springer-Verlag, 1999.
22. D. Boneh, A. Joux, and P. Q. Nguyen. Why textbook ElGamal and RSA encryption are insecure. In *Proc. of Asiacrypt '00*, volume 1976 of *LNCS*. IACR, Springer-Verlag, 2000.
23. D. Boneh and I. E. Shparlinski. Hard core bits for the elliptic curve Diffie-Hellman secret. In *Proc. of Crypto 2001*, LNCS. IACR, Springer-Verlag, 2001.
24. D. Boneh and R. Venkatesan. Hardness of computing the most significant bits of secret keys in Diffie-Hellman and related schemes. In *Proc. of Crypto'96*, LNCS. IACR, Springer-Verlag, 1996.
25. D. Boneh and R. Venkatesan. Breaking RSA may not be equivalent to factoring. In *Proc. of Eurocrypt'98*, volume 1233 of *LNCS*, pages 59–71. Springer-Verlag, 1998.
26. V. Boyko, M. Peinado, and R. Venkatesan. Speeding up discrete log and factoring based schemes via precomputations. In *Proc. of Eurocrypt'98*, volume 1403 of *LNCS*, pages 221–235. IACR, Springer-Verlag, 1998.

27. E. F. Brickell. Solving low density knapsacks. In *Proc. of Crypto '83*. Plenum Press, 1984.
28. E. F. Brickell. Breaking iterated knapsacks. In *Proc. of Crypto '84*, volume 196 of *LNCS*. Springer-Verlag, 1985.
29. E. F. Brickell and A. M. Odlyzko. Cryptanalysis: A survey of recent results. In G. J. Simmons, editor, *Contemporary Cryptology*, pages 501–540. IEEE Press, 1991.
30. J.-Y. Cai. Some recent progress on the complexity of lattice problems. In *Proc. of FCRC*, 1999. Available at [47] as TR99-006.
31. J.-Y. Cai. The complexity of some lattice problems. In *Proc. of ANTS-IV*, volume 1838 of *LNCS*. Springer-Verlag, 2000.
32. J.-Y. Cai and T. W. Cusick. A lattice-based public-key cryptosystem. *Information and Computation*, 151:17–31, 1999.
33. J.-Y. Cai and A. P. Nerurkar. An improved worst-case to average-case connection for lattice problems. In *Proc. of 38th FOCS*, pages 468–477. IEEE, 1997.
34. S. Cavallar, B. Dodson, A. K. Lenstra, W. Lioen, P. L. Montgomery, B. Murphy, H. te Riele, K. Aardal, J. Gilchrist, G. Guillerm, P. Leyland, J. Marchand, F. Morain, A. Muffett, C. Putnam, and P. Zimmermann. Factorization of 512-bit RSA key using the number field sieve. In *Proc. of Eurocrypt '00*, volume 1807 of *LNCS*. IACR, Springer-Verlag, 2000.
35. B. Chor and R.L. Rivest. A knapsack-type public key cryptosystem based on arithmetic in finite fields. *IEEE Trans. Inform. Theory*, 34, 1988.
36. H. Cohen. *A Course in Computational Algebraic Number Theory*. Springer-Verlag, 1995. Second edition.
37. J.H. Conway and N.J.A. Sloane. *Sphere Packings, Lattices and Groups*. Springer-Verlag, 1998. Third edition.
38. D. Coppersmith. Small solutions to polynomial equations, and low exponent RSA vulnerabilities. *J. of Cryptology*, 10(4):233–260, 1997. Revised version of two articles from Eurocrypt'96.
39. D. Coppersmith. Finding small solutions to small degree polynomials. In *Proc. of CALC 2001*, LNCS. Springer-Verlag, 2001.
40. D. Coppersmith and A. Shamir. Lattice attacks on NTRU. In *Proc. of Eurocrypt'97*, LNCS. IACR, Springer-Verlag, 1997.
41. M.J. Coster, A. Joux, B.A. LaMacchia, A.M. Odlyzko, C.-P. Schnorr, and J. Stern. Improved low-density subset sum algorithms. *Comput. Complexity*, 2:111–128, 1992.
42. C. Coupé, P. Q. Nguyen, and J. Stern. The effectiveness of lattice attacks against low-exponent RSA. In *Proc. of PKC'98*, volume 1431 of *LNCS*. Springer-Verlag, 1999.
43. W. Diffie and M. E. Hellman. New directions in cryptography. *IEEE Trans. Inform. Theory*, IT-22:644–654, Nov 1976.
44. I. Dinur. Approximating SVP_∞ to within almost-polynomial factors is NP-hard. Available at [47] as TR99-016.
45. I. Dinur, G. Kindler, and S. Safra. Approximating CVP to within almost-polynomial factors is NP-hard. In *Proc. of 39th FOCS*, pages 99–109. IEEE, 1998. Available at [47] as TR98-048.
46. G. Durfee and P. Q. Nguyen. Cryptanalysis of the RSA schemes with short secret exponent from Asiacrypt'99. In *Proc. of Asiacrypt '00*, volume 1976 of *LNCS*. IACR, Springer-Verlag, 2000.
47. ECCC. `http://www.eccc.uni-trier.de/eccc/`. The Electronic Colloquium on Computational Complexity.
48. E. El Mahassni, P. Q. Nguyen, and I. E. Shparlinski. The insecurity of Nyberg–Rueppel and other DSA-like signature schemes with partially known nonces. In *Proc. of CALC 2001*, LNCS. Springer-Verlag, 2001.

49. P. van Emde Boas. Another NP-complete problem and the complexity of computing short vectors in a lattice. Technical report, Mathematische Instituut, University of Amsterdam, 1981. Report 81-04.
Available at `http://turing.wins.uva.nl/~peter/`.
50. R. Fischlin and J.-P. Seifert. Tensor-based trapdoors for CVP and their application to public key cryptography. In *IMA Conference on Cryptography and Coding*, LNCS. Springer-Verlag, 1999.
51. A. M. Frieze. On the Lagarias-Odlyzko algorithm for the subset sum problem. *SIAM J. Comput*, 15(2):536–539, 1986.
52. A. M. Frieze, J. Håstad, R. Kannan, J. C. Lagarias, and A. Shamir. Reconstructing truncated integer variables satisfying linear congruences. *SIAM J. Comput.*, 17(2):262–280, 1988. Special issue on cryptography.
53. E. Fujisaki, T. Okamoto, D. Pointcheval, and J. Stern. RSA–OAEP is secure under the RSA assumption. In *Proc. of Crypto 2001*, LNCS. IACR, Springer-Verlag, 2001.
54. M. L. Furst and R. Kannan. Succinct certificates for almost all subset sum problems. *SIAM J. Comput*, 18(3):550–558, 1989.
55. C.F. Gauss. *Disquisitiones Arithmeticæ*. Leipzig, 1801.
56. C. Gentry. Key recovery and message attacks on NTRU-composite. In *Proc. of Eurocrypt 2001*, volume 2045 of *LNCS*. IACR, Springer-Verlag, 2001.
57. M. Girault and J.-F. Misarsky. Cryptanalysis of countermeasures proposed for repairing ISO 9796–1. In *Proc. of Eurocrypt '00*, volume 1807 of *LNCS*. IACR, Springer-Verlag, 2000.
58. O. Goldreich and S. Goldwasser. On the limits of non-approximability of lattice problems. In *Proc. of 30th STOC*. ACM, 1998. Available at [47] as TR97-031.
59. O. Goldreich, S. Goldwasser, and S. Halevi. Challenges for the GGH cryptosystem. Available at `http://theory.lcs.mit.edu/~shaih/challenge.html`.
60. O. Goldreich, S. Goldwasser, and S. Halevi. Eliminating decryption errors in the Ajtai-Dwork cryptosystem. In *Proc. of Crypto'97*, volume 1294 of *LNCS*, pages 105–111. IACR, Springer-Verlag, 1997. Available at [47] as TR97-018.
61. O. Goldreich, S. Goldwasser, and S. Halevi. Public-key cryptosystems from lattice reduction problems. In *Proc. of Crypto'97*, volume 1294 of *LNCS*, pages 112–131. IACR, Springer-Verlag, 1997. Available at [47] as TR96-056.
62. O. Goldreich, D. Micciancio, S. Safra, and J.-P. Seifert. Approximating shortest lattice vectors is not harder than approximating closest lattice vectors, 1999. Available at [47] as TR99-002.
63. M. I. González Vasco and I. E. Shparlinski. On the security of Diffie-Hellman bits. In K.-Y. Lam, I. E. Shparlinski, H. Wang, and C. Xing, editors, *Proc. Workshop on Cryptography and Comp. Number Theory (CCNT'99)*. Birkhauser, 2000.
64. M. Grötschel, L. Lovász, and A. Schrijver. *Geometric Algorithms and Combinatorial Optimization*. Springer-Verlag, 1993.
65. M. Gruber and C. G. Lekkerkerker. *Geometry of Numbers*. North-Holland, 1987.
66. J. Håstad. Solving simultaneous modular equations of low degree. *SIAM J. Comput.*, 17(2):336–341, April 1988. Preliminary version in Proc. of Crypto '85.
67. B. Helfrich. Algorithms to construct Minkowski reduced and Hermite reduced bases. *Theoretical Computer Science*, 41:125–139, 1985.
68. C. Hermite. Extraits de lettres de M. Hermite à M. Jacobi sur différents objets de la théorie des nombres, deuxième lettre. *J. Reine Angew. Math.*, 40:279–290, 1850. Also available in the first volume of Hermite's complete works, published by Gauthier-Villars.
69. J. Hoffstein, J. Pipher, and J.H. Silverman. NTRU: a ring based public key cryptosystem. In *Proc. of ANTS III*, volume 1423 of *LNCS*, pages 267–288. Springer-Verlag, 1998. Additional information at `http://www.ntru.com`.

70. N. A. Howgrave-Graham. Finding small roots of univariate modular equations revisited. In *Cryptography and Coding*, volume 1355 of *LNCS*, pages 131–142. Springer-Verlag, 1997.
71. N. A. Howgrave-Graham. *Computational Mathematics Inspired by RSA*. PhD thesis, University of Bath, 1998.
72. N. A. Howgrave-Graham. Approximate integer common divisors. In *Proc. of CALC 2001*, LNCS. Springer-Verlag, 2001.
73. N. A. Howgrave-Graham and N. P. Smart. Lattice attacks on digital signature schemes. Technical report, HP Labs, 1999. HPL-1999-90. To appear in *Designs, Codes and Cryptography.*
74. E. Jaulmes and A. Joux. A chosen ciphertext attack on NTRU. In *Proc. of Crypto 2000*, volume 1880 of *LNCS*. IACR, Springer-Verlag, 2000.
75. A. Joux and J. Stern. Lattice reduction: A toolbox for the cryptanalyst. *J. of Cryptology*, 11:161–185, 1998.
76. C. S. Jutla. On finding small solutions of modular multivariate polynomial equations. In *Proc. of Eurocrypt'98*, volume 1403 of *LNCS*, pages 158–170. IACR, Springer-Verlag, 1998.
77. R. Kannan. Improved algorithms for integer programming and related lattice problems. In *Proc. of 15th STOC*, pages 193–206. ACM, 1983.
78. R. Kannan. Algorithmic geometry of numbers. *Annual review of computer science*, 2:231–267, 1987.
79. R. Kannan. Minkowski's convex body theorem and integer programming. *Math. Oper. Res.*, 12(3):415–440, 1987.
80. P. Klein. Finding the closest lattice vector when it's unusually close. In *Proc. of SODA '00*. ACM–SIAM, 2000.
81. S. V. Konyagin and T. Seger. On polynomial congruences. *Mathematical Notes*, 55(6):596–600, 1994.
82. A. Korkine and G. Zolotareff. Sur les formes quadratiques positives ternaires. *Math. Ann.*, 5:581–583, 1872.
83. A. Korkine and G. Zolotareff. Sur les formes quadratiques. *Math. Ann.*, 6:336–389, 1873.
84. J. C. Lagarias. Point lattices. In R. Graham, M. Grötschel, and L. Lovász, editors, *Handbook of Combinatorics*, volume 1, chapter 19. Elsevier, 1995.
85. J. C. Lagarias and A. M. Odlyzko. Solving low-density subset sum problems. *Journal of the Association for Computing Machinery*, January 1985.
86. L. Lagrange. Recherches d'arithmétique. *Nouv. Mém. Acad.*, 1773.
87. A. K. Lenstra and H. W. Lenstra, Jr. *The Development of the Number Field Sieve*, volume 1554 of *Lecture Notes in Mathematics.* Springer-Verlag, 1993.
88. A. K. Lenstra, H. W. Lenstra, Jr., and L. Lovász. Factoring polynomials with rational coefficients. *Mathematische Ann.*, 261:513–534, 1982.
89. H. W. Lenstra, Jr. Integer programming with a fixed number of variables. Technical report, Mathematisch Instituut, Universiteit van Amsterdam, April 1981. Report 81-03.
90. H. W. Lenstra, Jr. Integer programming with a fixed number of variables. *Math. Oper. Res.*, 8(4):538–548, 1983.
91. L. Lovász. *An Algorithmic Theory of Numbers, Graphs and Convexity*, volume 50. SIAM Publications, 1986. CBMS-NSF Regional Conference Series in Applied Mathematics.
92. J. Martinet. *Les Réseaux Parfaits des Espaces Euclidiens*. Éditions Masson, 1996. English translation to appear at Springer-Verlag.
93. J. E. Mazo and A. M. Odlyzko. Lattice points in high-dimensional spheres. *Monatsh. Math.*, 110:47–61, 1990.

94. R.J. McEliece. A public-key cryptosystem based on algebraic number theory. Technical report, Jet Propulsion Laboratory, 1978. DSN Progress Report 42-44.
95. A. Menezes, P. Van Oorschot, and S. Vanstone. *Handbook of Applied Cryptography*. CRC Press, 1997.
96. R. Merkle and M. Hellman. Hiding information and signatures in trapdoor knapsacks. *IEEE Trans. Inform. Theory*, IT-24:525–530, September 1978.
97. D. Micciancio. *On the Hardness of the Shortest Vector Problem*. PhD thesis, Massachusetts Institute of Technology, 1998.
98. D. Micciancio. The shortest vector problem is NP-hard to approximate within some constant. In *Proc. of 39th FOCS*. IEEE, 1998. Available at [47] as TR98-016.
99. D. Micciancio. Lattice based cryptography: A global improvement. Technical report, Theory of Cryptography Library, 1999. Report 99-05.
100. D. Micciancio. The hardness of the closest vector problem with preprocessing. *IEEE Trans. Inform. Theory*, 47(3):1212–1215, 2001.
101. D. Micciancio. Improving lattice-based cryptosystems using the Hermite normal form. In *Proc. of CALC 2001*, LNCS. Springer-Verlag, 2001.
102. J. Milnor and D. Husemoller. *Symmetric Bilinear Forms*. Springer-Verlag, 1973.
103. H. Minkowski. *Geometrie der Zahlen*. Teubner-Verlag, Leipzig, 1896.
104. J.-F. Misarsky. A multiplicative attack using LLL algorithm on RSA signatures with redundancy. In *Proc. of Crypto'97*, volume 1294 of *LNCS*, pages 221–234. IACR, Springer-Verlag, 1997.
105. P. L. Montgomery. Square roots of products of algebraic numbers. In Walter Gautschi, editor, *Mathematics of Computation 1943-1993: a Half-Century of Computational Mathematics*, Proc. of Symposia in Applied Mathematics, pages 567–571. American Mathematical Society, 1994.
106. National Institute of Standards and Technology (NIST). *FIPS Publication 186: Digital Signature Standard*, May 1994.
107. P. Q. Nguyen. A Montgomery-like square root for the number field sieve. In *Algorithmic Number Theory – Proc. of ANTS-III*, volume 1423 of *LNCS*. Springer-Verlag, 1998.
108. P. Q. Nguyen. Cryptanalysis of the Goldreich-Goldwasser-Halevi cryptosystem from Crypto'97. In *Proc. of Crypto'99*, volume 1666 of *LNCS*, pages 288–304. IACR, Springer-Verlag, 1999.
109. P. Q. Nguyen. *La Géométrie des Nombres en Cryptologie*. PhD thesis, Université Paris 7, November 1999. Available at `http://www.di.ens.fr/~pnguyen/`.
110. P. Q. Nguyen. The dark side of the hidden number problem: Lattice attacks on DSA. In K.-Y. Lam, I. E. Shparlinski, H. Wang, and C. Xing, editors, *Proc. Workshop on Cryptography and Comp. Number Theory (CCNT'99)*. Birkhauser, 2000.
111. P. Q. Nguyen and I. E. Shparlinski. The insecurity of the Digital Signature Algorithm with partially known nonces. *J. of Cryptology*, 2001. To appear.
112. P. Q. Nguyen and I. E. Shparlinski. The insecurity of the elliptic curve Digital Signature Algorithm with partially known nonces. *Preprint*, 2001.
113. P. Q. Nguyen and J. Stern. Merkle-Hellman revisited: a cryptanalysis of the Qu-Vanstone cryptosystem based on group factorizations. In *Proc. of Crypto'97*, volume 1294 of *LNCS*, pages 198–212. IACR, Springer-Verlag, 1997.
114. P. Q. Nguyen and J. Stern. Cryptanalysis of a fast public key cryptosystem presented at SAC '97. In *Selected Areas in Cryptography – Proc. of SAC'98*, volume 1556 of *LNCS*. Springer-Verlag, 1998.
115. P. Q. Nguyen and J. Stern. Cryptanalysis of the Ajtai-Dwork cryptosystem. In *Proc. of Crypto'98*, volume 1462 of *LNCS*, pages 223–242. IACR, Springer-Verlag, 1998.

116. P. Q. Nguyen and J. Stern. The Béguin-Quisquater server-aided RSA protocol from Crypto '95 is not secure. In *Proc. of Asiacrypt'98*, volume 1514 of *LNCS*, pages 372–379. Springer-Verlag, 1998.
117. P. Q. Nguyen and J. Stern. The hardness of the hidden subset sum problem and its cryptographic implications. In *Proc. of Crypto'99*, volume 1666 of *LNCS*, pages 31–46. IACR, Springer-Verlag, 1999.
118. P. Q. Nguyen and J. Stern. Lattice reduction in cryptology: An update. In *Proc. of ANTS-IV*, volume 1838 of *LNCS*. Springer-Verlag, 2000.
119. A. M. Odlyzko. The rise and fall of knapsack cryptosystems. In *Cryptology and Computational Number Theory*, volume 42 of *Proc. of Symposia in Applied Mathematics*, pages 75–88. A.M.S., 1990.
120. R. L. Rivest, A. Shamir, and L. M. Adleman. A method for obtaining digital signatures and public-key cryptosystems. *Communications of the ACM*, 21(2):120–126, 1978.
121. C. P. Schnorr. A hierarchy of polynomial lattice basis reduction algorithms. *Theoretical Computer Science*, 53:201–224, 1987.
122. C. P. Schnorr. A more efficient algorithm for lattice basis reduction. *J. of algorithms*, 9(1):47–62, 1988.
123. C. P. Schnorr. Factoring integers and computing discrete logarithms via diophantine approximation. In *Proc. of Eurocrypt'91*, volume 547 of *LNCS*, pages 171–181. IACR, Springer-Verlag, 1991.
124. C. P. Schnorr and M. Euchner. Lattice basis reduction: improved practical algorithms and solving subset sum problems. *Math. Programming*, 66:181–199, 1994.
125. C. P. Schnorr and H. H. Hörner. Attacking the Chor-Rivest cryptosystem by improved lattice reduction. In *Proc. of Eurocrypt'95*, volume 921 of *LNCS*, pages 1–12. IACR, Springer-Verlag, 1995.
126. A. Shamir. A polynomial time algorithm for breaking the basic Merkle-Hellman cryptosystem. In *Proc. of 23rd FOCS*, pages 145–152. IEEE, 1982.
127. V. Shoup. Number Theory C++ Library (NTL) version 3.6. Available at `http://www.shoup.net/ntl/`.
128. V. Shoup. OAEP reconsidered. In *Proc. of Crypto 2001*, LNCS. IACR, Springer-Verlag, 2001.
129. I. E. Shparlinski. On the generalized hidden number problem and bit security of XTR. In *Proc. of 14th Symp. on Appl. Algebra, Algebraic Algorithms, and Error-Correcting Codes*, LNCS. Springer-Verlag, 2001.
130. I. E. Shparlinski. Sparse polynomial approximation in finite fields. In *Proc. 33rd STOC*. ACM, 2001.
131. C. L. Siegel. *Lectures on the Geometry of Numbers*. Springer-Verlag, 1989.
132. B. Vallée. La réduction des réseaux. autour de l'algorithme de Lenstra, Lenstra, Lovász. *RAIRO Inform. Théor. Appl*, 23(3):345–376, 1989.
133. B. Vallée, M. Girault, and P. Toffin. How to guess ℓ-th roots modulo n by reducing lattice bases. In *Proc. of AAEEC-6*, volume 357 of *LNCS*, pages 427–442. Springer-Verlag, 1988.
134. S. A. Vanstone and R. J. Zuccherato. Short RSA keys and their generation. *J. of Cryptology*, 8(2):101–114, 1995.
135. S. Vaudenay. Cryptanalysis of the Chor-Rivest cryptosystem. In *Proc. of Crypto'98*, volume 1462 of *LNCS*. IACR, Springer-Verlag, 1998.
136. E. R. Verheul. Certificates of recoverability with scalable recovery agent security. In *Proc. of PKC'00*, LNCS. Springer-Verlag, 2000.
137. M. Wiener. Cryptanalysis of short RSA secret exponents. *IEEE Trans. Inform. Theory*, 36(3):553–558, 1990.

A 3-Dimensional Lattice Reduction Algorithm

Igor Semaev

Moscow State University
Laboratory of Mathematical Problems of Cryptology
Vorobievy gori, 119899, Moscow, Russia
semaev@box.ru

Abstract. The aim of this paper is a reduction algorithm for a basis b_1, b_2, b_3 of a 3-dimensional lattice in $\mathbb{R}^n$ for fixed $n \geq 3$. We give a definition of the reduced basis which is equivalent to that of the Minkowski reduced basis of a 3-dimensional lattice. We prove that for $b_1, b_2, b_3 \in \mathbb{Z}^n$, $n \geq 3$ and $|b_1|, |b_2|, |b_3| \leq M$, our algorithm takes $O(\log^2 M)$ binary operations, without using fast integer arithmetic, to reduce this basis and so to find the shortest vector in the lattice. The definition and the algorithm can be extended to any dimension. Elementary steps of our algorithm are rather different from those of the LLL-algorithm, which works in $O(\log^3 M)$ binary operations without using fast integer arithmetic.
Keywords: 3-dimensional lattice, lattice reduction problem, shortest vector in a lattice, Gaussian algorithm, LLL-algorithm.

1 Introduction. The Gaussian Reduction Algorithm

Let $\mathbb{R}^n$ be n-dimensional space over the field $\mathbb{R}$ of real numbers. Let $b_1, b_2, \ldots, b_k$, $k \leq n$, be linearly independent vectors in $\mathbb{R}^n$. A lattice of dimension k in $\mathbb{R}^n$ is the set of vectors

$$L = \{z_1 b_1 + z_2 b_2 + \cdots + z_k b_k | z_i \in \mathbb{Z}\}.$$

The determinant of L is defined by $\Delta(L) = |\det BB'|^{1/2}$, where B is the $k \times n$ matrix whose rows are $b_1, b_2, \ldots, b_k$. For $k = n$ we have $\Delta(L) = |\det B|$. Let $|b|$ be the Euclidean length of the vector $b \in \mathbb{R}^n$. By $\lambda(L)$ we denote the length of a shortest nonzero vector in L. Finding the shortest vector of a lattice given by some basis is a difficult problem in high dimension. There exists the famous LLL-algorithm [4] for finding a nonzero $b \in L$ such that

$$|b| \leq 2^{(k-1)/2} \lambda(L).$$

For $L \subseteq \mathbb{Z}^n$ (n is fixed) and $|b_1|, |b_2|, \ldots, |b_k| \leq M$, the LLL-algorithm works in $O(\log^3 M)$ binary operations without using fast integer arithmetic. This algorithm performs the following operations:

1. exchanging vectors b_i and b_j for i, j such that $1 \leq i, j \leq k$;
2. replacement of some vector b_j by $b_j - r b_i$ for i, j such that $1 \leq i, j \leq k$ and an integer r.

J.H. Silverman (Ed.): CaLC 2001, LNCS 2146, pp. 181–193, 2001.

For $k = 2$ we have the Gaussian reduction algorithm [2]. We say that a pair $a, b \in \mathbb{R}^n$, $n \geq 2$, is *reduced* if

$$|a| \leq |b| \qquad \text{and} \qquad 2|(a,b)| \leq |a|^2.$$

Let b_1, b_2 be a basis of some 2-dimensional lattice L, where $|b_2| \leq |b_1|$. We compute $r = \lfloor (b_1, b_2)/|b_2|^2 \rceil$ and set $a = b_1 - rb_2$. (Here $\lfloor x \rceil$ denotes the closest integer to x.) If $|a| \geq |b_2|$, then b_2 is the shortest vector in L and the algorithm terminates. If $|a| < |b_2|$, then we set $b_3 = a$ and apply the above step to the pair b_2, b_3, and so on. Thus we get a reduced pair $b_s, a = G(b_1, b_2)$, where b_s is the shortest vector in L.

Let $b_1, b_2 \in \mathbb{Z}^n$, $n \geq 2$ and $|b_2| \leq |b_1| \leq M$. We suppose the basis is defined also by integers $|b_1|^2$, $|b_2|^2$, and (b_1, b_2). Let us prove that Gaussian reduction takes $O\left(\left(1 + \log \frac{|b_1|}{|b_s|}\right) \log |b_1|\right)$ binary operations for fixed n, where b_s is the shortest vector of the lattice spanned by b_1 and b_2. So it takes $O(\log^2 M)$ operations.

Consider the first step of the Gaussian algorithm. To accomplish this step one needs to divide (b_1, b_2) by $|b_2|^2$ with remainder. From the Cauchy-Schwarz inequality it follows $|(b_1, b_2)|/|b_2|^2 \leq |b_1|/|b_2|$. Then for $r = \lfloor (b_1, b_2)/|b_2|^2 \rceil$ we have to compute $|b_3|^2$ and (b_3, b_2) by the formulas

$$|b_3|^2 = |b_1|^2 - 2r(b_1, b_2) + r^2|b_2|^2 \qquad \text{and} \qquad (b_3, b_2) = (b_1, b_2) - r|b_2|^2.$$

This takes $O\left(\left(1 + \log \frac{|b_1|}{|b_2|}\right) \log |b_1|\right)$ binary operations. Generally, the algorithm takes

$$O\left(\sum_{i=2}^{s} \left(1 + \log \frac{|b_{i-1}|}{|b_i|}\right) \log |b_{i-1}|\right) = O\left(\left(s + \log \frac{|b_1|}{|b_s|}\right) \log |b_1|\right)$$

binary operations, where $s - 1$ is the number of steps. Let us show that $s = O\left(1 + \log \frac{|b_1|}{|b_s|}\right)$.

Let b_1, b_2 be a nonreduced pair, where $|b_2| \leq |b_1|$. Consider the first step of the algorithm. It is easy to prove that if $|(b_1, b_2)| \geq |b_2|^2$, then $|b_1|^2 \geq |b_3|^2 + |b_2|^2$; and if $|(b_1, b_2)| < |b_2|^2$, then the pair b_2, b_3 is reduced. So if b_2, b_3 is not reduced, then $|b_1|^2 \geq 2|b_3|^2$ and so on. Therefore $s = O\left(1 + \log \frac{|b_1|}{|b_s|}\right)$.

The aim of this paper is a reduction algorithm for a basis b_1, b_2, b_3 of a 3-dimensional lattice in $\mathbb{R}^n$, $n \geq 3$. We give a definition of reduced basis and prove that for $b_1, b_2, b_3 \in \mathbb{Z}^n$, $n \geq 3$ and $|b_1|, |b_2|, |b_3| \leq M$, our algorithm takes $O(\log^2 M)$ binary operations, without fast arithmetic, to reduce this basis and so to find the shortest vector in the lattice. The definition and the algorithm can be extended to any dimension. Elementary steps of our algorithm are rather different from those of the LLL-algorithm.

This paper contains four sections besides the introduction. In the second section, we derive some reduction algorithms based only on Gaussian pairwise reduction. Examples show that this algorithm can take $O(\log^3 M)$ binary operations and not give the shortest vector in some cases. In the third section, we give the definition of reduced basis which is equivalent to that of the Minkowski reduced basis for a 3-dimensional lattice. We give here a reduction algorithm, but it is still slow. In the fourth section we derive another reduction algorithm, and in the fifth section we finish the proof that it takes $O(\log^2 M)$ binary operations.

There is an algorithm for lattice reduction in dimension 3 due to Vallee [5]. It is geometric and deals with integer lattices in $\mathbb{R}^3$. While this algorithm is based essentially on the same ideas as ours, its complexity is bounded by

$$O(\mu(M)\log M) \text{ binary operations,}$$

as follows from Theorem 2 of [5], so $O(\log^3 M)$ binary operations without using fast arithmetic. Here $\mu(M)$ is the cost of a multiplication performed on two integers of absolute value less than M

Our algorithm can be slightly modified to reduce positive definite ternary quadratic forms in the same time. Recently a new algorithm for reducing such forms was derived by Eisenbrand [3]. It uses Gauss-Lagarias ideas and takes $O(\mu(M)\log\log M)$ binary operations, so $O((\log M)^2 \log\log M)$ binary operations without using fast arithmetic. With fast integer multiplication algorithms it may be faster than ours. But I think our algorithm is more practical for reasonable inputs and it is easy to implement.

Thanks to an anonymous referee for pointing to Vallee's paper [5], to Eisenbrand for sharing his paper [3] with me, and to the conference organizers for arranging for my paper to be retyped in TEX.

2 The Naive Approach

We first prove two useful lemmas.

Lemma 1. *Let b_1, b_2 be linearly independent vectors in $\mathbb{R}^n$, $n \geq 2$, such that $|b_1| \leq |b_2|$. Then the inequality $2|(b_1, b_2)| \leq |b_1|^2$ is equivalent to the inequality $|b_2 + xb_1| \geq |b_2|$ for all integers x.*

Proof. The proof of this lemma is obvious.

We denote by ϵ_{ij} the sign of the scalar product (b_i, b_j). Thus $\epsilon_{ij} = \operatorname{sgn}(b_i, b_j)$, where

$$\operatorname{sgn}(\alpha) = \begin{cases} 1 & \text{if } \alpha > 0, \\ -1 & \text{if } \alpha < 0, \\ 0 & \text{if } \alpha = 0. \end{cases}$$

Lemma 2. *Let b_1, b_2 be linearly independent vectors in $\mathbb{R}^n$, $n \geq 2$. If $\epsilon_{12} = 0$, then the pair b_1, b_2 is reduced; otherwise it is reduced if and only if*

$$|b_2 - \epsilon_{12} b_1| \geq \max\{|b_1|, |b_2|\}. \tag{1}$$

Proof. We need prove it only for $\epsilon_{12} = -1$; that is, we need to prove the inequality

$$|b_2 + b_1| \geq \max\{|b_1|, |b_2|\}. \tag{2}$$

Let $|b_1| \leq |b_2|$. Then $2|(b_1, b_2)| \leq |b_1|^2$ is equivalent to

$$0 \leq |b_1|^2 - 2|(b_1, b_2)| = |b_2 + b_1|^2 - |b_2|^2,$$

which is equivalent to (2). Thus the statement is proved.

Definition 1. *We say that a basis b_1, b_2, b_3 of a 3-dimensional lattice L in $\mathbb{R}^n$, $n \geq 3$, is* 2-reduced *if its vectors are pairwise reduced.*

Algorithm 1 (Algorithm for 2-Reducing).
Input: *A basis b_1, b_2, b_3 of a 3-dimensional lattice in $\mathbb{R}^n$, $n \geq 3$.*
Output: *A 2-reduced basis b_1, b_2, b_3 of this lattice.*
Step 1. *(find a nonreduced pair) If b_i, b_j is not reduced for some $i, j \in \{1, 2, 3\}$, then to to Step 2, else terminate.*
Step 2. *(reduce the pair) Evaluate $a, b = G(b_i, b_j)$, replace $b_i, b_j \leftarrow a, b$, and go to Step 1.*

Remark 1. We note that to evaluate the shortest vector, it is not enough to apply only the Gaussian pairwise reduction algorithm. Consider the basis:

$$b_1 = (1, 0, 0), \qquad b_2 = \left(\frac{1}{2}, \frac{\sqrt{3}}{2}, 0\right), \qquad b_3 = \left(\frac{1}{2}, -\frac{\sqrt{3}}{2}, \delta\right),$$

where $\delta \neq 0$. It is easy to see that

$$|b_1|^2 = |b_2|^2 = 1, \qquad |b_3|^2 = 1 + \delta^2,$$
$$(b_1, b_2) = (b_1, b_3) = \frac{1}{2}, \qquad (b_2, b_3) = -\frac{1}{2}.$$

Thus the basis is 2-reduced. But

$$a = b_3 - b_1 + b_2 = (0, 0, \delta) \quad \text{and} \quad |a| < |b_1|, |b_2|, |b_3| \quad \text{for } \delta < 1.$$

Remark 2. Algorithm 1 can take $O(\log^3 M)$ binary operations to find a 2-reduced basis for $b_1, b_2, b_3 \in \mathbb{Z}^n$, $n \geq 3$, where $|b_1|, |b_2|, |b_3| \leq M$. This is true for the following basis:

$$b_1 = (1, 1, 0), \qquad b_2 = (N, 0, 0), \qquad b_3 = (0, N^2, 1),$$

where M and N are integers satisfying $N^2 < M$.

3 Definition of Reduced Basis

Definition 2. *We say that a basis b_1, b_2, b_3 of a 3-dimensional lattice L in $\mathbb{R}^n$, $n \geq 3$, is* reduced *if its vectors satisfy*

1. $|b_1| \leq |b_2| \leq |b_3|$;
2. $|b_2 + x_1 b_1| \geq |b_2|$ *and* $|b_3 + x_2 b_2 + x_1 b_1| \geq |b_3|$ *for all integers* x_1, x_2.

Thus a reduced basis is 2-reduced. This definition is equivalent to that of the Minkowski reduced basis as follows from [1]. Thus b_1 is the shortest nonzero vector of L. It follows that

$$\lambda(L) = |b_1| \leq 2^{1/6} \Delta(L)^{1/3}.$$

Algorithm 2 (Algorithm for Reducing).
Input: *Basis vectors b_1, b_2, b_3 of a 3-dimensional lattice $\mathbb{R}^n$, $n \geq 3$.*
Output: *A reduced basis b_1, b_2, b_3 of this lattice.*
Step 1. *(find a nonreduced pair) If b_i, b_j is not reduced for some $i, j \in \{1, 2, 3\}$, then go to Step 2, otherwise go to Step 3.*
Step 2. *(reduce the pair) Evaluate $a, b = G(b_i, b_j)$, replace $b_i, b_j \leftarrow a, b$, and go to Step 1.*
Step 3. *(reduce the basis) If $\epsilon_{12}\epsilon_{13}\epsilon_{23} \neq -1$, then go to Step 4, else evaluate*

$$a = b_1 - \epsilon_{12} b_2 - \epsilon_{13} b_3.$$

If $|a| < \max\{|b_1|, |b_2|, |b_3|\} = |b_i|$, then replace $b_i \leftarrow a$.
Step 4. *(order the vectors) Order the vectors b_1, b_2, b_3 by increasing value of their norm.*

Remark 3. This algorithm uses the previous Algorithm 1. It can take as many as $O(\log^3 M)$ binary operations to find a reduced basis for $b_1, b_2, b_3 \in \mathbb{Z}^n$, $n \geq 3$, where $|b_1|, |b_2|, |b_3| \leq M$.

We are going to prove that Algorithm 2 really finds a reduced basis. We need only prove the following statements.

Proposition 1. *Let $b_1, b_2, b_3 \in \mathbb{R}^n$, $n \geq 3$, be a 2-reduced basis of some lattice. Then this basis is reduced or $\epsilon_{12}\epsilon_{13}\epsilon_{23} = -1$ and*

$$|b_1 - \epsilon_{12} b_2 - \epsilon_{13} b_3| < \max\{|b_1|, |b_2|, |b_3|\}.$$

Proposition 2. *Let $b_1, b_2, b_3 \in \mathbb{R}^n$, $n \geq 3$, be a 2-reduced basis of some lattice, and assume that*

$$|a| < \max\{|b_1|, |b_2|, |b_3|\} = |b_i|, \qquad \text{where} \quad a = b_1 - \epsilon_{12} b_2 - \epsilon_{13} b_3.$$

Then replacing b_i by a and ordering the vectors, we get a reduced basis.

These two propositions follow from Theorem 1.

Theorem 1. *Let $b_1, b_2, b_3 \in \mathbb{R}^n$, $n \geq 3$, be a 2-reduced basis of some lattice such that $|b_1| \leq |b_2| \leq |b_3|$. If for some integers x_1, x_2 we have*

$$|b_3 + x_1 b_1 + x_2 b_2| < |b_3|, \tag{3}$$

then

$$\epsilon_{12}\epsilon_{13}\epsilon_{23} = -1, \qquad x_1 = -\epsilon_{13}, \text{ and} \qquad x_2 = -\epsilon_{23}.$$

Proof. From (3) we get

$$|b_1|^2 x_1^2 + |b_2|^2 x_2^2 + 2(b_1, b_3)x_1 + 2(b_2, b_3)x_2 + 2(b_1, b_2)x_1 x_2 < 0. \tag{4}$$

To prove this theorem we need a lemma.

Lemma 3. *If at least one of the last three summands in the lefthand side of (4) is non-negative, then this inequality is impossible.*

Proof. Let $(b_1, b_3)x_1 \geq 0$. Then

$$\begin{aligned} 0 &> |b_1|^2 x_1^2 + |b_2|^2 x_2^2 - 2|(b_2, b_3)x_2| - 2|(b_1, b_2)x_1 x_2| \\ &\geq |b_1|^2 x_1^2 + |b_2|^2 x_2^2 - |b_2|^2 |x_2| - |b_1|^2 |x_1 x_2|. \end{aligned}$$

This inequality is impossible for $|x_1| \geq |x_2|$. Let $|x_1| < |x_2|$. Then

$$|b_1|^2 |x_1| \big(|x_2| - |x_1|\big) > |b_2|^2 |x_2| \big(|x_2| - 1\big).$$

Hence $|x_2| - |x_1| > |x_2| - 1$ and $|x_1| < 1$. So $x_1 = 0$. Therefore $|b_3 + x_2 b_2| < |b_3|$. This contradicts Lemma 1 and the theorem's condition. Similarly, we consider the case $(b_2, b_3)x_2 \geq 0$. Let now $(b_1, b_2)x_1 x_2 \geq 0$. Then

$$\begin{aligned} 0 &> |b_1|^2 x_1^2 + |b_2|^2 x_2^2 - 2|(b_2, b_3)x_2| - 2|(b_1, b_3)x_1| \\ &\geq |b_1|^2 (x_1^2 - |x_1|) + |b_2|^2 (x_2^2 - |x_2|). \end{aligned}$$

This inequality is impossible for integers x_1, x_2. Thus the lemma is proved.

Let us prove the theorem. If $\epsilon_{12}\epsilon_{13}\epsilon_{23} \neq -1$, then by changing signs of the vectors b_1, b_2, b_3, we get

$$(b_1, b_2) \geq 0, \qquad (b_1, b_3) \geq 0, \qquad (b_2, b_3) \geq 0.$$

According to Lemma 3, the inequality (4), and so (3), are impossible. Let $\epsilon_{12}\epsilon_{13}\epsilon_{23} = -1$. Then by changing signs of the vectors b_1, b_2, b_3, we get

$$(b_1, b_2) > 0, \qquad (b_1, b_3) > 0, \qquad (b_2, b_3) < 0.$$

Indeed one can replace b_2 by $\epsilon_{12} b_2$ and b_3 by $\epsilon_{13} b_3$. According to (3), we need consider only the case $x_1 < 0$ and $x_2 > 0$. From (4) it follows that

$$\begin{aligned} 0 &> |b_1|^2 x_1^2 + |b_2|^2 x_2^2 - 2(b_1, b_3)x_1 - 2|(b_2, b_3)|x_2 + 2(b_1, b_2)|x_1| x_2 \\ &\geq |b_1|^2 x_1^2 + |b_2|^2 x_2^2 - |b_1|^2 |x_1| - |b_2|^2 x_2 - |b_1|^2 |x_1| x_2 \\ &\geq |b_1|^2 \big(x_1^2 - |x_1| + x_2^2 - x_2 - |x_1| x_2\big). \end{aligned}$$

Hence

$$x_1^2 - |x_1| + x_2^2 - x_2 - |x_1|x_2 < 0 \qquad \text{and} \qquad \big(|x_1| - x_2\big)^2 + |x_1|x_2 < |x_1| + x_2.$$

The inequality $|x_1|x_1 < |x_1|+x_2$ is valid only for $|x_1| = 1$ or $x_2 = 1$. For example, in the last case $x_1^2 - 2|x_1| < 0$, so $|x_1| \le 1$. Therefore $x_1 = -1$ and $x_2 = 1$. We have in both cases

$$|b_3 - b_1 + b_2| < |b_3|.$$

Finally, for the initial basis we obtain

$$|b_3 - \epsilon_{13}b_1 - \epsilon_{23}b_2| < |b_3|.$$

This completes the proof of Theorem 1.

Let us prove Proposition 1. Suppose that the basis b_1, b_2, b_3 is not reduced. Then $\epsilon_{12}\epsilon_{13}\epsilon_{23} = -1$ and the basis b_1', b_2', b_3' is also not reduced, where $b_1' = b_1$, $b_2' = -\epsilon_{12}b_2$, and $b_3' = -\epsilon_{13}b_3$. We note that

$$(b_1', b_2') < 0, \quad (b_1', b_3') < 0, \quad (b_2', b_3') < 0.$$

Then by Theorem 1,

$$|b_1' + b_2' + b_3'| < \max\big\{|b_1'|, |b_2'|, |b_3'|\big\}.$$

Proposition 1 follows.

Let us prove Proposition 2. It suffices to deal with the case $\epsilon_{12} = \epsilon_{13} = \epsilon_{23} = -1$. Let $\{1,2,3\} = \{i,j,k\}$. Then $a = b_i + b_j + b_k$. We note that

$$\begin{aligned}(a, b_j) &= |b_j|^2 + (b_i, b_j) + (b_k, b_j)\\ &= \frac{1}{2}\big(|b_j|^2 + 2(b_i, b_j) + |b_j|^2 + 2(b_k, b_j)\big)\\ &\ge 0\end{aligned}$$

for all j with $1 \le j \le 3$, since b_i, b_j and b_k, b_j are reduced. If the pair a, b_j with $j \ne i$ is not reduced, then by Lemma 2,

$$|a - b_j| = |b_i + b_k| < \max\big\{|a|, |b_j|\big\} \le |b_i|.$$

This is a contradiction, since the pair b_i, b_k is reduced. So b_j, b_k, a is a 2-reduced basis. Let us show that it is reduced. If this is not the case, then according to Proposition 1,

$$|a - b_j - b_k| = |b_i| < \max\big\{|b_1|, |b_2|, |b_3|\big\} \le |b_i|.$$

This contradiction completes the proof of Proposition 2.

4 The Algorithm

Algorithm 3 (Algorithm for Reducing).
Input: *Basis vectors b_1, b_2, b_3 of a 3-dimensional lattice in $\mathbb{R}^n$, $n \geq 3$, ordered such that $|b_1| \leq |b_2| \leq |b_3|$.*
Output: *A reduced basis b_1, b_2, b_3 of this lattice.*
Step 1. *(reduce the pair b_1, b_2) Evaluate $a, b = G(b_1, b_2)$, replace $b_1, b_2 \leftarrow a, b$.*
Step 2. *(find a minimum of $|b_3 + x_2 b_2 + x_1 b_1|$) Compute integers x_1, x_2 such that $|b_3 + x_2 b_2 + x_1 b_1|$ is minimal. We have $|x_2 - y_2| \leq 1$ and $|x_1 - y_1| \leq 1$, where*

$$y_2 = -\frac{\dfrac{(b_2,b_3)}{|b_2|^2} - \dfrac{(b_1,b_2)}{|b_2|^2} \cdot \dfrac{(b_1,b_3)}{|b_1|^2}}{1 - \dfrac{(b_1,b_2)}{|b_1|^2} \cdot \dfrac{(b_1,b_2)}{|b_2|^2}}, \quad y_1 = -\frac{\dfrac{(b_1,b_3)}{|b_1|^2} - \dfrac{(b_1,b_2)}{|b_1|^2} \cdot \dfrac{(b_2,b_3)}{|b_2|^2}}{1 - \dfrac{(b_1,b_2)}{|b_1|^2} \cdot \dfrac{(b_1,b_2)}{|b_2|^2}}. \tag{5}$$

Set $a = b_3 + x_2 b_2 + x_1 b_1$.
Step 3. *(replace $b_3 \leftarrow a$) If $|a| \geq |b_3|$, then terminate, else replace $b_3 \leftarrow a$. Order b_1, b_2, b_3 such that $|b_1| \leq |b_2| \leq |b_3|$ and go to Step 1.*

Remark 4. We suppose that the basis b_1, b_2, b_3 is defined also by numbers

$$|b_1|^2, |b_2|^2, |b_3|^2, (b_1, b_2), (b_1, b_3), (b_2, b_3).$$

We change the numbers in each step when changing the basis.

Remark 5. Definition 2 and Algorithm 3 can be extended to higher dimension. Generally it is not clear if one could obtain the shortest vector of a lattice in so doing.

Lemma 4. *Let b_1, b_2, b_3 be linearly independent vectors in $\mathbb{R}^n$, $n \geq 3$, such that $|b_1| \leq |b_2| \leq |b_3|$ and $2|(b_1, b_2)| \leq |b_1|^2$. Let x_1, x_2 be integers such that $|b_3 + x_2 b_2 + x_1 b_1|$ is minimal. If y_1 and y_2 are defined by (5), then*

$$|x_2 - y_2| \leq 1 \qquad \textit{and} \qquad |x_1 - y_1| \leq 1.$$

Proof. It is easy to see that real y_1, y_2 such that $|b_3 + y_2 b_2 + y_1 b_1|$ is minimal are defined by (5). So for integers x_1, x_2 we have

$$|b_3 + x_2 b_2 + x_1 b_1|^2 = |b_3 + y_2 b_2 + y_1 b_1|^2 + |(x_2 - y_2) b_2 + (x_1 - y_1) b_1|^2.$$

Let $\lfloor y_i \rceil - y_i = \epsilon_i$ for $1 \leq i \leq 2$, so $|\epsilon_i| \leq \frac{1}{2}$. By assumption

$$|\epsilon_2 b_2 + \epsilon_1 b_1|^2 \geq |(x_2 - y_2) b_2 + (x_1 - y_1) b_1|^2.$$

Hence

$$0 \geq \left((x_2 - y_2)^2 - \epsilon_2^2\right)|b_2|^2 + 2\left((x_2 - y_2)(x_1 - y_1) - \epsilon_1 \epsilon_2\right)(b_1, b_2) + \left((x_1 - y_1)^2 - \epsilon_1^2\right)|b_1|^2.$$

By assumption $2|(b_1, b_2)| \le |b_1|^2$. Then

$$0 \ge \big((x_2-y_2)^2-\epsilon_2^2\big)|b_2|^2-\big(|x_2-y_2||x_1-y_1|+|\epsilon_1\epsilon_2|\big)|b_1|^2+\big((x_1-y_1)^2-\epsilon_1^2\big)|b_1|^2.$$

Hence

$$(x_2-y_2)^2-|x_2-y_2||x_1-y_1|+(x_1-y_1)^2 \le \epsilon_2^2+|\epsilon_1\epsilon_2|+\epsilon_1^2 \le \frac{3}{4},$$

since $(x_2-y_2)^2-\epsilon_2^2 \ge 0$ for any integer x_2 and $|b_2|^2 \ge |b_1|^2$. It follows that

$$\left(|x_2-y_2|-\frac{|x_1-y_1|}{2}\right)^2+\frac{3}{4}|x_1-y_1|^2 \le \frac{3}{4}.$$

Therefore $|x_1-y_1| \le 1$. Similarly $|x_2-y_2| \le 1$. The lemma is proved.

Lemma 5. *Let the basis $b_1, b_2, b_3 \in \mathbb{Z}^n$, $n \ge 3$, satisfy the assumptions of Lemma 4. Let r be the minimal natural number such that*

$$\frac{|(b_1,b_3)|}{|b_1|^2} \le r \qquad \textit{and} \qquad \frac{|(b_2,b_3)|}{|b_2|^2} \le r.$$

Then in

$$T = O\big((\log r+1)\log|b_3|\big) \textit{ binary operations,}$$

one can find y_1, y_2 with an accuracy of a few digits after the binary point.

Proof. It is easy to see that $|y_i| \le O(r)$, $1 \le i \le 2$. To find y_1 one computes integers r_1, r_2, s_1, s_2 such that

$$(b_1,b_3)=r_1|b_1|^2+s_1, \qquad 0 \le s_1 < |b_1|^2,$$
$$(b_2,b_3)=r_2|b_2|^2+s_2, \qquad 0 \le s_2 < |b_2|^2.$$

It takes T binary operations since $|r_1|, |r_2| \le r$. Then one computes integers r_3, s_3 such that

$$(b_1,b_2)r_2=r_3|b_1|^2+s_3, \qquad 0 \le s_3 < |b_1|^2.$$

This takes T binary operations as well. Thus we find the numerator for y_1 by

$$\begin{aligned}\frac{(b_1,b_3)}{|b_1|^2}-\frac{(b_1,b_2)}{|b_1|^2}\cdot\frac{(b_2,b_3)}{|b_2|^2} &= r_1+\frac{s_1}{|b_1|^2}-(b_1,b_2)\frac{r_2}{|b_1|^2}-(b_1,b_2)\frac{s_2}{|b_1|^2|b_2|^2}\\ &= r_1-r_3+\frac{s_1}{|b_1|^2}-\frac{s_3}{|b_1|^2}-(b_1,b_2)\frac{s_2}{|b_1|^2|b_2|^2}\\ &= r_1-r_3+\delta,\end{aligned}$$

where $|\delta| \le \frac{3}{2}$. One needs a few digits of δ after the binary point. Similarly, one finds the denominator for y_1 with an accuracy of $O(\log r+1)$ binary digits. This takes T binary operations. To find y_1 one has to again divide with remainder two numbers. This takes T binary operations also. Similarly we find y_2 in T binary operations. The lemma is proved.

Theorem 2. *Let $b_1, b_2, b_3 \in \mathbb{Z}^n$, $n \geq 3$, be linearly independent vectors such that $|b_1| \leq |b_2| \leq |b_3|$. Then in $O\left(\log \frac{|b_3|}{|b_1'|} + 1\right)$ steps, Algorithm 3 finds a basis b_1', b_2', b_3' which is reduced or satisfies*

$$|b_1'| \leq |b_2'| \leq |b_3'| \leq |b_1| \qquad \text{and} \qquad 2|b_3'|^2 \leq |b_3|^2.$$

This calculation takes $O\left(\left(\log \frac{|b_3|}{|b_1'|} + 1\right) \log |b_3|\right)$ binary operations.

We will prove Theorem 2 later in Section 5.

Corollary 1. *Let $b_1, b_2, b_3 \in \mathbb{Z}^n$, $n \geq 3$, satisfying $|b_1| \leq |b_2| \leq |b_3|$ be a basis of some 3-dimensional lattice. Let a be the shortest nonzero vector in this lattice. Then Algorithm 3 finds some reduced basis in $O\left(\left(\log \frac{|b_3|}{|a|} + 1\right) \log |b_3|\right)$ binary operations.*

Proof. Form Theorem 2 it follows that the algorithm does $s = O\left(\log \frac{|b_3|}{|a|} + 1\right)$ steps to find a reduced basis. To find a basis b_1', b_2', b_3' such that

$$|b_1'| \leq |b_2'| \leq |b_3'| \leq |b_1| \qquad \text{and} \qquad 2|b_3'|^2 \leq |b_3|^2,$$

one needs $O\left(\left(\log \frac{|b_3|}{|b_1|} + 1\right) \log |b_3|\right)$ binary operations. To find a basis b_1'', b_2'', b_3'' satisfying

$$|b_1''| \leq |b_2''| \leq |b_3''| \leq |b_1'| \qquad \text{and} \qquad 2|b_3''|^2 \leq |b_3'|^2$$

takes $O\left(\left(\log \frac{|b_3'|}{|b_1''|} + 1\right) \log |b_3|\right)$ binary operations, and so on. Generally it takes

$$\begin{aligned} O&\left(\left(s + \log \frac{|b_3|}{|b_1'|} + \log \frac{|b_3'|}{|b_1''|} + \cdots + \log \frac{|b_3^{(s-1)}|}{|a|}\right) \log |b_3|\right) \\ &= O\left(\left(s + \log \frac{|b_3|}{|a|} + \log \frac{|b_3'|}{|b_1'|} + \cdots + \log \frac{|b_3^{(s-1)}|}{|b_1^{(s-1)}|}\right) \log |b_3|\right) \\ &= O\left(\left(\log \frac{|b_3|}{|a|} + 1\right) \log |b_3|\right) \end{aligned}$$

binary operations. This concludes the proof.

5 Proof of Theorem 2

We begin with the proof of the following auxiliary proposition.

Theorem 3. *Let b_1, b_2, b_3 be linearly independent vectors in $\mathbb{R}^n$, $n \geq 3$, such that $|b_1| \leq |b_2| \leq |b_3|$ and*

$$0 \leq 2(b_1, b_2) \leq |b_1|^2, \qquad |(b_1, b_3)| < |b_1|^2, \qquad |(b_2, b_3)| < |b_2|^2.$$

Let x_1, x_2 be integers with

$$|b_3 + x_2 b_2 + x_1 b_1| < |b_3|. \tag{6}$$

(a) *If* $(b_1, b_3) \geq 0$ *and* $(b_2, b_3) \geq 0$, *then the integers* x_1, x_2 *are defined by*

$$x_1 = -1,\ -1,\ -2,\ \ 0,\ -1,\ \ 1,\ \ 1,\ -1,\ -2,$$
$$x_2 = -1,\ -2,\ -1,\ -1,\ \ 0,\ -2,\ -1,\ \ 1,\ \ 1,$$

or if not, then $2|b_1|^2 < |b_2|^2$ *and* $x_2 = -1$.

(b) *If* $(b_1, b_3) \geq 0$ *and* $(b_2, b_3) < 0$, *then the integers* x_1, x_2 *are defined by*

$$x_1 = 1,\ 0,\ -1,\ -1,\ -2,\ -3,\ -1,\ -2,\ -3,\ -1,\ -2,\ -3,\ -4,\ -3,$$
$$x_2 = 1,\ 1,\ \ 0,\ \ 1,\ \ 1,\ \ 1,\ \ 2,\ \ 2,\ \ 2,\ \ 3,\ \ 3,\ \ 3,\ \ 3,\ \ 4,$$

or if not, then $2|b_1|^2 < |b_2|^2$ *and* $x_2 = 1$.

Proof. From (6) it follows that

$$0 > x_2^2|b_2|^2 + x_1^2|b_1|^2 + 2x_2(b_2, b_3) + 2x_1(b_1, b_3) + 2x_1x_2(b_1, b_2). \tag{7}$$

Let us prove (a). If $x_1 = 0$, then

$$0 > x_2^2|b_2|^2 + 2x_2(b_2, b_3).$$

Hence $x_2 < 0$ and

$$0 > x_2^2|b_2|^2 - 2|x_2||b_2|^2 = (x_2^2 - 2|x_2|)|b_2|^2.$$

Therefore $x_2 = -1$. Similarly, if $x_2 = 0$, then $x_1 = -1$. Let $x_1, x_2 < 0$. It follows from (7) that

$$0 > \left(x_2^2 - 2|x_2|\right)|b_2|^2 + \left(x_1^2 - 2|x_1|\right)|b_1|^2.$$

Hence $0 > x_2^2 - 2|x_2|$ or $0 > x_1^2 - 2|x_1|$. In the first case $x_2 = -1$. Then

$$0 > -|b_2|^2 + \left(x_1^2 - 2|x_1|\right)|b_1|^2.$$

If $|x_1| \geq 3$, then $2|b_1|^2 < |b_2|^2$ and (a) is proved. Thus $x_1 = -1$ or $x_1 = -2$. Let $0 > x_1^2 - 2|x_1|$. Then $x_1 = -1$ and

$$0 > \left(x_2^2 - 2|x_2|\right)|b_2|^2 - |b_1|^2.$$

Then $|x_2| \leq 2$, so $x_2 = -1$ or $x_2 = -2$. If $x_1, x_2 > 0$, then (7) is impossible. Let $x_1 < 0$ and $x_2 > 0$. From (7) it follows that

$$0 > x_2^2|b_2|^2 + \left(x_1^2 - |x_1|(2 + x_2)\right)|b_1|^2.$$

So $|x_1| < 2 + x_2$. Since $|b_2|^2 \geq |b_1|^2$, we have $x_2 = 1$. Therefore $x_1 = -1$ or $x_1 = -2$. Let $x_1 > 0$ and $x_2 < 0$. From (7) it follows that

$$0 > \left(x_2^2 - 2|x_2|\right)|b_2|^2 + \left(x_1^2 - x_1|x_2|\right)|b_1|^2.$$

So $0 > x_2^2 - 2|x_2|$ or $0 > x_1^2 - x_1|x_2|$. In the first case $x_2 = -1$. If $x_1 \geq 2$, then from

$$0 > -|b_2|^2 + (x_1^2 - x_1)|b_1|^2$$

we have $2|b_1|^2 < |b_2|^2$ and (a) is proved. Thus $x_1 = 1$. Let $0 > x_1^2 - x_1|x_2|$, so $x_1 < |x_2|$. From $|x_2| \geq 2$, we have only $x_2 = -2$ and $x_1 = 1$. This completes the proof of the first case of the theorem. The proof of (b) is similar.

Let b_1, b_2, b_3 be linearly independent vectors in $\mathbb{R}^n$, $n \geq 3$, such that $|b_1| \leq |b_2| \leq |b_3|$. Let

$$M = \min_{y_1, y_2 \in \mathbb{Z}} \{|b_3 + y_2 b_2 + y_1 b_1|\}.$$

Corollary 2. *Under the conditions of Theorem 3 we have* $2|b_1|^2 < |b_2|^2$ *or one of the following conditions holds:*

(a) *If* $(b_1, b_3) \geq 0$ *and* $(b_2, b_3) \geq 0$, *then*

$$M = |b_3 - b_1| \quad or \quad M = |b_3 - b_2| \quad or \quad M = |b_3 - b_2 - b_1|.$$

(b) *If* $2|(b_3, b_1)| \leq |b_1|^2$, *then*

$$M = |b_3 - \epsilon_{23} b_2| \quad or \quad M = |b_3 - \epsilon_{23} b_2 - \epsilon_{13} b_1|.$$

(c) *If* $2|(b_3, b_2)| \leq |b_2|^2$, *then*

$$M = |b_3 - \epsilon_{13} b_1| \quad or \quad M = |b_3 - \epsilon_{23} b_2 - \epsilon_{13} b_1|.$$

Here $\epsilon_{ij} = \mathrm{sgn}(b_i, b_j)$.

Proof. Let $2|b_1|^2 \geq |b_2|^2$. Consider (a). Let $M = |b_3 + x_2 b_2 + x_1 b_1|$. Then x_1, x_2 are defined by the first case of Theorem 3. We note that

$$\begin{aligned} |b_3 - b_2 - b_1| &< |b_3 - 2b_2 - b_1|, |b_3 - b_2 - 2b_1|, \\ |b_3 - b_2| &\leq |b_3 - b_2 + b_1|, |b_3 - 2b_2 + b_1|, \\ |b_3 - b_1| &\leq |b_3 + b_2 - b_1|, |b_3 + b_2 - 2b_1|. \end{aligned}$$

Thus (a) is true. The proof of (b) and (c) is similar.

Proof (Outline of the proof of Theorem 2). In Step 1 of the algorithm we reduce the pair b_1, b_2. So we can consider only the case $2|(b_1, b_2)| \leq |b_1|^2$. In Step 2 we find integers x_1, x_2 such that

$$a = b_3 + x_2 b_2 + x_1 b_1 \qquad \text{and} \qquad |a| = \min_{y_1, y_2 \in \mathbb{Z}} \{|b_3 + y_2 b_2 + y_1 b_1|\}.$$

Using Corollary 2, we can reduce the proof to the case $|b_1| \leq |a|$. If $|b_1| \leq |b_2| \leq |a|$, then the basis b_1, b_2, a is reduced. Let $|b_1| \leq |a| < |b_2|$. Hence the pair b_1, a is reduced, so $2|(b_1, a)| \leq |b_1|^2$. In Step 2 we find integers x_1', x_2' such that

$$a' = b_2 + x_2' a + x_1' b_1 \qquad \text{and} \qquad |a'| = \min_{y_1, y_2 \in \mathbb{Z}} \{|b_2 + y_2 a + y_1 b_1|\}.$$

If $|a'| \geq |a|$, then the basis b_1, a, a' is reduced. Let $|a'| < |a|$. We note that if $|(b_2, a)| \geq |a|^2$,then

$$|a'|^2 \leq |b_2 - ra|^2 \leq \frac{1}{2}|b_2|^2 \qquad \text{for} \quad r = \left\lfloor \frac{(b_2, a)}{|a|^2} \right\rceil.$$

By Lemma 5, one can find a' in

$$O\big((\log|r|+1)\log|b_3|\big) = O\left(\left(\log\frac{|b_2|}{|a|}+1\right)\log|b_3|\right)$$

binary operations, since $|r| \le |b_2|/|a|$ by the Cauchy-Schwarz inequality. So the theorem can be proved in this case easily. Let $|(b_2, a)| < |a|^2$. Then by Theorem 3 and Corollary 2, there exist integers x_1'', x_2'', where $|x_2''| \le 1$, such that

$$a'' = b_2 + x_2''a + x_1''b_1 \qquad \text{and} \qquad |a''| = |a'|.$$

So $|a'| \ge |a|$. This is a contradiction, which completes the proof of the theorem.

References

1. J.W.S. Cassels, *Rational quadratic forms*, Academic Press, London, New York, 1978.
2. H. Cohen, *A course in computational algebraic number theory*, Springer-Verlag, Berlin, Heidelberg, 1993.
3. F. Eisenbrand, Fast reduction of ternary quadratic forms, this volume.
4. A.K. Lenstra, H.W. Lenstra, and L. Lovasz, Factoring polynomials with rational coefficients, *Math. Ann.* **261** (1982), 515–534.
5. B. Vallee, An affine point of view on minima finding in integer lattices of lower dimensions, *Proc. of EUROCAL'87*, LNCS 378, Springer-Verlag, Berlin, 1989, 376-378.

The Shortest Vector Problem in Lattices with Many Cycles

Mårten Trolin

Department of Numerical Analysis and Computer Science
Royal Institute of Technology, Stockholm, Sweden
marten@nada.kth.se

Abstract In this paper we investigate how the complexity of the shortest vector problem in a lattice Λ depends on the cycle structure of the additive group $\mathbb{Z}^n/\Lambda$. We give a proof that the shortest vector problem is **NP**-complete in the max-norm for n-dimensional lattices Λ where $\mathbb{Z}^n/\Lambda$ has $n-1$ cycles. We also give experimental data that show that the LLL algorithm does not perform significantly better on lattices with a high number of cycles.
Keywords: Lattices, LLL algorithm, shortest vector problem.

1 Introduction

Lattices were examined already in the middle of the 19th century, at that time mostly because of their connections to quadratic forms. The interest in algorithmic aspects of lattice problems started in the beginning of the 1980s.

In 1981, van Emde Boas [16] showed that finding the lattice point closest to a given point is **NP**-hard in ℓ_r-norm for any $r > 0$. In the following year, Lenstra, Lenstra and Lovász [7] published their lattice basis reduction algorithm, which is guaranteed to find a vector not more than an exponential factor longer than the shortest vector in polynomial time. This was a great achievement. Schnorr has improved approximation to a slightly sub-exponential factor [12].

The **NP**-hardness of the shortest vector problem in Euclidean norm was an open problem for a long time. It was proven to be **NP**-hard under randomized reductions by Ajtai in 1998 [2]. This result has been improved by several authors, and the strongest result today by Micciancio [8] is that the shortest vector problem is **NP**-hard to approximate within any factor smaller than $\sqrt{2}$ under randomized reductions. On the other hand, Goldreich and Goldwasser [4] have showed that this **NP**-hardness result cannot be extended to $\sqrt{n}$ unless the polynomial-time hierarchy collapses.

A lattice can be described either by a basis that spans the lattice, or as the solutions $\mathbf{x}$ of a set of modular equations $\langle \mathbf{a}_i, \mathbf{x} \rangle = 0 \pmod{k_i}$. Lattices can be classified after the cycle structure of the subgroup $\mathbb{Z}^n/\Lambda$. The number of cycles and the lengths of the cycles in this subgroup corresponds to the minimum number of equations and the moduli of the equations necessary to describe the lattice. Our main focus will be to investigate whether there is a difference in the complexity of computing short vectors between lattices with different number of cycles.

J.H. Silverman (Ed.): CaLC 2001, LNCS 2146, pp. 194–205, 2001.

In 1996, Ajtai [1] published a paper in which it is shown that a random lattice from a certain set of lattices is at least as hard as a certain shortest vector problem in the worst case. This result has been improved in [3]. These n-dimensional lattices have n/c cycles, whereas a random lattice usually has one cycle. On the other hand, Schnorr and Paz [11] have shown in a worst-case result that any lattice can be approximated arbitrarily well by a lattice with one cycle. Schnorr's and Paz's result indicates that the lattices with one cycles are the hardest, whereas Ajtai's result gives evidence that also lattices with n/c cycles are hard (although the problem studied by Ajtai is of a different kind than the common shortest vector problem). This gives rise to the question on whether or not lattices with more cycles are easier, or whether the number of cycles is of no importance to the shortest vector problem.

As far as we know, it has never previously been investigated how the cycle structure affects the complexity of lattice problems. Except for the lattice created by Ajtai [1], the previously published reductions that we know about [2,8,16] contain no analysis of the cycle structure of the lattices used. We will show that even for a large number of cycles the shortest vector problem is hard in the max-norm. We also give experimental data that indicate that the LLL lattice basis reduction algorithm does not perform significantly better when applied on lattices with a high number of cycles. However, we still lack a theoretical explanation for these results.

2 Definitions

A *lattice* Λ is the set $\{\sum_{i=1}^{n} \lambda_i \mathbf{b}_i \mid \lambda_i \in \mathbb{Z}\}$ where the vectors $\mathbf{b}_i \in \mathbb{Z}^n$ are linearly independent. The vectors $\mathbf{b}_i$'s are called a *basis* of the lattice, and the matrix $\mathbf{B}$ with the vectors $\mathbf{b}_i$'s as its rows is called a *basis matrix* for Λ. By $\Lambda(\mathbf{b}_1, \mathbf{b}_2, \ldots, \mathbf{b}_n)$ we mean the lattice spanned by the basis $\{\mathbf{b}_1, \mathbf{b}_2, \ldots, \mathbf{b}_n\}$. The determinant of a lattice is defined as $\det(\Lambda) = |\det(\mathbf{B})|$. The ℓ_k-*norm* of a vector $\mathbf{v}$ is defined as $||\mathbf{v}||_k = \left(\sum_{i=1}^{n} |v_i|^k\right)^{1/k}$. We also define the *max-norm*, ℓ_∞-norm, as $||\mathbf{v}||_\infty = \max_{i=1}^{n} |v_i|$. We can see that the ℓ_2-norm is the Euclidean norm. In this report we will mainly consider the ℓ_2-norm and the ℓ_∞-norm. When we leave out the index we mean the ℓ_2-norm.

A vector $\mathbf{v} \in \Lambda$ is called a *shortest lattice vector* if $||\mathbf{v}|| > 0$ and for every vector $\mathbf{u} \in \Lambda$ either $||\mathbf{u}|| \geq ||\mathbf{v}||$ or $||\mathbf{u}|| = 0$. Moreover we define the length of a shortest vector in Λ as $\lambda(\Lambda) = ||\mathbf{v}||$. We define the length of a basis as the length of the longest vector in the basis.

In the end of the 19th century, Minkowski [10] proved an upper bound on the length of the shortest vector in a lattice.

Theorem 1 (Minkowski's Inequality). *Let $\Lambda \in \mathbb{Z}^n$ be an n-dimensional lattice. Then there is a constant γ_n so that*

$$\lambda(\Lambda) \leq \sqrt{\gamma_n}(\det(\Lambda))^{1/n}.$$

The least constant γ_n is called Hermite's constant of rank n, and it has been proved that $\gamma_n \leq \frac{n}{\pi e}$ [5]. It is also known that $\gamma_n \geq \frac{n}{2\pi e}$.

An alternative way to describe a lattice is by giving a set of modular equation, that is, equations of the form $\langle \mathbf{a}_i, \mathbf{x} \rangle = 0 \pmod{k_i}$, where $\mathbf{a}_1, \mathbf{a}_2, \ldots, \mathbf{a}_m$ are n-dimensional vectors. Any lattice can be written on this form. We will say that a lattice described by m modular equations with moduli $k_1, k_2, \ldots, k_m$ has m cycles of lengths $k_1, k_2, \ldots, k_m$, provided that the equations are on as simple form as possible. More exactly, we demand that the coefficients and the modulus are relative prime within each equation, and that $k_i | k_{i+1}$, $i = 1, 2, \ldots, m-1$. If we construct a lattice by modular equations such that the moduli do not have this property, we can always combine the equations to become equations of this form, and this representation is easy to compute.

The Smith normal form of a matrix [14] gives us the relation that the lengths of the cycles of a lattice is given by the *determinant divisors* of the basis matrix:

Theorem 2. *Let Λ be a lattice and $\mathbf{B}$ its basis matrix. Then the lengths of the cycles of Λ, $k_1, k_2, \ldots, k_n$ are given by*

$$k_i = \frac{d_i}{d_{i-1}}$$

where d_i is gcd *of all i-minors of $\mathbf{B}$ and $d_0 = 1$.*

3 Background and Previous Results

3.1 Complexity of Finding Short Vectors

To the best of our knowledge, the first result of **NP**-hardness of calculating short vectors in a lattice was published by van Emde Boas in 1981 [16], where it is proved **NP**-hard to calculate the shortest vector in ℓ_∞-norm in a general lattice. The same problem for the ℓ_2-norm was long an open problem, until proven **NP**-hard under randomized reductions by Ajtai in 1998 [2]. Micciancio [8] improved this result by showing that it is **NP**-hard to approximate the shortest vector within a factor $\sqrt{2} - \varepsilon$ for any $\varepsilon > 0$ under randomized reductions.

3.2 On the Cycle Structure of lattices

We will now state a few theorems on the cycle structure. These are probably well known, and we will therefore omit the proofs. Please note that some of the cycle lengths k_i mentioned in the theorems may be 1.

Theorem 3. *Let $\Lambda \subseteq \mathbb{Z}^n$ be a lattice with cycle structure $\mathbb{Z}_{k_1} \times \mathbb{Z}_{k_2} \times \cdots \times \mathbb{Z}_{k_n}$. Then the lattice $d \cdot \Lambda$ has the cycle structure $\mathbb{Z}_{d \cdot k_1} \times \mathbb{Z}_{d \cdot k_2} \times \cdots \times \mathbb{Z}_{d \cdot k_n}$.*

The next theorem shows that we can always assume that the shortest cycle of a lattice has length 1.

Theorem 4. *Let $\Lambda \subseteq \mathbb{Z}^n$ be a lattice with cycle structure $\mathbb{Z}_{k_1} \times \mathbb{Z}_{k_2} \times \cdots \times \mathbb{Z}_{k_n}$, where $k_1 \leq k_i$, $i = 2, \ldots, n$. Then $\Lambda = k_1 \cdot \Lambda'$, where Λ' is a lattice with cycle structure $\mathbb{Z}_{k_2/k_1} \times \mathbb{Z}_{k_3/k_1} \times \cdots \times \mathbb{Z}_{k_n/k_1}$.*

3.3 Previous Results on the Cycle Structure

Paz and Schnorr [11] have shown the following theorem, which essentially says that any lattice can be approximated arbitrarily well by a lattice described by a single modular equation. We will call these lattices *cyclic*.

Theorem 5. *Let $\Lambda \in \mathbb{Z}^n$ be a lattice. Then for every $\varepsilon > 0$ we can efficiently construct a linear transformation $\sigma_{\Lambda,\varepsilon} : \Lambda \to \mathbb{Z}^n$ such that $\sigma_{\Lambda,\varepsilon}(\Lambda)$ is a lattice and for some integer k*

1. $\|\mathbf{u} - \sigma_{\Lambda,\varepsilon}(\mathbf{u})/k\| \leq \varepsilon\|\mathbf{u}\|$ *for all* $\mathbf{u} \in \Lambda$,
2. $\sigma_{\Lambda,\varepsilon}(\Lambda)$ *has one cycle.*

This theorem implies that the cyclic lattices, in some sense, are the hardest ones in the worst case. If we know a way of finding short vectors in cyclic lattices, this would give us a method of finding short vectors in any lattice.

The average case/worst case connection described by Ajtai [1], the class of hard lattices consist of lattices with n/c cycles, where n is the dimension and c some constant.

These first results show that lattices with just one cycle are hard, but the latter seems to indicate that also lattices with relatively many cycles are hard. Hence it is natural to investigate the role of the cycle structure in complexity questions.

4 The LLL Algorithm in Practice

In this section we will give data about the performance of the LLL algorithm in practice when applied to lattices with different number of cycles. The intention is to find out whether or not the result of LLL depends on the cycle structure of the lattice.

In all experiments, version 4.3 of the NTL library [13] was used.

4.1 Construction of Lattice Instances

For the experiments, we need to construct lattices in such a way that we have control over the number of cycles in the lattices. The idea is to create a set of linear modular equations and compute the null space of this set of equations.

To create an n-dimensional lattice with m cycles we create an $m \times n$ matrix $\mathbf{A}$ and set

$$\Lambda = \{\mathbf{x} \mid \mathbf{A}\mathbf{x} \equiv \mathbf{0} \pmod{q}\} .$$

The elements of $\mathbf{A}$ are given by a shift register generator. In this register we create a stream of bits using

$$x_{i+1} = \sum_{j=1}^{l} a_j x_{i-j} \bmod 2 .$$

The parameter $\mathbf{a} = (a_1, a_2, \ldots, a_l)$ is a constant to be chosen. Solving for the null space gives us $n - m$ basis vectors. To ensure the basis contains n vectors, m rows of the matrix $q\mathbf{I}_n$ are added to the basis.

The dimensions of $\mathbf{A}$ determine the cycle structure of the lattice. With m rows in $\mathbf{A}$, we get a lattice with m cycles of length q. To make it possible to compare the different lattices with each other, they were created in such a way that their determinants were equal. By Minkowski's inequality (theorem 1), this implies that the length of a shortest vector has the same upper bound. Also the expected length of the shortest vector is the same. More precisely, given the dimension n and the determinant d, the lattices were created as

$$\Lambda_m = \left\{\mathbf{x} \mid \mathbf{A}\mathbf{x} \equiv \mathbf{0} \ \left(\bmod p\left(d^{1/m}\right)\right)\right\}$$

where $p(x)$ is the smallest prime equal to or greater than x and $\mathbf{A}$ is an $m \times n$ matrix with random entries. Since the determinant is given by the product of the cycle lengths, we see that all the Λ_m have approximately the same determinant, which means that it makes sense to compare the results of the LLL algorithm on them.

An important factor is how the starting point for LLL is chosen. When we compute the basis matrix from the null space and add m rows of the form $q\mathbf{e}_k$ for unit vectors $\mathbf{e}_k$, we get a basis where the last rows are much shorter than the first ones. Micciancio [9] suggests the Hermite Normal Form (HNF) as a standard representation of a lattice. The HNF can be computed in polynomial time [15] and we can easily find an upper bound for the coordinates. The basis derived from the null space is in already in HNF, and for the results presented use this basis is used as starting point for LLL.

4.2 Result of the LLL Algorithm

In the experiments, the LLL algorithm was executed with 75-dimensional lattices created as explained above as input. The algorithm was executed at least four times for each number of cycles, and the length of the output vector was noted. The result is given in figures 1. The number of iterations needed by the algorithm to finish is given in figure 2.

As we can see in figure 1, the length of the vector produced by the LLL algorithm does not seem to depend on the cycle structure of the lattice examined. From figure 2 it seems that the number of iterations needed by the LLL algorithm to finish in our experiments decreases with the number of cycles.

Since the starting point for a lattice Λ of dimension n with m cycles contains vectors of length $q \approx \sqrt[m]{\det(\Lambda)}$, we get a better starting point for a higher number of cycles. Experimental data indicate that once the LLL algorithm has reached a point where the length of the shortest vector is that of the starting point for lattice with a higher number of cycles, the progress of the algorithm is similar for both lattices.

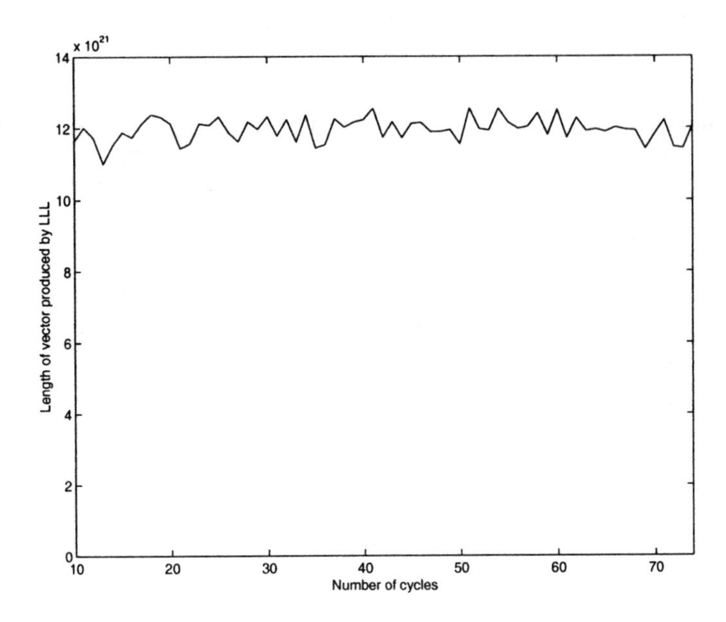

Figure 1. Shortest Vector Found by LLL in 75-Dimensional Lattices with Constant Determinant.

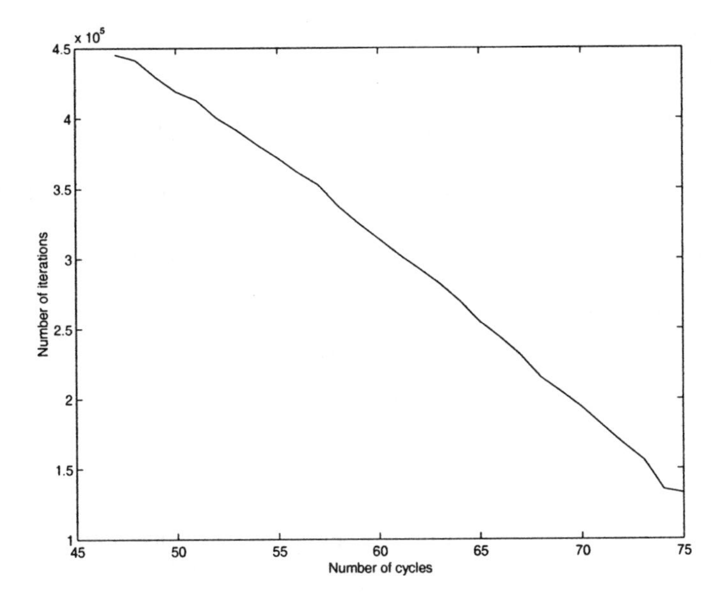

Figure 2. Number of Iterations as Function of the Number of Cycles for a 75-Dimensional Lattice.

5 Complexity of Computing Short Vectors in a Lattice with Many Cycles

We will now present an **NP**-completeness proof for lattices with a maximum number of cycles. We will prove that even if an n-dimensional lattice has $n-1$ cycles the problem of deciding whether there is a vector shorter than a given length in ℓ_∞-norm is **NP**-complete.

The problem that we will discuss is the following:

Definition 1. SVML$_\infty$ *is the problem of finding a short vector in a lattice with maximal number of cycles. Let $\Lambda \subseteq \mathbb{Z}^n$ be a lattice which has $n-1$ cycles of equal length q, and let $k \in \mathbb{Z}$. Then (Λ, k) is a* YES*-instance if there exists $\mathbf{v} \in \Lambda$ such that $\|\mathbf{v}\|_\infty \leq k$, and a* NO*-instance otherwise.*

We will prove the following theorem

Theorem 6. SVML$_\infty$ *is **NP**-complete.*

Proof. We first note that SVML$_\infty$ is in **NP**. Given a vector $\mathbf{v}$ we can in polynomial time verify that $\|\mathbf{v}\|_\infty \leq k$ and that $\mathbf{v} \in \Lambda$ by solving the system of linear equations $\mathbf{Bx} = \mathbf{v}$ where $\mathbf{B}$ is a basis matrix of Λ and check that $\mathbf{x}$ is integral. This can be done in polynomial time.

Before we continue the proof, we introduce some notation. For any $a \in \mathbb{R}$, define

$$\{a\} := |a \bmod \mathbb{Z}| = \min_{k \in \mathbb{Z}}(|a-k|) \ .$$

Informally, $\{a\}$ is the distance from a to the closest integer. We also introduce a related notation for vectors. For any $\mathbf{v} \in \mathbb{R}^n$, define

$$\{\{\mathbf{v}\}\} := \|\mathbf{v} \bmod \mathbb{Z}^n\|_\infty = \max_{i=1}^{n}(\{v_i\}) \ .$$

$\{\{\mathbf{v}\}\}$ can be seen as the distance between $\mathbf{v}$ and the closest integral vector, given that by distance we mean the max-norm.

We prove that SVML$_\infty$ is **NP**-hard by reducing from good simultaneous Diophantine approximation in ℓ_∞-norm, GDA$_\infty$, which was proven **NP**-hard by Lagarias [6]. GDA$_\infty$ is the following problem. Given a vector $\alpha \in \mathbb{Q}^n$ and integers N and s decide whether there exists an integer Q such that

$$1 \leq Q \leq N$$

and

$$\{\{Q\alpha\}\} \leq 1/s \ .$$

In other words, given a vector of rational numbers, we want to find good approximations to the components of this vector using rationals with a small common denominator.

We note that we can always assume that α is of the form

$$\alpha = \left(\frac{a_1}{b}, \frac{a_2}{b}, \ldots, \frac{a_n}{b}\right) \ .$$

Should α not be of this form, we can always rewrite all its components using the least common denominator.

We start by proving that a revision of GDA_∞, RGDA_∞, is **NP**-hard.

Definition 2. RGDA_∞ *is the following problem: Given integers s, q and N and a vector $\beta = (k_1, k_2, \ldots, k_n)/B \in \mathbb{Q}^n$ where $B = N^2 s(s-2)^q$, decide whether there exists an integer Q such that*

$$1 \leq Q \leq N^2(s-2)^q + \frac{N}{2}$$

and

$$\{\{Q\beta\}\} \leq \frac{1}{s} + \frac{1}{2Ns(s-2)^q} .$$

Lemma 1. RGDA_∞ *is **NP**-hard.*

Proof. We reduce GDA_∞ to RGDA_∞. Let α, N and s be an instance of GDA_∞. Assume b is the common denominator of α. Let c be the least integer such that $1/s < c/b$. Now choose q as the least integer for which

$$\frac{1}{s} + \frac{1}{Ns(s-2)^q} < \frac{c}{b}$$

and

$$(s-2)^q > N$$

and let $B = N^2 s(s-2)^q$. In other words, we choose q so that there is no multiple of $1/b$ in the interval $[1/s, 1/s + N/B)$.

Let the vector

$$\beta' = (k_1, k_2, \ldots, k_n)/B$$

with integral components $k_1, k_2, \ldots, k_n$ be such that $||\alpha - \beta'||_\infty$ is minimized. This can be done in polynomial time using ordinary division. It is easy to see that $||\alpha - \beta'||_\infty \leq 1/(2B)$.

We now define a new vector

$$\beta = \left(\beta', \frac{1}{Ns}, \frac{1}{Ns(s-2)}, \ldots, \frac{1}{Ns(s-2)^q}, \frac{1}{N^2 s}, \frac{1}{N^2 s(s-2)}, \ldots, \frac{1}{N^2 s(s-2)^q}\right),$$

that is, we append some new elements to the vector β'.

We see that β, N, q and s form an instance of RGDA_∞. Since q is logarithmic in N, β is not more than polynomially larger than α. Also the bit size of the common denominator does not grow more than polynomially.

We want to prove that Q is a solution of this RGDA_∞ problem if and only if it is a solution of this original GDA_∞ problem.

Let Q be a solution of the RGDA_∞ instance. We want to prove that Q also is a solution of the original GDA problem.

We first prove that $Q \leq N$. We know that $1/(Ns)$ is a component of β. Since Q is a solution,

$$\left\{Q\frac{1}{Ns}\right\} \leq \frac{1}{s} + \frac{1}{2s(s-2)^q} .$$

This implies that either

$$Q\frac{1}{Ns} \leq \frac{1}{s} + \frac{1}{2s(s-2)^q}$$

or

$$Q\frac{1}{Ns} \geq 1 - \left(\frac{1}{s} + \frac{1}{2s(s-2)^q}\right),$$

which can be rewritten as

$$Q \leq N + \frac{N}{2(s-2)^q}$$

or

$$Q \geq Ns - N - \frac{N}{2(s-2)^q} .$$

Since Q is integral, these two conditions imply that

$$Q \leq N$$

or

$$Q \geq N(s-1).$$

We also have that $\frac{1}{Ns(s-2)}$ is a component of β. The corresponding calculations for this component show that

$$Q \leq N(s-2) < N(s-1)$$

or

$$Q \geq N(s-2)(s-1).$$

We can use the same reasoning for the components β_{n+1} up to β_{n+q+1} (remember that $\beta_{n+q+1} = 1/\left(Ns(s-2)^q\right)$), which shows that either

$$Q \leq N$$

or

$$Q \geq N(s-2)^q(s-1) - \frac{N}{2} .$$

We do the same thing with $\beta_{n+q+2} = 1/(N^2 s)$, which gives us that either

$$Q \leq N^2$$

or

$$Q > N^2(s-2) .$$

Since $(s-2)^q > N$ this implies together with the previous results that $Q > N^2(s-2)$ unless $Q \le N$. Going through the remaining components we finally get that

$$Q \le N$$

or

$$Q \ge N^2(s-2)^q(s-1) - \frac{N}{2}\,.$$

Since we in the definition of the problem stated that $Q \le N^2(s-2)^q + N/2$, the only remaining possibility is that $Q \le N$.

We now prove that $\{\{Q\alpha\}\} \le 1/s$. We observe that $\{\{Q\alpha\}\} = k/b$ for some integer k (remember that b is the common denominator of α). We know that $\{Q\beta'\} \le \{Q\beta\} \le 1/s + 1/(2B)$. Since the distance between α and β' is at most $1/(2B)$ we can conclude that

$$\begin{aligned}\{\{Q\alpha\}\} &\le \{\{Q\beta'\}\} + Q\frac{1}{2B}\\ &\le \frac{1}{s} + \frac{1}{2Ns(s-2)^q} + \frac{1}{2Ns(s-2)^q}\\ &= \frac{1}{s} + \frac{1}{Ns(s-2)^q}.\end{aligned}$$

But, as we just stated, the approximation error in α is always a multiple of $1/b$ and since we have chosen q such that $1/s + 1/(Ns(s-2)^q)$ does not pass a $1/b$ boundary, this must indeed be

$$\{\{Q\alpha\}\} \le \frac{1}{s}\,.$$

This concludes the proof that Q is a solution of the GDA$_\infty$ instance if it is a solution of the RGDA$_\infty$ instance.

Now assume that Q is a solution of the GDA$_\infty$ instance. This means that $\{Q\alpha\} \le 1/s$ and $Q \le N$. We first note that $\{Q\beta_i\} \le Q\beta_i \le N\beta_i \le 1/s$ for $i = n+1, \ldots, n+2q+2$, i.e., the appended components. This means that we only need to consider β'. We know that $\|\alpha - \beta'\|_\infty \le 1/(2B)$, which means that

$$\{\{Q\beta'\}\} \le \{\{Q\alpha\}\} + Q\frac{1}{2B} \le \frac{1}{s} + \frac{1}{2Ns(s-2)^q}$$

and we can conclude that Q is a solution of the RGDA$_\infty$ problem.

This proves that the reduction is correct. We now turn to the proof of the **NP**-hardness of SVML$_\infty$. We do this by reducing from RGDA$_\infty$. Let β, N and s be an instance of RGDA$_\infty$, with $\beta = (\frac{k_1}{B}, \frac{k_2}{B}, \ldots, \frac{k_n}{B})$. We create the lattice with the following $(n+1) \times (n+1)$ matrix as its basis matrix (the basis vectors are the rows of the matrix)

$$\mathbf{A} = \begin{pmatrix} 1/B & k_1/B & k_2/B & \dots & k_n/B \\ 0 & 1 & 0 & \dots & 0 \\ 0 & 0 & 1 & \dots & 0 \\ \vdots & \vdots & \vdots & \ddots & \vdots \\ 0 & 0 & 0 & \dots & 1 \end{pmatrix} .$$

If we multiply this lattice by B we get an $(n+1)$-dimensional integral lattice. According to theorem 2, the lattice $B \cdot \mathbf{A}$ has n cycles of length B. This means that $B \cdot \mathbf{A}$ and $B/s + B/(2Ns(s-2)^q)$ is an instance of SVML_∞.

We now want to prove that this SVML_∞ instance is a YES instance if and only if the original RGDA_∞ is a YES instance.

Assume that the SVML_∞ instance is a YES instance, i.e., there is a Q such that

$$\max\left\{Q\frac{1}{B}, \{\{Q\beta\}\}\right\} \leq \frac{1}{s} + \frac{1}{2Ns(s-2)^q}$$

which implies that

$$\{\{Q\beta\}\} \leq \frac{1}{s} + \frac{1}{2Ns(s-2)^q}$$

and

$$Q \leq N^2(s-2)^q + \frac{N}{2}$$

so Q is a solution of the RGDA_∞ instance.

Assume that the RGDA_∞ instance is a YES instance. Then there is a $Q \leq N^2(s-2)^q + N/2$ such that

$$\{\{Q\beta\}\} \leq \frac{1}{s} + \frac{1}{2Ns(s-2)^q} .$$

We can calculate

$$Q\frac{1}{B} \leq \frac{1}{s} + \frac{1}{2Ns(s-2)^q}$$

which implies that

$$\max\left\{Q\frac{1}{B}, \{\{Q\beta\}\}\right\} \leq \frac{1}{s} + \frac{1}{2Ns(s-2)^q} ,$$

i.e., the SVML_∞ instance is a YES instance.

This concludes the proof that SVML_∞ is **NP**-complete.

Acknowledgements

I would like to thank Johan Håstad for valueable feedback and ideas during the preparation of this paper. I would also like to thank the anonymous referees for pointing out mistakes and possible improvements.

References

1. M. Ajtai. Generating Hard Instances of Lattice Problems. *Proc. 28th ACM Symposium on Theory of Computing*, pages 99–108, 1996.
2. M. Ajtai. The shortest vector problem in ℓ_2 is **NP**-hard for randomized reductions. *Proc. 30th ACM Symposium on the Theory of Computing*, pages 10–19, 1998.
3. J-Y. Cai and A. Nerurkar. An Improved Worst-Case to Average-Case Connection for Lattice Problems. *Proc. 38th IEEE Symposium on Foundations of Computer Science*, pages 468–477, 1997.
4. O. Goldreich and S. Goldwasser. On the limits of non-approximability of lattice problems. *Journal of Computer and System Sciences*, Academic Press, 60(3):540–563, 2000. Can be obtained from `http://www.eccc.uni-trier.de/eccc`.
5. Kabatjanskii and Levenshtein. Bounds for Packings on a Sphere and in Space. *Problems of Information Transmission 14*, 1:1–17, 1978.
6. J.C. Lagarias. The Computational Complexity of Simultanous Diophantine Approximation Problems. *SIAM Journal of Computing*, 14:196–209, 1985.
7. A.K. Lenstra, H.W. Lenstra and L. Lovász. Factoring Polynomials with Rational Coefficients. *Mathematische Annalen* 261:515–534, 1982.
8. D. Micciancio. The Shortest Vector in a Lattice is Hard to Approximate within Some Constant. *Proc. 39th IEEE Symposium on Foundations of Computer Science*, 1998, 92–98.
9. D. Micciancio. Lattice Based Cryptography: A Global Improvement. Technical report, Theory of Cryptography Library, 1999. Report 99-05. Can be obtained from `http://eprint.iacr.org`.
10. H. Minkowski. Über die positiven quadratischen Formen und über kettenbruchähnliche Algorithmen. *Crelles Journal für die Reine und Angewandte Mathematik*, 107:278–297, 1891.
11. A. Paz and C.P. Schnorr. Approximating Integer Lattices by Lattices with Cyclic Lattice Groups. *Automata, languages and programming (Karlsruhe)*, 1987, 386–393.
12. C.P. Schnorr. A Hierarchy of Polynomial Time Lattice Basis Reduction Algorithms. *Theoretical Computer Science*, 53:201–224, 1987.
13. V. Shoup. NTL: A Library for doing Number Theory. Can be obtained from `http://www.shoup.net`.
14. H.J.S. Smith. On Systems of Linear Indeterminate Equations and Congruences. *Philosophical Transactions of the Royal Society of London*, 151:293–326, 1861.
15. A. Storjohann and G. Labahn. Asymptotically Fast Computation of Hermite Normal Forms of Integer Matrices. *ISAAC'96*, 1996, 259–266.
16. P. van Emde Boas. Another **NP**-complete partition problem and the copmlexity of computing short vectors in lattices. Technical Report 81-04. Mathematics Department, University of Amsterdam, 1981. Can be obtained from `http://turing.wins.uva.nl/~peter`.

Multisequence Synthesis over an Integral Domain

Li-ping Wang[1] and Yue-fei Zhu[2*]

[1] State Key Laboratory of Information Security, Graduate School
University of Science and Technology of China, Beijing 100039, China
A9000@china.com

[2] Department of Network Engineering
Information Engineering University, Zhengzhou 450002, China
zyf0136@sina.com

Abstract. We first give an extension of $F[x]$-lattice basis reduction algorithm to the polynomial ring $R[x]$ where F is a field and R an arbitrary integral domain. So a new algorithm is presented for synthesizing minimum length linear recurrence (or minimal polynomials) for the given multiple sequences over R. Its computational complexity is $O(N^2)$ multiplications in R where N is the length of each sequence. A necessary and sufficient conditions for the uniqueness of minimal polynomials are given. The set of all minimal polynomials is also described.

1 Introduction

In this paper we investigate the problem of multisequence linear recurrence synthesis over an arbitrary integral domain R. It is of importance in coding theory, cryptography and signal processing field. Fitzpatrick and Norton gave an algorithm when R is a potential domain, i.e. both R and $R[[x]]$ are unique factorization domains [1]. Lu and Liu presented a method to find shortest linear recurrence over unique factorization domain by means of Gröbner basis theory [2]. Furthermore, Norton proposed an algorithm over an arbitrary integral domain [3]. However, the above algorithms are only used to solve the problem of single-sequence synthesis. By means of $F[x]$-lattice basis reduction algorithm [4], in [5] we gave a new multisequence synthesis algorithm over a field F. $F[x]$-lattice basis reduction algorithm can be consider as a generalization of famous LLL algorithm [6], which has been widely used in computational number theory and cryptography. Especially LLL algorithm is often used to attack some cryptosystems such as knapsack schemes [7] and RSA system [8]. Our results are an application to coding theory and cryptography about $F[x]$-lattice basis reduction algorithm. But it fails when applied to an integral domain. In the next section, we present a $R[x]$-lattice basis reduction algorithm and some properties about reduced basis. In Section 3, we give an algorithm to solve this synthesis problem. Furthermore, we develop a necessary and sufficient conditions for the

* Research supported by NSF under grants No. 19931010 and G 1999035803.

J.H. Silverman (Ed.): CaLC 2001, LNCS 2146, pp. 206–217, 2001.

uniqueness of minimal polynomials. When the solution is not unique, we describe the set of all minimal polynomials.

First we formulate the problem. Let R be an integral domain and $a = (a_1, \cdots, a_N)$ be a sequence with length N over R. A polynomial

$$q(x) = \sum_{i=0}^{d} c_i x^i \in R[x]$$

is an *annihilating polynomial* of a if

$$c_d a_k + c_{d-1} a_{k-1} + \cdots + c_0 a_{k-d} = 0 \qquad \text{for all } k,\ d < k \leq N. \tag{1}$$

A *minimal polynomial* of a is defined to be an annihilating polynomial with least degree. Its degree is called the *linear complexity* of a and indicates the minimum length of a linear recurrence capable of generating a.

Let $a^{(1)}, a^{(2)}, \cdots, a^{(m)}$ be m sequences over R. A polynomial $q(x)$ is an annihilating polynomial of $a^{(1)}, a^{(2)}, \cdots, a^{(m)}$ if $q(x)$ is an annihilating polynomial for each $a^{(i)}$, $1 \leq i \leq m$. A minimal polynomial of $a^{(1)}, a^{(2)}, \cdots, a^{(m)}$ is defined to be an annihilating polynomial with least degree. For brevity,

$$\mathrm{Ann}(a^{(1)}, a^{(2)}, \cdots, a^{(m)})$$

denotes the set of all annihilating polynomials of $a^{(1)}, a^{(2)}, \cdots, a^{(m)}$.

The *multisequence synthesis problem* of $a^{(1)}, a^{(2)}, \cdots, a^{(m)}$ is to find one of their minimal polynomials. Throughout the paper, we denote

$$a^{(i)} = (a_1^{(i)}, a_2^{(i)}, \cdots, a_N^{(i)}),$$

where $a_j^{(i)} \in R$, for $1 \leq i \leq m$ and $1 \leq j \leq N$.

Let

$$A = R((x^{-1})) = \left\{ \sum_{i=i_0}^{\infty} a_i x^{-i} \;\middle|\; i_0 \in Z,\ a_i \in R \right\}.$$

Obviously A is a formal power-negative series ring. Since R is an integral domain, then so is A. We define a map

$$\begin{array}{rcl} v : A & \longrightarrow & Z \cup \{\infty\} \\ a(x) = \sum_{i=i_0}^{\infty} a_i x^{-i} & \longmapsto & \begin{cases} \infty & \text{if } a(x) = 0 \\ \min\{i | a_i \neq 0\} & \text{otherwise} \end{cases} \end{array}$$

For arbitrary elements $a(x), b(x) \in A$, then v satisfies

1. $v(a(x)) = \infty$ if and only if $a(x) = 0$.
2. $v(a(x) \cdot b(x)) = v(a(x)) + v(b(x))$.
3. $v(a(x) + b(x)) \geq \min\{v(a(x)), v(b(x))\}$, with equality if $v(a(x)) \neq v(b(x))$.

Therefore we also call v a valuation on A. If $a = (a_1, \cdots, a_N)$, then $a(x) = \sum_{j=1}^{\infty} a_j \cdot x^{-j} \in A$ is the associated formal negative-power series of a. We give a lemma similar to Lemma 1 [5].

Lemma 1. *Let $a^{(1)}(x), a^{(2)}(x), \cdots, a^{(m)}(x)$ be the associated formal negative-power series of $a^{(1)}, a^{(2)}, \cdots, a^{(m)}$ respectively, and let $q(x)$ be a polynomial over R with degree d. Then $q(x) \in \mathrm{Ann}(a^{(1)}, a^{(2)}, \cdots, a^{(m)})$ if and only if for each i, $1 \leq i \leq m$, there exists a unique polynomial $p_i(x) \in R[x]$ such that*

$$v(q(x) \cdot a_i(x) - p_i(x)) > N - d \tag{2}$$

Lemma 1 reduces the multisequence synthesis problem of $a^{(1)}, a^{(2)}, \cdots, a^{(m)}$ to finding a polynomial of least degree satisfying (2), which can be solved by $R[x]$-lattice basis reduction algorithm.

2 $R[x]$-Lattice and Its Reduction Basis Algorithm

In this section we give some properties about $R[x]$−lattice and an $R[x]$-lattice basis reduction algorithm.

Let n be an arbitrary positive integer, then A^n denotes the free A−module $A \oplus \cdots \oplus A$ with rank n. The standard basis of A^n consists of the elements $\varepsilon_1 = (1_A, 0, \cdots, 0)$, $\varepsilon_2 = (0, 1_A, 0, \cdots, 0)$, $\varepsilon_n = (0, \cdots, 0, 1_A)$, where 1_A is the identity element of A. Apparently $1_A = 1_R$, and so is simply denoted 1 in the sequel. For any element in A^n, we use its coordinates under the standard basis to represent it and so also call an element in A^n a vector. A subset Λ of A^n is an $R[x]$-lattice if there exists a basis $\omega_1, \cdots, \omega_n$ of A^n such that

$$\Lambda = \Lambda(\omega_1, \cdots, \omega_n) = \sum_{i=1}^{n} R[x]\omega_i \ . \tag{3}$$

In this situation we say that $\omega_1, \cdots, \omega_n$ form a basis for Λ. Obviously, an $R[x]$-lattice may be also defined by a free $R[x]$-submodule of A^n with rank n. The determinant $d(\Lambda)$ of a lattice Λ is defined by

$$d(\Lambda) = v(\det(\omega_1, \cdots, \omega_n)) \ . \tag{4}$$

We define a map on A^n, denoted by V.

$$\begin{aligned} V : A^n &\longrightarrow Z \cup \{\infty\} \\ \beta = (b_i(x))_{1 \leq i \leq n} &\longmapsto \begin{cases} \infty & \beta = 0 \\ \min\{v(b_i(x)) | 1 \leq i \leq n\} & \text{otherwise} \end{cases} \end{aligned}$$

Let ϵ be a real number with $0 < \epsilon < 1$ and the length function $L : A^n \longrightarrow \mathcal{R}^{\geq 0}$ is given by $\beta \longmapsto \epsilon^{V(\beta)}$. Besides, we define a projection. For any integer k,

$$\begin{aligned} \theta_k : A^n &\longrightarrow R^n \\ \beta = (b_i(x))_{1 \leq i \leq n} &\longmapsto (b_{i,k})^T_{1 \leq i \leq n} \end{aligned}$$

where $b_i(x) = \sum_{j=j_0}^{\infty} b_{i,j} x^{-j}$, $1 \leq i \leq n$, and T denotes the transpose of a matrix.

Definition 1. *A basis $\omega_1, \cdots, \omega_n$ of a lattice Λ is reduced if the set*

$$\{\theta_{V(\omega_i)}(\omega_i) \in R^n \mid 1 \leq i \leq n\}$$

is R-linearly independent.

Definition 2. *A subset of a lattice is a sublattice if itself is also a lattice with rank n.*

Lemma 2. *Let $B \in \mathrm{Mat}_n(R)$, then the rows (columns) of B are R-linearly independent if and only if its determinant $|B| \neq 0$.*

Proof. Let K be the quotient field of the integral domain R. Then $|B| \neq 0$ over K if and only if the rows (columns) of B are K-linearly independent. Since R is an integral domain, the rows (columns) of B are R-linearly independent if and only if they are K- linearly independent. In addition, $|B| \neq 0$ over R if and only if $|B| \neq 0$ over K. Hence the conclusion is proved. □

Lemma 3. *Let $\omega_1, \cdots, \omega_n$ be a basis of a lattice Λ and*

$$\Delta(\omega_1, \cdots, \omega_n) = d(\Lambda) - \sum_{i=1}^{n} V(\omega_i) \quad . \tag{5}$$

Then $\Delta(\omega_1, \cdots, \omega_n) \geq 0$. The equality holds if and only if the basis $\omega_1, \cdots, \omega_n$ is reduced.

Proof. We use the notation

$$\omega_i = \theta_{V(\omega_i)}(\omega_i) x^{-V(\omega_i)} + \cdots \qquad (i = 1, \cdots, n) \ . \tag{6}$$

So we have

$$\det(\omega_1, \cdots, \omega_n) = x^{-(V(\omega_1)+\cdots+V(\omega_n))} \det(\theta_{V(\omega_1)}(\omega_1), \cdots, \theta_{V(\omega_n)}(\omega_n)) + \cdots \tag{7}$$

where $\cdots$ at the right indicates terms with x^{-1} raised to exponents larger than $V(\omega_1) + \cdots + V(\omega_n)$. Thus $d(\Lambda) \geq \sum_{i=1}^{n} V(\omega_i)$. By Lemma 2, the equality holds if and only if the basis $\omega_1, \cdots, \omega_n$ is reduced. □

Lemma 4. *Let $\omega_1, \cdots, \omega_n$ be a basis of a lattice Λ and $u_i \in R^*$, for all i, $1 \leq i \leq n$, where R^* denotes $R/\{0\}$. Then*

1. *$u_1\omega_1, \cdots, u_n\omega_n$ are also a basis of A^n.*
2. *The lattice $\Lambda(u_1\omega_1, \cdots, u_n\omega_n)$ is a sublattice of lattice $\Lambda(\omega_1, \cdots, \omega_n)$ and $\Lambda(u_1\omega_1, \cdots, u_n\omega_n) = \Lambda(\omega_1, \cdots, \omega_n)$ if and only if $u_1 u_2 \cdots u_n$ is a unit in R.*
3. *$d(\Lambda(u_1\omega_1, \cdots, u_n\omega_n)) = d(\Lambda(\omega_1, \cdots, \omega_n))$.*

Proof. Properties 1. and 2. are easily obtained. As to 3., it follows from the identity $\det(u_1\omega_1, \cdots, u_n\omega_n) = u_1 \cdots u_n \det(\omega_1, \cdots, \omega_n)$. □

We give an algorithm as follows.

Algorithm 1

Input: $\omega_1, \cdots, \omega_n$ are a basis of a lattice and set $(\phi_1, \cdots, \phi_n) := (\omega_1, \cdots, \omega_n)$.
Output: a reduced basis for some sublattice of a lattice $\Lambda(\omega_1, \cdots, \omega_n)$.

1. For $i := 1$ to n do begin Set $u_i := 1$ end;
2. Set $\Delta := d(\Lambda) - \sum_{i=1}^{n} V(\phi_i)$ and $r := 0$;
3. While $\Delta > 0$ Do begin
   - Set $r := r + 1$;
   - Put

$$(\phi_1, \cdots, \phi_n) := (\phi_{\rho(1)}, \cdots, \phi_{\rho(n)})$$
$$(u_1, \cdots, u_n) := (u_{\rho(1)}, \cdots, u_{\rho(n)}),$$

   where some permutation $\rho \in S_n$ such that $V(\phi_1) \geq \cdots \geq V(\phi_n)$;
   - Set $M := (\theta_{V(\phi_1)}(\phi_1), \cdots, \theta_{V(\phi_n)}(\phi_n))$;
   - (reduction step) Find a nonzero solution $(c_1, \cdots, c_n)^T$ of the equation system $Mx = 0$ and determine j such that $2 \leq j \leq n$, $c_j \neq 0$ and $c_i = 0$ for $i > j$, and so set

$$\phi_j := c_j \phi_j + \sum_{i=1}^{j-1} c_i x^{V(\phi_i) - V(\phi_j)} \cdot \phi_i \quad \text{and} \quad u_j := c_j u_j;$$

   - Set $\Delta := d(\Lambda) - \sum_{i=1}^{n} V(\phi_i)$;

   end;
4. Output $r, u_1, \cdots, u_n$ and $\phi_1, \cdots, \phi_n$.

Remark 1.

1. By algorithm 1, we get a reduced basis for a sublattice $\Lambda(u_1\omega_1, \cdots, u_n\omega_n)$ of lattice $\Lambda(\omega_1, \cdots, \omega_n)$, where $u_1, \cdots, u_n$ are the output. Therefore it possibly is not a basis of the lattice $\Lambda(\omega_1, \cdots, \omega_n)$. Especially if R is a finite field, it is also a reduced basis of lattice $\Lambda(\omega_1, \cdots, \omega_m)$ by Lemma 4.
2. We define

$$\psi : \{(\phi_1, \cdots, \phi_n) | \phi_1, \cdots, \phi_n \text{ are a basis in Algorithm 1}\} \longrightarrow Z$$
$$(\phi_1, \cdots, \phi_n) \longmapsto d(\Lambda(\phi_1, \cdots, \phi_n)) - \sum_{i=1}^{n} V(\phi_i)$$

   Assume $\phi_1, \cdots, \phi_n$ are a current basis for a sublattice $\Lambda(u_1\omega_1, \cdots, u_n\omega_n)$ yielded at the beginning of the each while-loop. By Lemma 4, we know

$$d(\Lambda(\phi_1, \cdots, \phi_n)) = d(\Lambda(\omega_, \cdots, \omega_n)).$$

   Thus in Algorithm 1 the value Δ strictly decreases and $\Delta = 0$ implies $\phi_1, \cdots, \phi_n$ becomes reduced by Lemma 3. So the algorithm terminates after at most $\psi(\omega_1, \cdots, \omega_n)$ steps.

3. In Algorithm 1, we have to find a solution of the equation system $Mx = 0$ over R. We solve it over K, where K is the quotient field of R. Thus we get a solution in R by multiplying denominators.

In the following discussion we develop some properties about reduced basis. To simplify mathematical expressions, we denote π_n the n−th component of a vector in R^n, Lcf($f(x)$) the leading coefficient of $f(x)$ and deg$f(x)$ the degree of $f(x)$. Define deg$f(x) = -\infty$ if $f(x) = 0$. Let $\omega_1, \cdots, \omega_n$ be a reduced basis of a lattice Λ and $V(\omega_1) \geq \cdots \geq V(\omega_n)$. If any vector $\beta \in \Lambda(\omega_1, \cdots, \omega_n)$, then β is a linear combination $\beta = f_1(x)\omega_1 + \cdots + f_n(x)\omega_n$ with $f_i(x) \in R[x]$ for $1 \leq i \leq n$. Set

$$\begin{aligned} d_i &= \deg(f_i(x)), \\ c_i &= \mathrm{Lcf}(f_i(x)), \\ d(\beta) &= \min\{V(\omega_i) - d_i | 1 \leq i \leq n\}, \\ I(\beta) &= \{i|\ 1 \leq i \leq n, V(\omega_i) - d_i = d(\beta)\}. \end{aligned}$$

Since the following proofs are similar to those in [5], we omit them here.

Lemma 5. *With above notation, then* $d(\beta) = V(\beta)$.

Lemma 6. *With above notation, if* $V(\beta) > V(\omega_t)$ *with some* t, $1 \leq t \leq n$, *then* $f_t(x) = \cdots = f_n(x) = 0$.

Note 1. Especially in the case $t = 1$, we know $\beta = 0$, i.e. ω_1 is a shortest length vector in all non-zero vectors of Λ according to Lemma 6.

Theorem 1. *Let* $\omega_1, \cdots, \omega_n$ *be a reduced basis of a lattice* Λ *with* $V(\omega_1) \geq \cdots \geq V(\omega_n)$ *and*

$$S(\Lambda) = \{\beta \in \Lambda(\omega_1, \cdots, \omega_n)|\ \pi_n(\theta_{V(\beta)}(\beta)) \neq 0\}\ . \tag{8}$$

Let s *satisfy* $1 \leq s \leq n$, $\omega_s \in S(\Lambda)$ *and* $\omega_j \notin S(\Lambda)$ *for* $j < s$. *Then* ω_s *is a vector with shortest length in* $S(\Lambda)$.

3 Multisequence Synthesis Algorithm

In this section we present a multisequence synthesis algorithm based on the above $R[x]$-lattice basis reduction algorithm. Furthermore, we consider the uniqueness of minimal polynomials.

Let $a^{(i)}(x)$, $1 \leq i \leq m$, be the associated formal negative-power series of the sequences $a^{(i)}$. We construct $m + 1$ vectors in A^{m+1} and set

$$\begin{aligned} e_1 &= (-1, 0, \cdots, 0)_{m+1}, \\ &\vdots \\ e_m &= (0, \cdots, 0, -1, 0)_{m+1}, \\ \alpha &= (a^{(1)}(x), \cdots, a^{(m)}(x), x^{-N-1})_{m+1}. \end{aligned}$$

Obviously $e_1, \cdots, e_m, \alpha$ are A-linearly independent. Therefore they span an $R[x]$-lattice Λ with rank $m+1$. In detail,

$$\Lambda(e_1, \cdots, e_m, \alpha) = \left\{ q(x) \cdot \alpha + \sum_{i=1}^{m} p_i(x) e_i | \; p_i(x), q(x) \in R[x], 1 \le i \le m \right\}.$$

Then we have $d(\Lambda) = N + 1$. Set

$$\Omega = S(\Lambda(e_1, \cdots, e_m, \alpha)).$$
$$\Gamma = \mathrm{Ann}(a^{(1)}, a^{(2)}, \cdots, a^{(m)}).$$

For any $q(x) \in \Gamma$, it follows from Lemma 1 that for each $i, 1 \le i \le m$, there exists a unique polynomial $p_i(x) \in R[x]$ such that $v(q(x) \cdot a_i(x) - p_i(x)) > N - d$, where d is the degree of $q(x)$. However, $v(q(x) \cdot x^{-(N+1)}) = N + 1 - d$ and hence we obtain a unique vector β in Ω. Write

$$\beta = (q(x) \cdot a^{(1)}(x) - p_1(x), \cdots, q(x) \cdot a^{(m)}(x) - p_m(x), q(x) \cdot x^{-(N+1)}).$$

So a map $\varphi : \Gamma \longrightarrow \Omega$ is well defined by $q(x) \longmapsto \beta$. Define two natural total orderings, i.e. Ω is ordered by length of a vector and Γ by degree of a polynomial.

Theorem 2. *The map $\varphi : \Gamma \longrightarrow \Omega$ is an order-preserving bijection.*

The proof of Theorem 2 is omitted because of the similarity to Theorem 2 [5].

By means of Algorithm 1, we get a reduced basis $\omega_1, \cdots, \omega_{m+1}$ for the sublattice $\Lambda(u_1 e_1, \cdots, u_m e_m, u_{m+1}\alpha)$ of a lattice $\Lambda(e_1, \cdots, e_m, \alpha)$ with $V(\omega_1) \ge \cdots \ge V(\omega_{m+1})$, where $u_1, \cdots, u_{m+1}$ is the output. Let s satisfy $1 \le s \le m+1$, $\pi_{m+1}(\theta_{V(\omega_s)}(\omega_s)) \ne 0$ and $\pi_{m+1}(\theta_{V(\omega_i)}(\omega_i)) = 0$, for $i < s$. From Theorem 1, we know ω_s is a vector with shortest length in $S(\Lambda(u_1 e_1, \cdots, u_m e_m, u_{m+1}\alpha))$. For convenience, set $S = S(\Lambda(u_1 e_1, \cdots, u_m e_m, u_{m+1}\alpha))$. Furthermore we have:

Theorem 3. *The vector ω_s is also a vector with shortest length in Ω.*

Proof. Let β be a vector in Ω with shortest length. It is obvious that $V(\omega_s) \ge V(\beta)$. Setting $\gamma = u_1 \cdots u_{m+1}\beta$, then γ is in S and $V(\gamma) = V(\beta)$. Thus $V(\gamma) \ge V(\omega_s)$ and so $V(\omega_s) = V(\beta)$. □

By Theorem 2, there is a one-to-one correspondence between annihilating polynomials in Γ and vectors in Ω. We get a reduced basis $\omega_1, \cdots, \omega_{m+1}$ of the lattice $\Lambda(u_1 e_1, \cdots, u_m e_m, u_{m+1}\alpha)$ by Algorithm 1. It easily follows from Theorem 3 that ω_s is also a vector with shortest length in Ω and hence $\varphi^{-1}(\omega_s)$ is a minimal polynomial of $a^{(1)}, a^{(2)}, \cdots, a^{(m)}$. The above ideas lead to Algorithm 2.

Algorithm 2
Input: m sequences $a^{(1)}, a^{(2)}, \cdots, a^{(m)}$, each of length N, over R.
Output: a minimal polynomial of $a^{(1)}, a^{(2)}, \cdots, a^{(m)}$.

1. Compute the associated formal negative-power series, i.e.

$$a^{(i)}(x) := \sum_{j=1}^{N} a_j^{(i)} x^{-j}, \quad 1 \le i \le m;$$

2. Set $n := m+1$ and $\omega_1 := e_1, \cdots, \omega_{n-1} := e_m, \omega_n := \alpha$;
3. Execute Algorithm 1 and output r, $u_1, \cdots, u_n$ and $\phi_1, \cdots, \phi_n$.
4. Let s satisfy $1 \le s \le n$ and $\pi_n(\theta_{V(\phi_s)}(\phi_s)) \ne 0$;
5. Compute $q(x) := \varphi^{-1}(\phi_s)$ and output $q(x)$.

Remark 2. It takes at most N reduction steps according to Remark 1. And in each reduction step at most N multiplications in R are needed. So the computational complexity is $O(N^2)$ operations in R.

We know r is the total number of the reduction steps. Let each k, $0 \le k \le r$, be a superscript number such that the basis

$$\omega_1^{(k)}, \omega_2^{(k)}, \cdots, \omega_{m+1}^{(k)}$$

for some sublattice denotes the current basis yielded in the k-th while-loop. Analogous to Lemma 4 [5], we also have:

Lemma 7. *Assume $0 \le k \le r$. Then the set*

$$\left\{\theta_{V(\omega_i^{(k)})}(\omega_i^{(k)}) | 1 \le i \le m+1\right\}$$

is R-linearly dependent, but there exist m vectors in it which are R-linearly independent and their $m+1$-th components are 0.

From Lemma 7, we easily get the following Lemma.

Lemma 8. *Let $\omega_1, \omega_2, \cdots, \omega_{m+1}$ be a reduced basis for the sublattice*

$$\Lambda(u_1 e_1, \cdots, u_m e_m, u_{m+1}\alpha)$$

by Algorithm 1. Then there is a unique integer s such that $1 \le s \le m+1$ and $\pi_{m+1}(\theta_{V(\omega_s)}(\omega_s)) \ne 0$.

The subsequent discussion is devoted to the uniqueness about minimal polynomials. Recall that $a, b \in R^*$ are associates if there is a unit $u \in R^*$ such that $a = ub$. This defines an equivalence relation: $[b]$ denotes the class of associates of $b \in R^*$ and $[R^*]$ denotes the set of associate classes of R^*. Therefore, two polynomials $f_1(x)$ and $f_2(x)$ are equivalent if and only if there exists a unit $u \in R^*$ such that $f_1(x) = uf_2(x)$. The vectors β , $\gamma \in \Lambda(e_1, \cdots, e_m, \alpha)$ are equivalent if and only if $\beta = u\gamma$, where u is a unit in R. From now on we take the equivalent elements as one.

Theorem 4. *Let $\omega_1, \omega_2, \cdots, \omega_{m+1}$ be a reduced basis of the sublattice*

$$\Lambda(u_1 e_1, \cdots, u_m e_m, u_{m+1}\alpha)$$

by Algorithm 1 and $V(\omega_1) \ge \cdots \ge V(\omega_{m+1})$. Write $c = \mathrm{Lcf}(\varphi^{-1}(\omega_s))$, where s is defined same as in Lemma 8. Then the shortest length vector in S is unique if and only if $s = 1$, $V(\omega_1) > V(\omega_2)$ and the equation $cx = b$ with any element $b \in R^*$ and $[b] \ne [c]$ has no solution in R.

Proof. Suppose a vector $\beta \in S$ with $V(\beta) = V(\omega_1)$. Then β is a linear combination $\beta = f_1(x)\omega_1 + \cdots + f_{m+1}(x)\omega_{m+1}$ with $f_i(x) \in R[x]$ for $1 \le i \le m+1$. Since $V(\beta) > V(\omega_2)$, we have $f_2(x) = \cdots = f_{m+1}(x) = 0$ according to Lemma 6 and $\beta = f_1(x) \cdot \omega_1$. Because of $V(\beta) = V(\omega_1)$, we know $f_1(x) = a \in R^*$. Let $b = \text{Lcf}(\varphi^{-1}(\beta))$, then $b = ac$ and so $[b] = [c]$. Thus a is a unit in R and β and ω_s belong to same equivalent class, i.e. the shortest length vector in S is unique. Conversely, suppose $s \neq 1$. Setting $\beta = x^{V(\omega_1)-V(\omega_s)}\omega_1 + \omega_s$, then $\beta \in S$, $V(\beta) = V(\omega_s)$ and $\text{Lcf}(\varphi^{-1}(\beta)) = c$. Since $\beta \neq \omega_s$, then $\varphi^{-1}(\beta) \neq \varphi^{-1}(\omega_s)$, a contradiction to the uniqueness. Suppose $s = 1$ and $V(\omega_1) = V(\omega_2)$. From Lemma 8, we know $\omega_2 \notin S$. Putting $\beta = \omega_1 + \omega_2$, then $\beta \in S$, $V(\beta) = V(\omega_1)$ and $\text{Lcf}(\varphi^{-1}(\beta)) = c$. But $\beta \neq \omega_1$ and $\varphi^{-1}(\beta) \neq \varphi^{-1}(\omega_1)$, which also contradicts with the uniqueness. Suppose there exists $b \in R^*$ and $[b] \neq [c]$ such that the equation $cx = b$ has a solution in R. Setting $\beta = b\,\omega_1$, then $V(\beta) = V(\omega_1)$, but β and ω_1 are not same equivalent class. □

Theorem 5. *Let $\omega_1, \omega_2, \cdots, \omega_{m+1}$ be a reduced basis of the sublattice*

$$\Lambda(u_1e_1, \cdots, u_me_m, u_{m+1}\alpha)$$

by Algorithm 1 and $V(\omega_1) \ge \cdots \ge V(\omega_{m+1})$. Write $c = Lcf(\varphi^{-1}(\omega_s))$, where s is defined same as in Lemma 8. Then the minimal polynomial of $a^{(1)}, a^{(2)}, \cdots, a^{(m)}$ is unique if and only if $s = 1$, $V(\omega_1) > V(\omega_2)$, the equation $cx = b$ with any element $b \in [R^*]$ and $[b] \neq [c]$ has no solution in R and $u_1 \cdot u_2 \cdots u_{m+1}$ is a unit in R.

Proof. Since $S \subseteq \Omega$, it suffices to show the shortest length vector in Ω is unique if and only if the shortest length vector in S is unique and $u_1 \cdots u_{m+1}$ is a unit in R from Theorem 2, Theorem 3 and Theorem 4. Suppose $u_1 \cdots u_{m+1}$ is not a unit in R. Then $u_1 \cdots u_{m+1}\omega_s$ and ω_s are not same equivalent class, a contradiction to the uniqueness. Conversely, Suppose there exists a vector $\beta \in \Omega$ and $\beta \neq u\omega_s$, for all u is a unit in R. Then $u_1 \cdots u_{m+1}\beta \in S$, also a contradiction to the uniqueness of the shortest vector in S. □

Theorem 6. *Let $\omega_1, \omega_2, \cdots, \omega_{m+1}$ be a reduced basis of the sublattice*

$$\Lambda(u_1e_1, \cdots, u_me_m, u_{m+1}\alpha)$$

by Algorithm 1 and $V(\omega_1) \ge \cdots \ge V(\omega_{m+1})$. Assume

$$V(\omega_s) = V(\omega_{s+1}) = \cdots = V(\omega_t) > V(\omega_{t+1}),$$

where s is defined same as the above and $s \le t \le m+1$. Then the set W which contains all shortest length vectors in S is

$$W = \left\{ \sum_{i=1}^{s-1} f_i(x)\omega_i + c_s\omega_s + \sum_{j=s+1}^{t} c_j\omega_j \;\middle|\; \begin{array}{l} \deg(f_i) \le V(\omega_i) - V(\omega_s) \\ f_i(x) \in R[x] \textit{ for all } 1 \le i \le s-1 \\ c_j \textit{ runs all of } R \\ s+1 \le j \le t \\ c_s c \textit{ runs all of } [R^*] \end{array} \right\}$$

Proof. Assume β is a vector in S with shortest length. Then β is a linear combination

$$\beta = f_1(x)\omega_1 + \cdots + f_{m+1}(x)\omega_{m+1}$$

with $f_i(x) \in R[x]$ for $1 \le i \le m+1$. Since $V(\beta) = V(\omega_s) > V(\omega_{t+1})$, we have

$$f_{t+1}(x) = \cdots = f_{m+1}(x) = 0$$

by Lemma 6 and $d(\beta) = V(\beta)$ by Lemma 5. If $1 \le i \le s-1$, we have

$$V(\omega_i) - \deg(f_i(x)) \ge d(\beta) = V(\omega_s).$$

If $s \le i \le t$, we obtain

$$V(\omega_i) - \deg(f_i(x)) \ge d(\beta) = V(\omega_s).$$

Thus $\deg(f_i(x)) = 0$ and so write

$$\beta = f_1(x)\omega_1 + \cdots + f_{s-1}(x)\omega_{s-1} + c_s\omega_s + \cdots + c_t\omega_t$$

with $c_i \in R$ for all i, $s \le i \le t$. Setting

$$w = f_1(x)\omega_1 + \cdots + f_{s-1}(x)\omega_{s-1} \qquad \text{and} \qquad u = \sum_{j=s+1}^{t} c_j\omega_j,$$

it follows that $c_s c \in [R^*]$ from

$$\pi_{m+1}(\theta_{V(\beta)}(\beta)) \ne 0, \quad \pi_{m+1}(\theta_{V(w)}(w)) = 0, \quad \text{and} \quad \pi_{m+1}(\theta_{V(u)}(u)) = 0.$$

Conversely, for any vector $\beta \in W$, then $V(\beta) = V(\omega_s)$ and $\beta \in S$. □

Corollary 1. *The set U of all shortest length vectors in Ω is given by*

$$U = \{\beta \in \Omega | u_1 \cdots u_{m+1}\beta = \gamma, \text{ where } \gamma \text{ runs all } W\}$$

From Corollary 1 and Theorem 2 we easily get all minimal polynomials of $a^{(1)}, a^{(2)}, \cdots, a^{(m)}$.

Note 2. If R is a finite field, for any element $b \in R^*$, then $[b] = [1]$ and so there are no such equation $cx = b$ with $c = \text{Lcf}(\varphi^{-1}(\omega_s))$ and $[c] \ne [b]$. Therefore our results are in accord with those in [5].

4 An Example

In this section a concrete example is presented.

Given $a^{(1)} = (1, -2, 0, 4, -4)$ and $a^{(2)} = (0, 1, -1, -1, 3)$ two sequences over integers ring Z. Find a minimal polynomial.

First we calculate the formal-negative series of $a^{(1)}$ and $a^{(2)}$, i.e.,

$$a^{(1)}(x) = x^{-1} - 2x^{-2} + 4x^{-4} - 4x^{-5},$$
$$a^{(2)}(x) = x^{-2} - x^{-3} - x^{-4} + 3x^{-5}.$$

Setting

$$\omega_1 = (a^{(1)}(x), a^{(2)}(x), x^{-6}), \quad \omega_2 = (-1, 0, 0), \quad \text{and} \quad \omega_3 = (0, -1, 0),$$

we have

$$V(\omega_1) = 1, \quad V(\omega_2) = V(\omega_3) = 0, \quad \Delta \leftarrow 5,$$
$$\theta_{V(\omega_1)}(\omega_1) = (1, 0, 0)^T, \quad \theta_{V(\omega_2)}(\omega_2) = (-1, 0, 0)^T, \quad \theta_{V(\omega_3)}(\omega_3) = (0, -1, 0)^T.$$

Putting $M \leftarrow (\theta_{V(\omega_1)}(\omega_1), \cdots, \theta_{V(\omega_3)}(\omega_3))$, then $(1, 1, 0)^T$ is a solution of the equation system $Mx = 0$ and so set

$$\omega_2 \leftarrow \omega_2 + x\omega_1 = (-2x^{-1} + 4x^{-3} - 4x^{-4}, x^{-1} - x^{-2} - x^{-3} + 3x^{-4}, x^{-5}).$$

Thus

$$V(\omega_1) = 1, \quad V(\omega_2) = 1, V(\omega_3) = 0, \quad \Delta \leftarrow 4,$$
$$\theta_{V(\omega_1)}(\omega_1) = (1, 0, 0)^T, \quad \theta_{V(\omega_2)}(\omega_2) = (-2, 1, 0)^T, \quad \theta_{V(\omega_3)}(\omega_3) = (0, -1, 0)^T.$$

Then $(2, 1, 1)^T$ is a solution of new equation system $Mx = 0$ and so set

$$\begin{aligned}\omega_3 &\leftarrow \omega_3 + x\omega_2 + 2x\omega_1 \\ &= (-4x^{-1} + 4x^{-2} + 4x^{-3} - 8x^{-4}, x^{-1} - 3x^{-2} + x^{-3} + 6x^{-4}, x^{-4} + 2x^{-5}).\end{aligned}$$

We get

$$V(\omega_1) = 1, \quad V(\omega_2) = 1, V(\omega_3) = 1, \quad \Delta \leftarrow 3,$$
$$\theta_{V(\omega_1)}(\omega_1) = (1, 0, 0)^T, \quad \theta_{V(\omega_2)}(\omega_2) = (-2, 1, 0)^T, \quad \theta_{V(\omega_3)}(\omega_3) = (-4, 1, 0)^T.$$

Thus $(2, -1, 1)^T$ is a solution of $Mx = 0$ and hence set

$$\omega_3 \leftarrow \omega_3 - \omega_2 + 2\omega_1 = (4x^{-4} - 8x^{-5}, x^{-4} + 6x^{-5}, x^{-4} + x^{-5} + 2x^{-6}).$$

We obtain

$$V(\omega_1) = 1, \quad V(\omega_2) = 1, \quad V(\omega_3) = 4, \quad \Delta \leftarrow 0.$$

So that we get a reduced basis and $q(x) = x^2 + x + 2$ is a minimal polynomial.

5 Conclusion

When applied to single sequence synthesis, we can derive an algorithm equivalent to Norton algorithm [3]. Because of space limitations, we omit it in this paper.

References

[1] Fitzpatrick, P., Norton, G. H.: The Berlekamp-Massey algorithm and linear recurring sequences over a factorial domain. AAECC, **6** (1995) 309–323

[2] Lu, P. Z., Liu, M. L.: Gröbner basis for characteristic ideal of linear recurrence sequence over UFD. Science in China (Series A), vol.**28**, No. 6. 508–519 (1998)

[3] Norton, G. H.: On shortest linear recurrences. J. Symb. Comp. **27** (1999) 325–349

[4] Schmidt, W. M.: Construction and estimation of bases in function fields. J. Number Theory **39**, no. 2, 181–224 (1991)

[5] Wang, L. P., Zhu, Y. F.: $F[x]$-lattice basis reduction algorithm and multisequence synthesis. Science in China, in press

[6] Lenstra, A. K., Lenstra, H. W., Lovasz, L.: Factoring polynomials with rational coefficients. Math. Ann., vol. **261** (1982) 515–534

[7] Lagarias, J. C. , Odlyzko, A. M.: Soluting low density subset problems. Pro. 24th Annual IEEE Symp. on Found of Comp. Science (1983) 1–10

[8] Coppersmith, D.: Small solutions to polynomial equations and low exponent RSA vulnerabilities. Journal of Cryptology **10** (1997) 233–260

[9] Berlekamp, E. R.: Algebraic Coding Theory. New York: McGrawHill (1968)

[10] Massey, J. L.: Shift-register synthesis and BCH decoding. IEEE Trans. Inform. Theory, vol. IT-15, no. 1, 122-127 (1969)

Author Index

Lecture Notes in Computer Science

For information about Vols. 1–2048
please contact your bookseller or Springer-Verlag

Vol. 1982: S. Näher, D. Wagner (Eds.), Algorithm Engineering. Proceedings, 2000. VIII, 243 pages. 2001.

Vol. 1996: P.L. Lanzi, W. Stolzmann, S.W. Wilson (Eds.), Advances in Learning Classifier Systems. Proceedings, 2000. VIII, 273 pages. 2001. (Subseries LNAI).

Vol. 2049: G. Paliouras, V. Karkaletsis, C.D. Spyropoulos (Eds.), Machine Learning and Its Applications. VIII, 325 pages. 2001. (Subseries LNAI).

Vol. 2060: T. Böhme, H. Unger (Eds.), Innovative Internet Computing Systems. Proceedings, 2001. VIII, 183 pages. 2001.

Vol. 2062: A. Nareyek, Constraint-Based Agents. XIV, 178 pages. 2001. (Subseries LNAI).

Vol. 2064: J. Blanck, V. Brattka, P. Hertling (Eds.), Computability and Complexity in Analysis. Proceedings, 2000. VIII, 395 pages. 2001.

Vol. 2065: H. Balster, B. de Brock, S. Conrad (Eds.), Database Schema Evolution and Meta-Modeling. Proceedings, 2000. X, 245 pages. 2001.

Vol. 2066: O. Gascuel, M.-F. Sagot (Eds.), Computational Biology. Proceedings, 2000. X, 165 pages. 2001.

Vol. 2068: K.R. Dittrich, A. Geppert, M.C. Norrie (Eds.), Advanced Information Systems Engineering. Proceedings, 2001. XII, 484 pages. 2001.

Vol. 2069: C. Peters (Ed.), Cross-Language Information Retrieval and Evaluation. Proceedings, 2000. IX, 389 pages. 2001.

Vol. 2070: L. Monostori, J. Váncza, M. Ali (Eds.), Engineering of Intelligent Systems. Proceedings, 2001. XVIII, 951 pages. 2001. (Subseries LNAI).

Vol. 2071: R. Harper (Ed.), Types in Compilation. Proceedings, 2000. IX, 207 pages. 2001.

Vol. 2072: J. Lindskov Knudsen (Ed.), ECOOP 2001 – Object-Oriented Programming. Proceedings, 2001. XIII, 429 pages. 2001.

Vol. 2073: V.N. Alexandrov, J.J. Dongarra, B.A. Juliano, R.S. Renner, C.J.K. Tan (Eds.), Computational Science – ICCS 2001. Part I. Proceedings, 2001. XXVIII, 1306 pages. 2001.

Vol. 2074: V.N. Alexandrov, J.J. Dongarra, B.A. Juliano, R.S. Renner, C.J.K. Tan (Eds.), Computational Science – ICCS 2001. Part II. Proceedings, 2001. XXVIII, 1076 pages. 2001.

Vol. 2075: J.-M. Colom, M. Koutny (Eds.), Applications and Theory of Petri Nets 2001. Proceedings, 2001. XII, 403 pages. 2001.

Vol. 2076: F. Orejas, P.G. Spirakis, J. van Leeuwen (Eds.), Automata, Languages and Programming. Proceedings, 2001. XIV, 1083 pages. 2001.

Vol. 2077: V. Ambriola (Ed.), Software Process Technology. Proceedings, 2001. VIII, 247 pages. 2001.

Vol. 2078: R. Reed, J. Reed (Eds.), SDL 2001: Meeting UML. Proceedings, 2001. XI, 439 pages. 2001.

Vol. 2079: E. Burke, W. Erben (Eds.), Practice and Theory of Automated Timetabling III. Proceedings, 2001. XII, 359 pages. 2001.

Vol. 2080: D.W. Aha, I. Watson (Eds.), Case-Based Reasoning Research and Development. Proceedings, 2001. XII, 758 pages. 2001. (Subseries LNAI).

Vol. 2081: K. Aardal, B. Gerards (Eds.), Integer Programming and Combinatorial Optimization. Proceedings, 2001. XI, 423 pages. 2001.

Vol. 2082: M.F. Insana, R.M. Leahy (Eds.), Information Processing in Medical Imaging. Proceedings, 2001. XVI, 537 pages. 2001.

Vol. 2083: R. Goré, A. Leitsch, T. Nipkow (Eds.), Automated Reasoning. Proceedings, 2001. XV, 708 pages. 2001. (Subseries LNAI).

Vol. 2084: J. Mira, A. Prieto (Eds.), Connectionist Models of Neurons, Learning Processes, and Artificial Intelligence. Proceedings, 2001. Part I. XXVII, 836 pages. 2001.

Vol. 2085: J. Mira, A. Prieto (Eds.), Bio-Inspired Applications of Connectionism. Proceedings, 2001. Part II. XXVII, 848 pages. 2001.

Vol. 2086: M. Luck, V. Mařík, O. Štěpánková, R. Trappl (Eds.), Multi-Agent Systems and Applications. Proceedings, 2001. X, 437 pages. 2001. (Subseries LNAI).

Vol. 2087: G. Kern-Isberner, Conditionals in Nonmonotonic Reasoning and Belief Revision. X, 190 pages. 2001. (Subseries LNAI).

Vol. 2088: S. Yu, A. Păun (Eds.), Implementation and Application of Automata. Proceedings, 2000. XI, 343 pages. 2001.

Vol. 2089: A. Amir, G.M. Landau (Eds.), Combinatorial Pattern Matching. Proceedings, 2001. VIII, 273 pages. 2001.

Vol. 2090: E. Brinksma, H. Hermanns, J.-P. Katoen (Eds.), Lectures on Formal Methods and Performance Analysis. Proceedings, 2000. VII, 431 pages. 2001.

Vol. 2091: J. Bigun, F. Smeraldi (Eds.), Audio- and Video-Based Biometric Person Authentication. Proceedings, 2001. XIII, 374 pages. 2001.

Vol. 2092: L. Wolf, D. Hutchison, R. Steinmetz (Eds.), Quality of Service – IWQoS 2001. Proceedings, 2001. XII, 435 pages. 2001.

Vol. 2093: P. Lorenz (Ed.), Networking – ICN 2001. Proceedings, 2001. Part I. XXV, 843 pages. 2001.

Vol. 2094: P. Lorenz (Ed.), Networking – ICN 2001. Proceedings, 2001. Part II. XXV, 899 pages. 2001.

Vol. 2095: B. Schiele, G. Sagerer (Eds.), Computer Vision Systems. Proceedings, 2001. X, 313 pages. 2001.

Vol. 2096: J. Kittler, F. Roli (Eds.), Multiple Classifier Systems. Proceedings, 2001. XII, 456 pages. 2001.

Vol. 2097: B. Read (Ed.), Advances in Databases. Proceedings, 2001. X, 219 pages. 2001.

Vol. 2098: J. Akiyama, M. Kano, M. Urabe (Eds.), Discrete and Computational Geometry. Proceedings, 2000. XI, 381 pages. 2001.

Vol. 2099: P. de Groote, G. Morrill, C. Retoré (Eds.), Logical Aspects of Computational Linguistics. Proceedings, 2001. VIII, 311 pages. 2001. (Subseries LNAI).

Vol. 2100: R. Küsters, Non-Standard Inferences in Description Logocs. X, 250 pages. 2001. (Subseries LNAI).

Vol. 2101: S. Quaglini, P. Barahona, S. Andreassen (Eds.), Artificial Intelligence in Medicine. Proceedings, 2001. XIV, 469 pages. 2001. (Subseries LNAI).

Vol. 2102: G. Berry, H. Comon, A. Finkel (Eds.), Computer-Aided Verification. Proceedings, 2001. XIII, 520 pages. 2001.

Vol. 2103: M. Hannebauer, J. Wendler, E. Pagello (Eds.), Balancing Reactivity and Social Deliberation in Multi-Agent Systems. VIII, 237 pages. 2001. (Subseries LNAI).

Vol. 2104: R. Eigenmann, M.J. Voss (Eds.), OpenMP Shared Memory Parallel Programming. Proceedings, 2001. X, 185 pages. 2001.

Vol. 2105: W. Kim, T.-W. Ling, Y-J. Lee, S.-S. Park (Eds.), The Human Society and the Internet. Proceedings, 2001. XVI, 470 pages. 2001.

Vol. 2106: M. Kerckhove (Ed.), Scale-Space and Morphology in Computer Vision. Proceedings, 2001. XI, 435 pages. 2001.

Vol. 2107: F.T. Chong, C. Kozyrakis, M. Oskin (Eds.), Intelligent Memory Systems. Proceedings, 2000. VIII, 193 pages. 2001.

Vol. 2108: J. Wang (Ed.), Computing and Combinatorics. Proceedings, 2001. XIII, 602 pages. 2001.

Vol. 2109: M. Bauer, P.J. Gymtrasiewicz, J. Vassileva (Eds.), User Modelind 2001. Proceedings, 2001. XIII, 318 pages. 2001. (Subseries LNAI).

Vol. 2110: B. Hertzberger, A. Hoekstra, R. Williams (Eds.), High-Performance Computing and Networking. Proceedings, 2001. XVII, 733 pages. 2001.

Vol. 2111: D. Helmbold, B. Williamson (Eds.), Computational Learning Theory. Proceedings, 2001. IX, 631 pages. 2001. (Subseries LNAI).

Vol. 2116: V. Akman, P. Bouquet, R. Thomason, R.A. Young (Eds.), Modeling and Using Context. Proceedings, 2001. XII, 472 pages. 2001. (Subseries LNAI).

Vol. 2117: M. Beynon, C.L. Nehaniv, K. Dautenhahn (Eds.), Cognitive Technology: Instruments of Mind. Proceedings, 2001. XV, 522 pages. 2001. (Subseries LNAI).

Vol. 2118: X.S. Wang, G. Yu, H. Lu (Eds.), Advances in Web-Age Information Management. Proceedings, 2001. XV, 418 pages. 2001.

Vol. 2119: V. Varadharajan, Y. Mu (Eds.), Information Security and Privacy. Proceedings, 2001. XI, 522 pages. 2001.

Vol. 2120: H.S. Delugach, G. Stumme (Eds.), Conceptual Structures: Broadening the Base. Proceedings, 2001. X, 377 pages. 2001. (Subseries LNAI).

Vol. 2121: C.S. Jensen, M. Schneider, B. Seeger, V.J. Tsotras (Eds.), Advances in Spatial and Temporal Databases. Proceedings, 2001. XI, 543 pages. 2001.

Vol. 2123: P. Perner (Ed.), Machine Learning and Data Mining in Pattern Recognition. Proceedings, 2001. XI, 363 pages. 2001. (Subseries LNAI).

Vol. 2124: W. Skarbek (Ed.), Computer Analysis of Images and Patterns. Proceedings, 2001. XV, 743 pages. 2001.

Vol. 2125: F. Dehne, J.-R. Sack, R. Tamassia (Eds.), Algorithms and Data Structures. Proceedings, 2001. XII, 484 pages. 2001.

Vol. 2126: P. Cousot (Ed.), Static Analysis. Proceedings, 2001. XI, 439 pages. 2001.

Vol. 2129: M. Goemans, K. Jansen, J.D.P. Rolim, L. Trevisan (Eds.), Approximation, Randomization, and Combinatorial Optimization. Proceedings, 2001. IX, 297 pages. 2001.

Vol. 2130: G. Dorffner, H. Bischof, K. Hornik (Eds.), Artificial Neural Networks – ICANN 2001. Proceedings, 2001. XXII, 1259 pages. 2001.

Vol. 2132: S.-T. Yuan, M. Yokoo (Eds.), Intelligent Agents. Specification. Modeling, and Application. Proceedings, 2001. X, 237 pages. 2001. (Subseries LNAI).

Vol. 2136: J. Sgall, A. Pultr, P. Kolman (Eds.), Mathematical Foundations of Computer Science 2001. Proceedings, 2001. XII, 716 pages. 2001.

Vol. 2138: R. Freivalds (Ed.), Fundamentals of Computation Theory. Proceedings, 2001. XIII, 542 pages. 2001.

Vol. 2139: J. Kilian (Ed.), Advances in Cryptology – CRYPTO 2001. Proceedings, 2001. XI, 599 pages. 2001.

Vol. 2141: G.S. Brodal, D. Frigioni, A. Marchetti-Spaccamela (Eds.), Algorithm Engineering. Proceedings, 2001. X, 199 pages. 2001.

Vol. 2143: S. Benferhat, P. Besnard (Eds.), Symbolic and Quantitative Approaches to Reasoning with Uncertainty. Proceedings, 2001. XIV, 818 pages. 2001. (Subseries LNAI).

Vol. 2146: J.H. Silverman (Eds.), Cryptography and Lattices. Proceedings, 2001. VII, 219 pages. 2001.

Vol. 2147: G. Brebner, R. Woods (Eds.), Field-Programmable Logic and Applications. Proceedings, 2001. XV, 665 pages. 2001.

Vol. 2149: O. Gascuel, B.M.E. Moret (Eds.), Algorithms in Bioinformatics. Proceedings, 2001. X, 307 pages. 2001.

Vol. 2150: R. Sakellariou, J. Keane, J. Gurd, L. Freeman (Eds.), Euro-Par 2001 Parallel Processing. Proceedings, 2001. XXX, 943 pages. 2001.

Vol. 2154: K.G. Larsen, M. Nielsen (Eds.), CONCUR 2001 – Concurrency Theory. Proceedings, 2001. XI, 583 pages. 2001.

Vol. 2161: F. Meyer auf der Heide (Ed.), Algorithms – ESA 2001. Proceedings, 2001. XII, 538 pages. 2001.

Vol. 2164: S. Pierre, R. Glitho (Eds.), Mobile Agents for Telecommunication Applications. Proceedings, 2001. XI, 292 pages. 2001.